WAR UNDERGROUND

War Underground

A History of Military Mining in Siege Warfare

EARL J. HESS

 *University Press of Kansas*

Published by the University Press of Kansas (Lawrence, Kansas 66045), which was organized by the Kansas Board of Regents and is operated and funded by Emporia State University, Fort Hays State University, Kansas State University, Pittsburg State University, the University of Kansas, and Wichita State University.

Library of Congress Cataloging-in-Publication Data

Names: Hess, Earl J., author.
Title: War underground : a history of military mining in siege warfare / Earl J. Hess.
Other titles: History of military mining in siege warfare
Description: [Lawrence, Kansas] : [University Press of Kansas], [2025] | Includes
    bibliographical references and index.
Identifiers: LCCN 2024028931 (print) | LCCN 2024028932 (ebook) | ISBN 9780700638413 (cloth)
    | ISBN 9780700638420 (ebook)
Subjects: LCSH: Tunnel warfare—History. | Mines (Military explosives)—History. | Siege
    warfare—History. | BISAC: HISTORY / Military / General | HISTORY / World
Classification: LCC UG340 .H47 2025 (print) | LCC UG340 (ebook) | DDC 355.4/4—dc23/
    eng/20240922
LC record available at https://lccn.loc.gov/2024028931.
LC ebook record available at https://lccn.loc.gov/2024028932.

British Library Cataloguing-in-Publication Data is available.

For Pratibha and Julie, with love

CONTENTS

Introduction

Sieges have played an important role in military operations ever since the earliest recorded history. Whenever a commander wanted to reduce a fortified position, a number of options presented themselves. Among these were bribing enemy personnel to betray the garrison, investing the place to cut it off from outside aid and starve it into submission, or conducting an attack designed to scale defensive walls. In addition, every commander had the option of military mining, that is, digging underground approaches called galleries or tunnels beneath the target to literally as well as figuratively "reduce" the fortified position.

This study focuses on military mining as an element in siege warfare and in open-field campaigns from the earliest documented use of this tactic in 880 BCE to the ongoing war in Ukraine. Its focus is on more than three hundred sieges in which the besieger dug underground approaches to an enemy target while the defender made corresponding underground attempts to intercept them. The 2,900-year time span, plus looking at developments in as many countries and across as many continents as possible, gives this study a wide scope missing in previous works on individual sieges and wars. It thus traces the trajectory of several trends in military mining over time.

An allied type of siege approach, from the ninth century BCE until the seventeenth century CE, was called sapping. It consisted of working above ground in an attempt to break out individual stones of a defending masonry wall, using the protection of a cover usually made of wood and often called a penthouse, a sow, a cat, or a tortoise. When this operation did not involve any tunneling, it was a separate type of siege approach. But sometimes tunneling was combined with sapping in the same operation. Thus, in addition to chipping away at the wall above ground, miners used the protection afforded by the penthouse to dig a tunnel in an effort to reach the wall's underground foundation. They could excavate a cavity, propping up the wall with wood posts; fill the cavity with combustibles; and in burning the props, bring down a section of the wall. This process often was called prop mining.

Mining was, in short, not the same type of operation as sapping even though both could take place under the same penthouse. If opportunity and need arose, miners could also start their tunnel at a distance from the defending wall and dig a longer underground approach to it. In the seventeenth century, with the advent of Sébastien Le Prestre de Vauban's system of siege warfare, the definition of the term "sapping" changed dramatically, but underground tunneling continued to be associated with it. Rather than denoting the chipping away at a masonry wall above ground, sapping now denoted the digging of an approach trench, called a sap, across ground covered by enemy artillery fire. Sappers often placed wood covers over the trench to protect themselves from incoming rounds, or sometimes they dug the sap in a zig-zag fashion to reduce the danger if the defender managed to fire down the length of the trench.

When close to the enemy fortification, miners then took over and started to dig underground galleries from the end of the sap and toward the foundation of the defensive structure. By that time, gunpowder had replaced wooden props as military mining entered the era of explosive mines. The new, professionally trained engineer troops being added to standing armies in Europe during the seventeenth century were appropriately called sappers and miners. They replaced the civilian ore miners typically hired for short-term service by besieging armies since the classical era.

As a distinct type of siege approach that was conducted underground, mining had a set of circumstances, possibilities, and problems that were unique in the history of warfare. A distinctly technical military service, mining was heavily influenced by the technology and methods of the civilian mining industry in every era. And it very often borrowed experienced civilian miners to dig the tunnels.

All of this aimed at trying to achieve success in static warfare, whether it was against a castle or a fortified city in the classical, medieval, and early modern eras or a long line of enemy earthwork defenses in the nineteenth and twentieth centuries. In other words, the purpose was to reduce the target in a campaign-winning stroke. But to do this, more than just technical success in the mine operation was required. The underground attack had to be coordinated with follow-up action on the surface to take advantage of the technical success. That was the rub that often bedeviled commanders from the start to the finish of military mining history. The problem was never truly solved in any era, but that did not prevent commanders from investing the time, resources, and energy needed to prosecute underground warfare. The history of military mining is one of great continuity up to the introduction of gunpowder in the fifteenth and sixteenth centuries. After that, military mining changed more rapidly so that. by the early twentieth century, it was brought to its ultimate expression on the Western Front of World War I (1914–1918). There, more men were employed in digging tunnels than in any previous war, all the modern technologies and methods were fully employed, and more mine explosions were set off than in all previous conflicts combined.

Siege mining's long reach in history has garnered relatively little attention from historians as far as detailed study is concerned. This book aims to fill the need for an in-depth examination of the topic. Its main objective is to document occurrences of military mining with details concerning the process. In addition, it examines mining's tactical significance as far as the outcome of a specific siege is concerned. A third contribution is to include evidence from an interdisciplinary perspective. For some operations, archaeological findings offer more information on siege mining than do extant written sources. Because of language limitations, most of the evidence comes from English sources. Despite this restriction, however, the comprehensive time span and the global perspective here remain unique in the literature devoted to siege mining to date.

The first and longest period in the history of military mining began in 880 BCE and extended to the introduction of gunpowder in the fifteenth century. Tactics and methods hardly changed during that period, maintaining a tight focus on prop mining by besiegers and the digging of countermines by the besieged. Technically called galleries, the tunnels began either at the bottom of vertical excavations called shafts or by digging directly either into a vertical bank of earth or from the basement of a building. Commanders needed the expertise of experienced civilian miners who knew how to navigate underground and how to shore up the sides and ceilings of galleries to prevent cave-ins. Besiegers also needed to understand how to provide a supply of breathable air; when galleries exceeded about sixty feet (18.28 meters) of length, it became virtually impossible to breathe without some ventilation scheme. As a result, most tunneling during this first period was at shallow depths and over relatively short distances.

The introduction of gunpowder started the second period of military mining history, which extended to the middle of the nineteenth century. Gunpowder produced a range of problems. Learning how to set the delicate charge, how to detonate it properly, and how to prevent the force of the blast from being wasted by traveling along the open tunnel all had to be painfully worked out through experience over several decades. Thanks to the development of the mechanical printing press in the middle of the fifteenth century and the onset of the Scientific Revolution two centuries later, the results of experiments and practice in the field could be disseminated throughout Western cultures in the form of technical manuals. Exactly how a gunpowder charge affected the ground around it and created craters on the surface became an object of intense scrutiny, as an inexact craft slowly evolved into a scientific process.

Vauban's work offered a doctrine for modern siege approaches that seemed to work in many sieges of the era. It involved sapping forward on the surface from a line of earthworks established parallel to the defensive wall but several hundred yards away from it. Sappers dug an approach trench across no-man's-land, after which miners tunneled from the end of the sap. They could explode a charge designed to fill in a section of the defensive ditch to prepare the way for an infantry attack or collapse the defensive wall itself by undermining its foundation. Miners could also explode a charge designed to blow out a crater in no-man's-land, then start a new mine from the bottom of the crater so that they would not need to rely on one long gallery requiring ventilation. As long as they were started from the head of the sap, offensive galleries continued to be relatively short during this second period of mining history. But on occasions when miners started their work from another location, gallery length increased with consequent complications for underground navigation and ventilation.

Counterminers learned from the scientific experiments of the eighteenth century that they could destroy the underground galleries of the besiegers by tunneling forward as close as possible to them and exploding small mines. These countermine charges, termed "camouflets," were not designed to create craters on the surface but to shatter nearby galleries within a globe of compression (later called the zone of destruction). Whether exploding an offensive mine or a defensive camouflet, miners had to block up their own tunnel to prevent the blast from escaping along its length rather than pushing out in a circle from the center of the charge. This process was called tamping the gallery. At times during this period, as during the preceding centuries of mining history, it was possible for the counterminers to break into the gallery of the besiegers and fight hand to hand in the dark and confined space of the tunnels.

The third and final period of military mining history began in the late nineteenth century with the introduction of new high explosives, which gradually replaced gunpowder, and electrical detonators to set off the charge. A wide variety of high explosives became available to the military miner and to the civilian miner at this time, all having varied advantages and disadvantages. Electricity gradually replaced the use of powder troughs (wooden boxes containing powder), powder hoses (powder encased in linen tubes), and safety fuzes (powder encased in rubberlike substances) to set off the mine.[1]

During World War I, the Western Front became the ultimate venue for the military miner. The armies were locked in static and sophisticated trench defenses for four years, resulting in the most prolonged and intense positional warfare in global history. Now armed with a wide array of high explosives, miles of electrical leads, and sophisticated detonators to connect the leads to the explosives, it was possible for miners to push their military art as far as it could go. All the armies heavily recruited civilian miners to supplement their engineer troops, borrowed electric-powered pumps for ventilation and water drainage, used modern drilling machines to supplement hand-powered digging, and organized for the first time mine rescue systems utilizing oxygen-generating rescue packs to deal with cave-ins and gas-poisoning underground. Although this intense activity failed to achieve a breakthrough in the static confrontation that drained all the armies, the Great War provided the most fruitful venue for the development of mining operations in world history.

The Western Front also provided the venue for digging long galleries that were completely separate from sapping operations on the surface. In fact, there were very few examples of sapping in France and Belgium, while tunneling proliferated to an unprecedented degree. Those tunnels came in many lengths, long and short, but the strong tendency was to plan and dig them starting from friendly trench lines. The comparatively huge numbers of miners, the adaptation of modern surveying equipment to old methods of navigating underground, and the employment of academically trained geologists to study the geomorphology of the battlefield enabled Great War engineers to dig long galleries. A survey of available and reliable reports on tunnel length indicates that on average they were dug 125 feet (38.10 meters) out before 1914 in contrast to the Western Front, where they were 568 feet (173.12 meters) long on average. The longest gallery recorded before the Great War was 510 feet (155.44 meters), while some galleries on the Western Front extended 1,810 feet (551.68 meters).

After 1918, interest in underground approaches withered so much as to almost disappear. The primary reason for this was that a range of new military equipment and a dramatic change in doctrine during the 1930s restored mobility to modern armies. The tendency to focus on capturing strongpoints throughout history now shifted to give primary weight to speed and movement as ways to achieve strategic goals. It is true that even during and after World War II military operations sometimes produced positional warfare, but those instances seldom resulted in mining projects. Very soon, even the knowledge of how to dig tunnels and set charges disappeared from military manuals. Underground approaches became relics of the past.

But that past is a rich, interesting, and important history. While mining appears in a marginal way in virtually every study of siege warfare, it has not yet been treated with depth or comprehensiveness as a distinct subject for book-length analysis. Kenneth Wiggins provided a brief overview in his booklet *Siege Mines and Underground Warfare* (2003), and Simon Jones has dealt in depth with the mining and countermining along the Western Front in *Underground Warfare, 1914–1918* (2010). But for most military historians, mining plays only a slim role in studies that range across the many characteristics of siege operations throughout history.

A comprehensive study of siege mining across time and space leaves one with the impression that there was a core area of the world where it developed and flourished. One can find examples of military mining in Asia and Africa, but a fuller view of it in Chinese military history is difficult to obtain because of a shortage of scholarly literature in English on that particular subject. As it is, the most attainable view we have of underground warfare in English-language sources is within the Near East–Mediterranean–European region. The concept of siege mining originated in the Near East, as far as we know, and migrated to the Greek city-states. The Romans brought it to a further point of repeated use and sophistication, bequeathing it to medieval commanders, who failed to advance the craft much farther. But after the onset of the gunpowder revolution, European technicians and military leaders advanced the precepts of military mining into the modern era.

Middle Eastern military cultures continued to take mining seriously, especially during the crusading era. Great Britain can be seen as on the periphery of the intense military mining culture of continental Europe, sharing in it but not pursuing it quite as vigorously as others (except during World War I). North America can also be viewed as part of this periphery. In fact, far less mining has taken place in the Western Hemisphere than even in Britain. But to a greater or lesser degree, the practice of digging underground approaches against established defenses has resonated throughout the world.

In addition to a wide coverage, incorporating at least three hundred examples of mining in action across international boundaries from the ancient Near East to twenty-first century Europe, this study follows a close link to civilian mining techniques, methods, and material. Military mining grew out of this culture, adopting its basic processes. Just as civilian miners dug vertical shafts from which they advanced horizontal galleries into lode-bearing strata, military miners worked similarly to approach enemy targets. That was a major reason why commanders tended to hire or impress civilian miners to do much of the planning and digging, usually improvising this recruitment when a decision was made to mine. For the most part, such men remained civilians and were not incorporated into the military force on a real or permanent basis. This trend weakened only during the time from the seventeenth century on, when the engineering arm of major Western armies began to professionalize. These militaries created professional corps of sappers and miners and trained them in how to plan, construct, and explode mines without the aid of civilians.

Therefore it is important to view the history of military mining within the context of civilian mining history. Commercial mining has drawn a great deal more study from historians than has its military counterpart, and its details and contours are well developed in the scholarly and popular literature. Essentially, there have been two major strains of commercial mining: the first was oriented toward accessing precious minerals and ores, and the second was meant to create transportation routes underground.[2]

Military mining has drawn from both major civilian traditions, although it has more in common with ore mining than with the digging of transit tunnels. Military mines were designed for even shorter-term use than commercial ore mines, useful for as long as it took to prop and burn or to explode them in order to bring down a defensive barrier. At times military miners in the classical and medieval eras used galleries as transit tunnels to insert troops into enemy enclosures, but even then they were useful only for the duration of the siege. It was only during World War I that military mines of both types—explosive and transit—were needed for extended periods due to the long and static nature of operations on the Western Front. Otherwise, military mining has been a comparatively infrequent process of quickly digging, then burning props or exploding gunpowder that lasted days, weeks, or a few months.

The ephemeral nature and the relatively sporadic application of military mining are the chief reasons why it usually trailed behind the commercial realm in terms of new developments. Commercial mining and tunneling were relatively constant and widespread components of civilizations around the world, placing great demands on those involved to develop new methods of digging, ventilating, and navigating underground as well as maintaining shafts and galleries. Before World War I, military mining had always played catch up with the progress to be seen in commercial mining, and because the tactic was dropped as soon as a commander felt it did not pay, whatever advances might have taken place in a short campaign also tended to disappear. It is astonishing how much soldiers had to relearn every time they wanted to employ mining as an operational choice, which also explains why armies have historically been so ready to hire commercial miners to help them dig military galleries.

This pattern began to change a bit with the onset of gunpowder, scientific study, the mechanical printing press, and the creation of professional military miners in the fifteenth through seventeenth centuries. Those developments eventually created a consistent stream of knowledge about theory and technique within an army's culture. But those advantages still failed to push the military miner ahead of their commercial peers because civilian mining industries in the most advanced countries also experienced advances in technology, technique, and knowledge during the same time period. And despite repeated wars and a marked increase in the use of mines to attack fortified positions in the seventeenth and eighteenth centuries, civilian mining continued to plow ahead. New concepts and machines always were used first in civilian mines, and many military miners failed to keep up with these latest developments.

The only real exception to this rule took place on the Western Front of World War I. There, military mining increased so sharply that, for the first and only time, one could say it caught up with civilian mining in intensity, innovative practices, and new material. An influx of hundreds of professional miners from around the British Empire and Commonwealth nations brought in a new spirit and new methods such as clay-kicking—a faster technique to dig a tunnel through the right kind of clay strata. But even as the military miner caught up with his civilian counterpart, the war came to an end, and with it, the long history of military mining also nearly ended.

No matter what time period in this long history one chooses to study, it is important to understand the terminology associated with military mining. Many historians have failed to differentiate between sapping and mining when discussing the time period before the introduction of gunpowder. Others conflate military mining with the use of landmines, which are antipersonnel devices planted just below the surface of the ground. Still others confuse the mid-twentieth century use of tunnel complexes as shelter and supply systems for insurgency warfare with mines that are designed to attack enemy strongpoints.[3]

But planting landmines has nothing to do with digging tunnels to undermine an enemy structure. In contrast, the construction of elaborate underground shelters in which to base guerrilla fighters has much more affinity with military mining. This practice actually started on the Western Front when armies used their mining companies to dig deep underground shelters to protect infantrymen from heavy shelling. But that was a diversion for the mining personnel, whose primary job was to dig attack tunnels to undermine enemy strongpoints. For the purposes of this study, the focus is maintained on military mining, defined as the digging of underground passages, technically termed "galleries," for the purpose of breaching enemy strongholds. Included in this definition are efforts by the defender to counter that underground approach. Kenneth Wiggins has suggested terms such as "underground mining" or "gallery mining" to capture the basic concept of military mining. World War I British writers liked the word "tunneling" to capture it.[4] "Siege mining" also is a term that is highly relevant for this operation. "Offensive galleries" and "countermine galleries" are other terms one can use to refer to military mining. Any underground digging to undercut a defensive structure or to prevent someone from accomplishing that goal is military mining, whether the tunnel is very short and linked with above-ground sapping operations or very long with no such connection. The concept behind military mining is consistent and quite distinct from other types of military digging, and it has a long pedigree throughout global history.

In a work that sweeps across such a length of time as this one, the quality of sources becomes acute. The work of all historians is empowered as well as limited by the primary sources available. In this case the collective work of many men who lived and wrote in the classical and medieval periods is largely what we have regarding the history of military mining. The digging that took place under the foundation of masonry walls was known to them, but it rarely took center stage in their narrative accounts of sieges. Technical details especially escape them, and the interested student all too often wishes for more information than is typically provided about gallery depths and lengths in ancient and medieval sources.

For example, when Appian wrote about the Roman civil wars, he devoted much detail to Octavian's siege of Lucius Antonius at Perusia in the winter of 41–40 BCE. Appian gave statistics on the size and dimension of Octavian's fortified line of circumvallation, how many towers were constructed, and how far apart they were placed along the line. He went on to detail activity such as the method of attack on the city's defenses. In fact, virtually everything that took place on the surface of the siege area drew Appian's attention, but when it came to the underground war, that was a different story. In an entirely offhand way, the historian merely noted that some of the defenders tried to dig a tunnel to collapse part of Octavian's line of circumvallation. One might forgive him for not providing the type of detail for mining that he devoted to surface events because the underground war was largely a secret operation. It took place in an environment where only those immediately involved in it knew what was happening, and they were comparatively few in number when considering the large armies operating on the surface.[5]

This is why archaeology can help the military historian a great deal, such as the discoveries of shaft and gallery remnants at Dura-Europos and a handful of other siege sites of the classical and medieval eras. Afterward, with the many cultural and intellectual developments in the early modern era and beyond, textual sources begin to pile up and often contain rich detail. Much information is available for salient sieges like that at Schweidnitz in the Seven Years' War (1756–1763), at Sebastopol during the Crimean War (1853–1856), at Petersburg during the American Civil War (1861–1865), and of course for the Western Front of World War I, from which a mother lode of detailed information on mining has emerged both in published sources and in rich archaeological remnants.

But we do have a serious lack of balance in the long sweep of mining history. The interested student has to squeeze kernels of information from a variety of sources that provide scant details for the long periods of the classical and medieval eras, but fortunately military mining did not change very much during those centuries. In contrast, when important elements did begin to change a lot and quite rapidly from the early modern era and beyond, an ever-increasing body of rich resources document that transformation. The end result is that enough information on the story of military mining exists to save it from the oblivion often associated with any activity conducted underground.

Much information about some aspects of military mining simply cannot be found in traditionally published academic books and articles. For example, there is almost no information about permanent countermine systems at various European towns in traditional sources. Thus, if the reader notices the more than two dozen online sources cited in this book, mostly to Wikipedia articles, please be aware that I could not find a better source for those subjects. Moreover, I used the information on those websites with discrimination and care in the same way that I evaluated the reliability of information found in Sir John Froissart's *Chronicles*, for example. No matter the source, the historian must determine how far to rely on it.

Taking a multidisciplinary approach has contributed much to this book. In addition to consulting archaeological reports, the writing of scholars in the geosciences has helped in understanding a process that took soldiers into the surface layers of the earth, especially during World War I. In addition, the work of film historians has illuminated the cultural history of military mining, as that unique element of war history has found its way into a handful of feature films of the twentieth and twenty-first centuries. Not only over time and across international borders but also across the disciplines, I have "mined" many sources of information to make this book as comprehensive as possible.

I wish to thank Kenneth Wiggins for sharing photos of his work at King John's Castle, Limerick, Ireland, where the most important remnants of underground timber shoring were found. Ken also read chapter 5 and offered helpful comments for its improvement. He and editors Helen Dunne and Nick Maxwell at Wordwell Books allowed the use of a map from his book *Anatomy of a Siege*. Adrian Stewart at Ninety Six National Historic Site was very helpful in pointing me toward information on the mine remnants at that place. My thanks also go to Clifford J. Rogers for reading chapters 1–3 and offering very helpful suggestions in addition to providing additional primary material. Many thanks also to Sarah Melville for important advice about the Assyrians and for alerting me to two wall-panel reliefs that were very helpful. Tonio Andrade also made thoughtful suggestions about sources. I also thank Jeremy Black and Samuel J. Watson for reading the entire manuscript and offering very helpful advice about improving it. Kevin Brock did his usual thorough and thoughtful copyediting of the manuscript, for which I am very grateful.

And, as always, thanks to my wife Pratibha for all she does for me.

I

The Near East and Greece

The civilian practice of mining for precious ores and the creation of walled cities were the fundamental preconditions for the development of military mining. Siege mining drew its concepts, technology, and procedures from civilian activities but always remained a bit behind the latest developments used by civilian miners. The creation of walled cities provided the purpose of and the rational for siege mining. From its earliest appearance in the Near East through its spread into the Greek and Mediterranean world, siege mining was a highly technical military craft unlike the other methods of capturing cities. But given the time, patience, labor, and resources required, military mining was not universally employed in siege warfare. Nevertheless, it became one of many tactics available to a besieging force to capture the target city and succeeded about as often as most of the other methods employed, except starvation.

The earliest known civilian mining of any type in the archaeological record was at the Lion Cavern in African Swaziland, dating to 43,000 years ago. From that time onward, the most common way to access ores was through mining on the surface. Modern terms such as "open pit," "quarrying," or "strip pit" refer to this practice. Underground working, also called drift mining, slope mining, or shaft mining, has been less common but more visible in the public imagination of mining history.[1]

Military mining bears no relationship with surface commercial mining but owes most of its techniques and principles to civilian subsurface operations. Going underground to seek usable ores or precious minerals became more common with the passage of time and the growing sophistication of cultures. In Greece surface deposits began to be exhausted during the Bronze Age (3,000–1,000 BCE). Little is known of mining during the Dark Age of 1,000–700 BCE, but the shafts and galleries of subsurface mining became more common during the Archaic Era (700–480 BCE). Slaves and animals increasingly performed the work, digging in search of "gold, silver, copper, iron, lead and—to a lesser extent—tin." Also during the Archaic Era, miners began to shore up shafts (vertical diggings) and galleries (horizontal diggings), which they drained of water and ventilated to remove noxious fumes, smoke, and gases. In short, by the end of the Archaic Era in the fifth century BCE, all the basic components, methods, problems, and advantages of subsurface gallery mining had appeared in the civilian economy.[2]

The largest workings in Greece were located at Laurion near Athens. Initially opened before the sixth century BCE, they were worked for some five hundred years. At its height of production, there were about 2,000 shafts dug by a large labor force of slaves. The shafts there were either rectangular or square in shape and averaged 1.90 meters (6.23 feet) by 1.30 meters (4.26 feet) wide, with some estimated going as deep as 119 meters (390.42 feet). The galleries were rectangular, square, or trapezoidal in profile and not over 1 meter (3.28 feet) tall and 60–90 centimeters (23.62–35.43 inches) wide. One could not stand up straight in them, so miners would lay on their backs or sides to work. It would have taken nine to ten hours to advance a gallery only 10–12 centimeters (3.93–4.72 inches), or about a month to dig 12 meters (39.37 feet) in the schist and limestone at Laurion. Shoring the sides and tops of galleries was little needed considering their size and the surrounding type of ground and rock. But when needed, the wooden shoring employed mortise-and-tenon joints to link upright props on the sides with the lintels that supported the roof. Archaeologists have found fragments of shoring, identifying them as made of olive wood.[3]

Historian John F. Healy estimates that from 10,000 to 20,000 slaves worked in mines around the Greek states. Their work environment often was atrocious, with small, hot galleries filled with bad air and insufficient lighting. They used a few simple tools, including chisels, hammers, mattocks, and shovels. From earliest times, fire setting was a common technique to crack open rock formations. This involved building a fire in a fissure to heat the surrounding rock and then rapidly cooling it with water to produce further cracks. Gases, mineral dust, miners' exhalations, decaying timber, and the smoke and heat of lamps used for illumination contributed to bad air in mines. Efforts to deal with this problem focused mostly on natural methods, using draught and convection to create air currents within the system of shafts and galleries, all linked by cross tunnels. Artificial means of ventilation were quite limited, mostly consisting of setting up linen strips and shaking them in an attempt to force fresh air into a shaft or gallery. Wind scoops to catch surface breezes and divert them into shafts or galleries were also possible. The bones of miners found at several locations indicate that ventilation was not only a persistent but also a deadly problem.[4]

Bringing light to the utter darkness of underground workings also had limited effect. Lighting methods used in the home were transferred to the galleries in the form of torches made of resinous wood. Also used were skins soaked in oil and set alight or terracotta vessels containing oil and a floating wick; such a lamp found in the gold mines of Thrace dated to around 350 BCE. All these methods of illumination would have produced what Healy calls a "smoky dense atmosphere" in the galleries.[5]

Military miners would have encountered much the same environment as these slave miners of Greece, and they would have used the same techniques, designs, tools, ventilation, and illumination. An important difference, however, was that they would have endured these conditions for a far shorter period of time because of the ephemeral nature of siege mining. Because military shafts and galleries were dug at a shallow depth, the problems of working them would also have lessened. But galleries at any depth and more than about sixty feet (18.28 meters) long would have needed some degree of ventilation, and even short ones required illumination.

While military mining was most closely related to commercial gallery mining, it also bore a relation to transportation tunneling. If a commander wanted to insert soldiers inside the enclosure of the fortified town he was attacking, rather than propping and burning to collapse a section of the city wall, he would imitate the civilian practice of digging tunnels for transportation purposes. That civilian construction has been documented to about 1200 BCE in Syria and Palestine. These tunnels were passages aimed at accessing underground springs and thus were meant to be permanent and frequently used. Wherever dug in the relatively dry climate of the Mediterranean and Near East, they often were converted into aqueducts. The Greeks dug such a tunnel through limestone for 3,300 feet (1005.84 meters) at Samos around 535–522 BCE. It was 5.5 feet (1.67 meters) high by 5.5 feet wide. Of course, even if a commander wanted to insert troops into an enclosure, the resulting tunnel would be for a one-time use rather than a permanent fixture. But the purpose had to influence design; even a temporary transit tunnel had to be big enough to allow soldiers to move relatively quickly through it. The small size of civilian galleries would be a hindrance to the success of an underground attack. While a gallery something like the dimensions of the Samos tunnel would be ideal, it would take much longer to dig than one the size of commercial mining galleries.[6]

Civilian mining appeared much earlier than the second precondition for siege mining. The earliest recovered city wall is at Jericho and dates to about 7000 BCE, but the evidence for sieges greatly predates that time period. Siege techniques developed with the spread of fortified cities and the desire by ambitious polities to acquire them. Along with a blockade to starve out the defender, escalading to overcome the walls, and treachery, sapping the walls constituted one of the major ways to deal with a fortified city. Egyptian reliefs dating to the twenty-fourth and twenty-third centuries BCE depict various siege techniques, but sapping did not appear in this evidence until the Twelfth Dynasty (1991–1786 BCE). Based on the images on numerous reliefs, the Assyrians (2500–605 BCE) developed siege craft to its highest state yet. Their techniques included sapping and tunneling in addition to siege ramps and towers. A wall-panel relief dated to the reign of Ashurnasirpal II in the ninth century BCE shows two Assyrian soldiers digging out a cavity under a city wall in a clear illustration of mining. Another wall panel dated to the reign of Ashurbanipal, 669–627 BCE, shows Assyrian soldiers sapping the wall of the Elamite city of Hamanu. The Assyrians also provided the earliest example of a military mine, documented by evidence other than relief depictions, dating to 880 BCE. They tunneled under a defending wall, excavated a cavity while bracing the foundation, filled the hole with combustible material, and set it afire to collapse the structure.[7]

The Persians (550–330 BCE) employed the full range of siege techniques to create the largest empire yet known in the Near East. The involvement of mining in their siege of Barca, on what is today the eastern coast of Libya, is well documented by Herodotus. Amasis, the Persian commander, laid siege to this Greek colony around 512 BCE by digging mines. A metalworker in the city used a bronze shield to detect the diggings. According to Herodotus, he "struck the shield" on the ground in various spots. It usually produced "a dull sound," but where it "rang out" the metalworker assumed there was a cavity underground. When the Barcans dug at this spot, they intercepted the Persian gallery and killed the miners. This is the first recorded example of detection of an offensive mine and of successful countermining. It failed, however, to save the colony. After a nine-month siege, Amasis tricked the colonists by arranging for a truce and then violating it to gain entry into the city.[8]

Countermining techniques developed along with the advancement of siege techniques. In the Persian siege of Paphos on the southwest coast of Cyprus in 498 BCE, the Greek defenders countermined a siege ramp constructed near the city wall. At what archaeologists call Tunnel 1, the Greeks dug a gallery 4 meters (13.12 feet) below the surface of the ground, from 1.2 meters (3.93 feet) to 1.7 meters (5.57 feet) wide, and from 1.7 meters (5.57 feet) to 2.3 meters (7.54 feet) high. The diggers filled a part of their gallery with combustibles and set it afire, which collapsed a section of the ramp and prevented the attackers from using it. Tunnel 4 had smaller dimensions: 1.1 meters (3.60 feet) to 1.5 meters (4.92 feet) wide and 1.4 meters (4.59 feet) to 1.8 meters (5.90 feet) high. It also had "nearly vertical side walls and a roughly concave roof," with niches in the walls for lamps. Tunnel 2 apparently was never finished because of a partial collapse of the roof. No timber shoring was found in any of the galleries even though three of them were dug through "soft conglomerate rock"; the last one went through solid rock. The flammable material used to burn the props in Tunnel 1 created an intense fire that turned the ramp above it "into a compact cone-shaped mass of calcined stone and lime," according to archaeologists. Despite the Greeks' success at neutralizing the siege ramp, the Persians managed to capture Paphos. Nearby, they also besieged Soli and took it by bringing down its wall either through sapping or mining.[9]

Ancient sources fail to indicate how long galleries tended to be. According to the Greek historian Polyaenus, writing in the second century CE, the Persian satrap Otones besieged the city of Chalcedon on the north coast of the Sea of Marmora, near the later location of Constantinople, during the sixth century BCE. His men started a gallery on the slope of Aphasius Hill fifteen stades from the city wall. A stade is typically considered to be about 600 feet (185.62 meters), making this projected gallery 9,000 feet (2743.20 meters), or 3,000 yards, long. This length, one and three-quarter miles, is virtually unbelievable for this time period. Even on the Western Front of World War I, which had the most sophisticated mining galleries in history, this length would have been a tall order.[10]

Mining techniques disseminated throughout the Mediterranean world. The Phoenicians took them to Sicily, where the Greeks passed them on to the Romans. But according to historian Josiah Ober, the Greeks did not readily take to military tunneling. They had little direct connection with commercial mining, a specialized industry conducted by a small and marginal element of society. Tunneling was "hot, nasty work," as Ober reminds us. "Unlike rock cutting and hauling, tunneling was not the sort of labor that a peasant soldier would undertake as part of his normal agricultural round." Tunneling tended to be "prototypically the work of the slave, and thus was regarded as labor beneath the dignity of the free citizen," at least in Athens.[11]

And yet military mining played a prominent role in a few Greek operations. In the Spartan siege of Platea in 429–427 BCE during the Peloponnesian

War, the besiegers conducted a multifaceted attack on the city that lasted for many months. They constructed a masonry wall to blockade the city and made a large siege ramp to mount rams in an attempt to break through its defensive wall. The Plateans countermined the ramp not once, but twice. The first gallery was blocked by the Spartans, who discovered it and filled the opening with mud wrapped in reed mats. But the Plateans dug a second gallery from a different direction and managed to extract enough material from the underside of the ramp to cause that section to subside. This second operation neutralized the siege ramp, but in the end the city fell due to food shortages.[12]

Another Spartan force tried to employ mining while besieging the city of Larissa, an ally of Athens, in 401–399 BCE. The aggressors dug a shaft and tried to extend a gallery toward the water supply of the city to cut it off. But the defenders stole out at night and filled in the shaft with stones and pieces of wood. When the Spartans erected a wooden shed over the shaft, the Larissans set it on fire. The attacking force later raised the siege and moved on to a softer target.[13]

Despite employing a wide range of siege tactics, the Greeks resorted to mining only occasionally, which was not necessarily true of the Carthaginians. A force from that North African city-state besieged Himera in Sicily during 409 BCE and constructed an offensive gallery that collapsed a section of the defending wall. The Himerans managed to blunt the follow-up assault by fighting on the rubble. That night they moved most of their troops out of the city and attacked the enemy camp but were defeated. When the Carthaginians renewed their attack through the gap in the wall the next day, they captured Himera.[14]

Mining techniques began to appear in tactical manuals about this time. Aineias the Tactician, writing between 360 and 346 BCE, included one of the first discussions of siege-mining practice. He actually concentrated on protecting sappers, who worked at the base of a masonry wall, rather than on tunneling. Aineias also discussed various countermeasures and was aware of the metalworker at the siege of Barca who had detected the offensive mine there by using a shield. He also discussed countermining siege ramps in the manner already tried at Platea. But Aineias also had learned that in past sieges the defender had released wasps and bees into enemy galleries to annoy the miners. He further suggested piling wood shavings at the opening of an enemy mine, setting the pile afire, and trying to blow the smoke into the cavity.[15]

In the far-reaching operations of Alexander the Great, mining played little role. Although conducting numerous sieges, he preferred the quicker tactics of scaling walls or collapsing sections of them by ramming. One of the few examples of an Alexandrian siege that involved mining took place at Gaza in 332 BCE. The Macedonian force brought down one section of its wall through mining and another section by ramming, then captured the city with an attack.[16]

Although never universally successful, siege mining contributed to changes in the design of city defenses. While the increasing use of artillery and other siege tactics later played larger roles in this process, aggressive mining contributed to the digging of defensive ditches in front of masonry walls. Sources have documented this as a new feature in Greek fortification design by the onset of the post-Alexandrian period. To be truly effective against offensive mining, however, the ditch needed to be as deep as the wall's foundation.[17]

During the series of wars that led to the division of Alexander's empire after his death, mining failed to bring victory to the forces of Polyperchon at the siege of Megalopolis in 317 BCE. Even though the attackers managed to undermine the wall and, through burning combustibles, bring down three towers and the curtains between them, the defenders fought at every breach to prevent the follow-up attack from entering the city. Polyperchon eventually lifted the siege after inadvertently proving that technical success with a mining operation did not necessarily lead to tactical or strategic victory. The use of mines also failed Demetrius Poliorcetes in his siege of Rhodes during 305–304 BCE.[18]

Despite high-profile failures such as these, mining remained a suitable tactic for commanders to use in siege work. Philo of Byzantium's manual on fortifications and siege craft, *Poliorketika*, written about the 240s BCE, asserts its place in the tactical list of choices. "It is necessary to have recourse to tunneling secretly under walls, in precisely the manner which miners (of ore) now employ," he writes.[19]

In the many wars of Philip V of Macedon, according to Polybius, sieges were common, and mining seemed to play a larger role than usual. During the Social War of 220–217 BCE, in which Philip led several Greek city-states against others, he besieged the city of Palus on an island off the northwest coast of the Peloponnese in 218 BCE. Mining undercut two hundred feet (60.96 meters) of the city wall through the burning of props, but the follow-up attack failed due to treachery on the part of the men who led the assault. A year later, Philip laid siege to Thebes. Here, his miners encountered difficult ground, taking them nine days to reach the foundation of the city wall. Then, digging in shifts, they undercut two hundred feet of the wall, setting up props to keep the section in place. But the supports gave way before they had prepared the combustibles, and the wall only partly collapsed. The Macedonians worked to clear away what they could of the rubble in order to make a passage capable for assaulting parties to use, but the city gave up before they managed to mount an attack.[20]

If Polybius's report of the time it took for Philip's miners to reach the foundation of the Theban wall is accurate, it indicates that they dug a short gallery, especially if the ground proved to be difficult for their work. He noted that it took three days to undercut and prop two hundred feet of wall, which seems probable, especially if the men were working in shifts with an air of urgency. These details point to highly skilled and professional miners, probably recruited directly from the civilian workforce.

Philip's operations against Echinus during the First Macedonian War (214–205 BCE) proved to be the highest-profile siege of his reign. His forces constructed elaborate siege works, including two towers as high as the city wall with a connecting work between them for safe communication. The connecting structure was built in three stories. Philip also advanced what Polybius calls "two trenches" from this connecting gallery toward the city wall. Historians have tended to refer to these "trenches" as mines, but if they truly were trenches, they should be considered ground-level saps or approaches rather than gallery or underground mines. Regardless, the city surrendered before Philip completed his siege operations.[21]

During the Cretan War of 205–200 BCE, Philip's siege of Prinassus in Asia Minor hit an apparently insurmountable block to mining. When his miners began work in 202 BCE, they encountered so much underground rock that it proved impossible to proceed. But an ingenious ploy suggested itself. The men created noises as if they were digging during the day and at night dirt from elsewhere was hauled in and piled at the mouth of the mine gallery. After a few days of this, Philip informed the garrison that a long stretch of its wall had been underpinned and was ready to tumble down. The city therefore gave up.[22]

Philip's next siege, of Abydos on the Asian coast of the Hellespont, produced a successful mining operation in 200 BCE. But while it breached the city wall, the defenders had constructed a secondary wall at that spot and continued to resist. Macedonian miners then started another gallery against this secondary wall. The city offered to surrender, but Philip refused terms. This led the defenders to adopt extreme measures. The fighting men swore to resist to the death, and all agreed that elderly men should be responsible for killing the women and children and hiding the precious metals before the city fell. The Macedonians breached the secondary wall, but Philip's attack was stalled in the breach. At this critical moment, however, city leaders lost their nerve to continue resisting, yet the plan of self-destruction still took place with disastrous results. Philip captured Abydos but at a sickening human cost to its inhabitants.[23]

The higher profile of mining in the wars of Philip V of Macedon may indicate that this siege technique was rising in importance by the turn of the second century BCE. Yet it may simply be that Polybius, the historian of those wars, was particularly interested in the history of siege craft and was more ready to record its details than previous chroniclers. Still, Philip's mining operations were successful as many times as they failed, but the failures were not due to lack of planning, effort, or ingenuity on the part of the Macedonian diggers. If a commander wanted to employ mining on a regular basis, he could hardly have bettered Philip's record in its use.

According to historian Arnold W. Lawrence, the frequency with which commanders employed mining in Greek sieges was very low. While Lawrence's tabulation of eighty-two prominent sieges from 432 to 189 BCE is far from complete, it indicates that mining took place in only five of those operations. Other tactics, including ramming, escalade, or the use of towers, ramps, and mounds, were employed in roughly similar proportion, but the most common tactic was blockade and starvation.[24] In comparison with this tabulation, the frequency with which Philip employed mining in his various wars is striking.

From the earliest evidence of military mining in the ancient Near East to the writings of Polybius in the post-Alexandrian period, the rise, development, and spreading of siege mining across the ancient world is evident. Adopting the techniques, procedures, tools, and even personnel from civilian operations, gallery mining took its place alongside many other siege tactics, with a roughly equal chance of success as most of them except for blockade and starvation. Unlike all the others, mining was a highly technical solution to breaking into fortified cities, thus the need to find experienced miners to conduct it. The central irony of siege mining, that even a highly successful effort at breaching city walls could be followed by a failed assault through the breach, was well established during this era. But that was a universal fact of life no matter who, where, or how siege miners did their work.

2

Rome

With its emphasis on supporting large forces in the field over long periods of time and its tendency to favor hard preparatory work, the Roman army took mining seriously as an element of siege warfare. Engineers used the principles inherited from others, only more efficiently than their predecessors. Just as important, their work was highlighted by a growing number of surviving historical accounts and archaeological finds. With the Roman period, military mining came into its own a bit more prominently as both an element of siege warfare and as a historical phenomenon.

The Romans inherited civilian mining techniques from their Greek and Egyptian predecessors and added some new ones as well. They did surface as well as gallery mining and standardized the process across their vast empire so that similar pick work and dimensions of galleries can be found from one corner of its vast territory to another. When constructing shafts, Romans tended to use wood to line and support rectangular constructions but stone to support circular ones. For climbing up and down the shaft, they cut into the sides either foot holes or holes for baulks to build a wooden ladder. They used criminals, slaves, and later Christians as mineworkers. Archaeologists have found in galleries chains that were used to keep these enforced laborers on the job. Members of the local population were sometimes pressed as workers in the mines as needed, and there is evidence that soldiers also were used in the galleries.[1]

Introducing iron tools was one of the innovations to be seen in the Roman era of commercial mining. In terms of lighting, however, they typically used old technology. Pine splints or tapers have been found in Roman gold mines in Romania dating to the second and third centuries CE. These tapers were bunched together to provide dim light in the perpetual darkness underground. But as with the Greeks, oil lamps resting in niches or ledges of gallery walls also were used. Romans also used the Greek methods of ventilation that were based mostly on the natural tendencies of air to circulate if one created a sufficient draft. When digging near or just below the water level, drainage became a problem. The Romans ameliorated this by creating drains if the circumstances permitted or by using extensive labor to manually bale water out of the galleries. If the problem was too big for either of these methods, they created mechanical waterwheels or Archimedean screws, both in use by the first century CE.[2]

As mining historian John F. Healy has put it, the Romans in some ways were "more ambitious than the Greeks." At their extensive Rio Tinto mines in Spain, they constructed four levels of galleries, each one about twenty-five feet (7.62 meters) above or below its nearest neighbor. Like the Greeks, however, Roman galleries tended to be small in dimension. They were either rectangular in shape or trapezoidal and narrower at the top than the bottom. Archaeologists have found some arched roofs and enlarged width at about shoulder height. In fact, in some galleries the chisel marks have been smoothed away "by constant wear as they [miners] passed through the gallery." The most difficult problem Roman miners encountered related to underground navigation. They simply did not have the kind of instruments necessary to keep deep galleries straight and thus often sank additional shafts vertically from the surface along the line of the gallery to guide the ongoing horizontal digging.[3]

The most advanced stage of Roman commercial mining occurred during the late republican and early imperial period, but their armies employed mining in many sieges from early in the history of the city-state until the end of the empire. They frequently were successful in a technical sense, undermining and collapsing walls, but there is no evidence that follow-up attacks were any more successful on the whole than in the past. Roman commanders more consistently utilized the underground siege approach, but they could never guarantee it would result in the fall of the target city.

Livy records details of a Roman mining operation during the siege of Veii, the climax of a ten-year war between the two city-states in 396 BCE. Marcus Furius Camillus pushed a gallery through, the miners digging in six-hour shifts day and night. The intention was not to undermine the wall but to dig underneath it and emerge within the defended perimeter. In other words, the object was to use the gallery as a transit for soldiers. To facilitate the operation, Camillus attacked the wall to draw attention away from the miners when they opened the forward end of their gallery. It was a complete surprise. A force emerging from the gallery opened the gates and attacked the defenders as the city fell to the Romans.[4]

Mining played a role in the major Roman siege of Carthaginian Lilybaeum on the west coast of Sicily during the First Punic War (264–241 BCE). Initially, when the siege began in 250 BCE, the Romans made a concentrated effort to take the city, advancing siege approaches toward six towers, employing battering rams, and cutting off all land approaches. Livy provides no details of the mining operations except to note that the Carthaginians countermined the Roman galleries to neutralize their effectiveness. After the initial phase ended in stalemate, the siege lingered for nine years until a Roman naval victory cut off Lilybaeum's link with Carthage. Soon afterward a general peace concluded the fighting across Sicily, in which the Carthaginians agreed to evacuate the city. Livy also notes that the Carthaginian army under Hannibal employed mining in its siege of Casilinum in 216 BCE during the Second Punic War (218–202 BCE). The defending Romans foiled the effort through countermining.[5]

The most detailed description of mining operations to date comes from Polybius and relates to the Roman siege of Ambracia, the modern city of Arta in Greece, in 189 BCE. Describing the effort as "siege operations on an extensive scale," Polybius records the use of several battering rams, long-handled sickles to tear off battlements, and intensive mining. The Romans dug a trench parallel along about one hundred yards (91.44 meters) of the wall, covering it with wattle screens. From this they advanced at least one tunnel toward the wall, working in shifts night and day. This activity remained unknown to the defenders because of the wattle screens until the Romans became careless about disposing of the spoil from the gallery. When the Ambraciots noticed the fresh dirt, they began countermeasures. Preparing their own parallel trench inside the wall opposite the screened Roman work, they dug it as deep as the wall's foundation. The defenders lined the outside wall of this trench with very thin plates of brass so they could place their ears close to it at any location and listen for sounds of digging. By the time they pinpointed the active area and dug a countermine, the Romans were propping up a cavity to collapse the wall.[6]

At this point the first reliably recorded combat underground in military mines occurred. The Ambraciots initially tried to deal with the Roman miners by using pikes, but this proved ineffective in the confined space. The stalemate was broken when the Ambraciots came up with an ingenious idea. They found a large corn jar nearly as big as the tunnel dimensions and bored a hole through its bottom. Inserting a hollow tube through it nearly to the top, they filled the jar with feathers and placed pieces of smoldering charcoal around the feathers along the sides. Covering the jar with an iron top punctured with holes, the Ambraciots then inserted it top first into the tunnel as near the Romans as possible. They then sealed the passage as much as possible, leaving room only for two spears to go through to keep the Romans away from the jar. Attaching a blacksmith bellows to the end of the tube sticking out from the base, they were able to blow air into the jar to encourage the charcoal to burn the feathers, pulling the iron tube back to ignite more feathers as needed. The scheme worked and is the first recorded use of a chemical weapon in military mining. This countermeasure saved Ambracia at the time, but a negotiated truce took place soon after in which the city agreed to give up on terms.[7]

The Roman siege of Ambracia and the spirited reaction of the Ambraciots highlights a complicated and sophisticated underground attack and defense. We can assume that at least some previous sieges in different cultures were probably as sophisticated and spirited yet lack firm documentation. Ambracia provides the first solid evidence of underground combat between besieger and besieged as well as the first documented use of what could be termed a chemical solution to the problem of neutralizing an offensive mine gallery. But it is reasonable to assume that similar actions were more rare than frequent across the long sweep of previous mining history.

There is much evidence that mining and countermining of what might be termed an ordinary sort compared to that of Ambracia was very common in Roman wars. Sulla's siege of Athens during the First Mithridatic War (89–85 BCE) employed several different siege tactics and lasted from 87 to 86 BCE. Relying primarily on starvation to bring down the city and its anti-Roman Greek coalition led by Mithradates VI of Pontus, Sulla concentrated active siege approaches against the port city of Piraeus. Defenders undermined the siege mound and quietly excavated a cavity that caused it to subside, but Roman workers managed to fill it up again. Then Sulla's forces began mining operations, which were met with a countermine. According to Appian, underground combat ensued when the two sides' galleries met, soldiers fighting "with swords and spears as well as they could in the darkness." The Romans tipped the scales in this strange fighting and continued pushing their gallery forward until under the walls of the city. They propped a long section with wooden beams and used "sulphur, hemp, and pitch" as combustibles. Once set afire, the effect was prolonged but successful. "The walls fell," writes Appian, "now here, now there—carrying the defenders down with them. This great and unexpected crash demoralized the forces guarding the walls everywhere, as each one expected that the ground would sink under him next. Fear and loss of confidence kept them turning this way and that way, so that they offered only a feeble resistance to the enemy."[8]

Despite this technical and emotional triumph, mining failed to deliver a clear victory for Sulla. The Greeks had already constructed secondary works in this sector, and the warriors fought well enough to stall the enemy attack. The siege continued, as Sulla's troops conducted extended operations against the city's secondary line by sapping under the cover of penthouses, bringing down part of its still moist and uncured wall. The Romans wore down the Greeks and eventually captured Piraeus.[9]

While events transpiring above ground in a siege tend to be recorded in detail, what occurred underground typically remains murky in the pages of ancient historians. None of those underground events are more subject to vague reports than the topic of combat in mines. Appian relates improbable stories about gallery fighting during Lucius Lucullus's siege of Themiscyra during the Third Mithridatic War of 73–63 BCE. Offensive galleries were so large "that great subterranean battles were fought in them." He continues: "The inhabitants cut openings into these tunnels from above and thrust bears and other wild animals and swarms of bees into them against the workers."[10]

While Appian provides more coverage than is usually found in the vague references to underground combat, his details are not convincing. How the defenders could gain access to the Roman gallery from above and manage to convince bears and other wild animals to enter it are not explained. We must consign such reports to the realm of imaginative war stories rather than history.

The Gallic Wars of Julius Caesar from 58 to 50 BCE provide some information on the relative use of mining compared to the employment of other siege techniques and field fortifications during an intensive war of territorial conquest late in the republican period. Although written primarily for propaganda purposes and with grossly exaggerated reports of numbers of troops, Caesar's book on the subject contains a great deal of detail about the military operations that can be considered reliable. The Romans pursued this series of campaigns with heavy reliance on fortified camps and other types of field fortifications and only rarely resorted to siege approaches. Even more rarely did they attempt mining.

Interestingly, Caesar recorded several instances of Gallic tribes employing mining and countermining. Belgic tribes, for example, in attacking a Roman garrison at the town of Remi in 57 BCE, attempted to sap the wall, using shields as an overhead cover. Caesar notes that this was a typical technique among his enemy while trying to enter a fortified city. When Publius Licinius Crassus invaded Aquitania in 56 BCE, he besieged the capital city of the Sotiates. The defenders countermined a siege ramp, using civilian copper miners to do the work, but the Romans neutralized the effort and captured the city. During Caesar's siege of Avaricum, a city of the Bituriges Cubi, during 52 BCE, the Roman siege mound was countermined as well. The Bituriges "[have] extensive iron mines in their country and are thoroughly familiar with every kind of underground working," Caesar reports. He goes on to praise the wide array of countermeasures taken by his Gallic adversaries, noting that when his men began to dig an offensive mine, it was met by effective countermining. Caesar writes that the Gauls planted sharpened stakes and dropped rocks and boiling pitch into the Roman gallery, although he does not provide details about how they accomplished this.[11]

At the Roman siege of Uxellodunum in 51 BCE, Caesar directed a mine gallery dug not to undermine a wall, but to tap into the source of a spring that provided water for the Carducis and Senones who defended the site. Centuries afterward, an exploration team sponsored by Napoleon III of France discovered this gallery in 1865. It was 1.8 meters (5.90 feet) tall and 1.45 meters (4.75 feet) wide and filled almost to the top with mud washed in by water. The gallery followed a stratum of workable material until encountering tough bedrock and diverted in a different direction before reaching its target. A modern estimate has indicated that the gallery probably could have been dug at a rate of 0.75 meters (2.46 feet) per day, which would have required a total of fifteen days to complete. That would amount to 11.25 meters (36.90 feet) in length. The vertical wooden props and horizontal roof bracing remained intact. It is the only Roman offensive gallery that has been recovered to date. And it proved successful in its purpose of cutting off the defenders' water supply.[12]

Caesar's detailed operational history of his Gallic Wars indicates that mining and countermining took place in relatively few places compared to the intense and widespread operations the Romans embarked upon during this turbulent period. Moreover, the rate of success of these mining efforts was not exceptionally high. Countermeasures often worked, and it is interesting to note how often the Gauls conducted mining and countermining operations of their own, banking on their experienced copper and iron miners to lend their skills to military purposes.

A century later, during the massive Jewish Revolt that broke out in 66 CE, Roman siege operations centered on Jerusalem. They serve as a good case study in mining and countermining. But even before the revolt, the city had been the site of several siege and countermining efforts. When Herod the Great besieged the city in his Roman-supported effort to assert claims to being king of Judaea in 37 BCE, it was protected by two wall circuits. The older one circled the Davidic city and the newer one the Upper City. The

Jewish defenders constructed secondary walls at the point of Herod's siege approaches and conducted sallies. They also countermined, according to Flavius Josephus, who unfortunately does not provide details of the mining operations. After capturing Jerusalem, Herod used it as his capital. He strengthened its defenses by constructing two citadels, reinforced the wall circuits, and rebuilt the temple. One of the citadels, Antonia, was square, with four corner towers that also were square. In 41 CE Herod Agrippa I began the construction of a third wall, intended to enclose the New City, but stopped when the Romans protested.[13]

The uprising that occurred in 66 CE led to the quick takeover of most of Jerusalem by the insurgents, known as Zealots. What was left of the Roman garrison and some loyal Jewish allies took refuge in the old palace of Herod the Great in August. The insurgents started a gallery mine some distance away from this site because they could not get closer, although Josephus does not indicate exactly how far this gallery had to go. The miners aimed at a tower of the palace and succeeded in digging out a cavity under the foundation. They shored it up, filled it with combustibles, and set it afire. By the time the tower collapsed, however, the Romans already had constructed a secondary line of defense. Despite this, the garrison evacuated the palace and took refuge in the remaining three towers but soon were forced to give up. After winning control of all Jerusalem, the Zealots completed the construction of the third wall around the New City, making it a massive fortification.[14]

The Romans brought three legions of troops under Titus to initiate the most famous siege of Jerusalem, occurring in 70 CE. Working quickly, within fifteen days of the start, a Roman tower had breached the outer wall, and the defenders retired to the second circuit. Here the Romans advanced two siege ramps with towers toward the Antonia Tower; at least one of them was made mostly of timber. The Zealots countermined and filled a cavity under the ramp with "pitch and bitumen," according to Josephus. The resulting fire consumed all the wood from the ramp that fell into the cavity, and the Romans did not even try to put it out.[15]

Two more ramps appeared, also aimed at the Antonia Tower, as siege approaches continued in relentless fashion. The Jews had not made the effort to fill in their countermine, and that now came to haunt them. It was so shallow that the wall of the tower began to crack as its weight bore down on the abandoned tunnel. The Romans noticed this and started sapping the wall under the cover of a testudo, and during the night, a section of the wall collapsed. Initial attacks failed because a secondary wall was now in place at this location. A second night attack also failed, but two days later the third effort captured this secondary wall, allowing the Romans to penetrate to the temple. Eventually, and in a very ironic twist, Jewish countermining had played an important and unintended part in the Roman capture of Jerusalem by September 8, 70 CE.[16]

While mining and countermining played a significant role in high-profile operations such as the Gallic Wars and the 70 CE siege of Jerusalem, we know far more about its role in the siege of the relatively obscure city of Dura-Europos on the west bank of the Euphrates River in 256 CE. The Romans had been involved in this area since about 165 CE and began to greatly strengthen the city walls after 210. They constructed a massive rampart along the inside of the west wall, which was the only sector not protected by gullies and ravines or the river itself, and in the process buried a number of houses and other buildings. The rampart was fifty feet wide (15.24 meters) at the base and rose up to the top of the thirty-foot-tall (9.14 meters) city wall.[17]

A Sasanian Persian force besieged Dura-Europos in 256. Even though there are scant textual accounts, the well-preserved site of the old city has yielded a rich archaeological record that surpasses that of any other siege in ancient history. The most valuable part of these findings is physical evidence of intense underground combat in the mine and countermine galleries discovered by a joint French-American expedition headed by Robert du Mesnil du Buisson in the early 1930s. This evidence was not fully interpreted until eighty years later, yielding a dramatic story of military mining.[18]

The Persians constructed several approaches to the west wall of Dura-Europos. One mine gallery aimed at Tower 19, which was located midway along the wall, which ran 856 meters (936.13 yards). A second mine gallery aimed at Tower 14, located at the southern end of this wall. These two galleries were designed to create a breach in the wall, while a third gallery was clearly designed for the movement of troops. This one was much bigger than the other two, 10 feet (3.04 meters) wide and 67 inches (170.18 centimeters) high, dimensions that compel us to assume it was meant to gain entry into the fortified enclosure. The Persians also constructed a siege ramp between Tower 14 and Tower 15, though much closer to the latter. They started it about 100 feet (30.48 meters) from the wall. Although judged to have been finished by the builders, this ramp does not appear to have a leveled place for siege artillery on it and probably was meant for use only by foot soldiers. The large transit gallery was dug directly under the ramp to lessen the chance that the Romans would discover it. The gallery was barely deep enough to go under the wall's foundation and aimed at emerging just beyond the interior edge of the rampart. The Romans were aware of the first two mines and of course the ramp, but there is no evidence that they suspected the third (transit) gallery's existence.[19]

The mine aimed at Tower 14 began in a ravine about 130 feet (39.62

meters) to the west. It was "a narrow, twisting tunnel, unsupported by wood and scarcely the height of a man," according to Clark Hopkins, who worked with du Mesnil in the excavations. Digging through soft rock, the miners pushed through two burial chambers on the way to Tower 14, where the gallery rose closer to the surface and widened. The Persians then dug a branch gallery to the right toward a ravine in order to create an air draft to ventilate the main tunnel. They would have needed ventilation, for typically galleries could not extend more than about 60 feet (18.28 meters) before the air became unbearable. The Persians excavated under the tower, propped the gallery, and fired it, but the effort only resulted in the slumping of the tower rather than its collapse.[20]

At Tower 19, however, the most dramatic physical evidence of mine warfare emerged. Du Mesnil, a military officer and self-taught archaeologist, and his team found the remains of nineteen Roman soldiers and one Persian soldier in the gallery. They recorded the evidence and then covered it up again, with du Mesnil correctly assuming an underground battle had taken place. Not until decades later did other archaeologists examine the archives of the expedition and argue convincingly that the evidence denotes the use of a chemical agent in the mine gallery. In short, the Persians gassed the Romans in order to stop their countermining effort.[21]

The Persians started the mine aimed at Tower 19 in a burial mound at the necropolis located west of the city wall. A mound of earth hid the entrance about 40 meters (131.23 feet) from the tower. This would have allowed them to gain access to the softer soil below the layer of limestone, about 1 meter (3.28 feet) thick, that lay just below the surface in this area. The gallery went under the northern half of the tower and then veered toward the north under a section of the curtain. The Romans began a countermine against this gallery. The limestone level was deep enough here to allow the countermine to be dug above it, with the miners having to shore up the gallery with timber posts and lintels. The countermine was therefore about 3 meters (9.84 feet) higher than the Persian gallery. French excavators found the shoring intact 2,188 years later and photographed it. They also found the shoring of the Persian offensive gallery at Tower 19. It consisted of "two lines of round hardwood posts 0.10–0.11 m [0.32–0.36 feet] in diameter and about 2 m [6.56 feet] in length sawed straight at the two ends." Planks up to four centimeters (1.57 inches) thick were laid on top of the posts to cover the roof of the gallery.[22]

Du Mesnil uncovered only two-thirds of the Roman countermine gallery because he placed higher priority on civilian sites in the city rather than on the military history of the siege. But his team uncovered a complicated set of evidence where the countermine neared the Persian gallery. Here, the

Shoring Timbers of a Roman Countermine, Dura-Europos. Uncovered by
a joint French-American team in the 1932–1933 season, this Roman countermine aimed at intercepting a Sasanian Persian offensive mine headed
toward Tower 19 during the siege of 256 CE. The shoring timbers, 2,188
years old, were well preserved. The large block of stone in the foreground
was placed there by the Persians in an effort to block further Roman
progress. Dura-Europos is the earliest siege in history for which physical
remains of underground workings has been discovered. Yale University
Art Gallery, Dura-Europos Collection, neg. F-IX-43.

Roman gallery was 1.2 meters (3.93 feet) wide between the timber posts and
only 1.65–1.75 meters (5.41–5.74 feet) high, "barely enough to permit many
adult males to stand upright," in the words of Simon James, who has ana-
lyzed the expedition archives. Just east of its western end the team found a
burned zone scorched by fire. Du Mesnil discovered in the 1932–1933 season
the remains of nineteen Roman soldiers piled up just east of this area. He
also found a body that he identified as that of a Persian during his work in
the next season; it also was located just east of the burn zone. This skeleton

Bones of a Persian Soldier, Dura-Europos. Convincing evidence indicates that these are the remains of a Persian killed in a successful use of chemical warfare to stop the progress of a Roman countermine near Tower 19 during the siege of 256 CE. The remains were uncovered by a joint French-American expedition in the 1932–1933 season. Yale University Art Gallery, Dura-Europos Collection, neg. G-908.

had its legs burned off, while the Roman remains were "incompletely decomposed," with some hair, linen, and leather intact. Coins, nails from the timber shoring, and burned pieces of wood lay about. The western part of the Roman body stack was heavily burned, partially burned in the middle, and not at all burned on the eastern side. Excavators found a desiccated brain in one skull on the east side, and "the bodies here still reeked" of odor. There were shields for only half of the Romans found; James speculates that the other half were not soldiers but civilians or perhaps slaves who were there to pass buckets of dirt forward to extinguish the fire that had obviously occurred.[23]

Du Mesnil acknowledged that the Roman countermine was stopped just before intersecting the Persian offensive gallery and assumed the Persians had tunneled into it. They then set fire to the timber shoring to block Roman access to their offensive mine. He also assumed that the Romans responded to this by collapsing their end of the countermine to deny their enemy its use. The bodies, Du Mesnil speculated, were those of men who had not been able to evacuate the countermine before it was collapsed in haste, the Persians killing them either by the sword or by firing the timber shoring.[24]

James, however, came to a different conclusion after examining the detailed description of the body stack provided by du Mesnil. He believes the bodies were deliberately piled up by the Persians, with shields placed on top, to serve as a physical barrier to Roman movement forward toward the junction of the countermine and the offensive gallery. James then speculates that the burn zone had been created by a chemical agent. The famous example of inserting smoke at the siege of Ambracia served as an inspiration for James's conclusion. Sulfur crystals and bitumen creates hydrocarbon smoke (containing carbon dioxide, carbon monoxide, and sulfur dioxide gases) as well as sulfurous acid when making contact with the moisture of the eyes and nasal mucus. In this case, however, James assumes the Romans were killed in hand-to-hand combat before the burn zone was created, although he wonders how many actually could have been dispatched in the narrow confines of this gallery. The Persian skeleton with burned-off legs found east of the burn zone apparently was that of the man who set off the chemical device that started the fire; he must have made a mistake and could not get away fast enough. The chemical device was used, then, not to kill the Romans, but to start a fire that would destroy the timber shoring and collapse the countermine gallery. Du Mesnil also found some stone balls in the countermine gallery, and James speculates that the Romans, before collapsing their end of it, shot artillery through the gallery at the Persians after they broke through at the western end.[25]

James's interpretation is intriguing and mostly convincing, although it also rests to a degree on speculation. In the end, the result of du Mesnil's efforts leaves us with some unanswered questions. Was a chemical weapon actually employed by the Persians at Dura-Europos? Could they have burned the timber shoring just as easily with traditional methods? Du Mesnil and his team reburied the remains he found but did not record whether this was in the covered-up gallery or in a nearby cemetery. James speculates that modern osteoarchaeological examination of the remains might reveal much more of what happened inside the countermine gallery. For example, it could verify that the Romans were killed by edged weapons, which would support James's conclusion. We also do not know exactly how the Persians succeeded in breaking into the city, although there is no doubt that they captured it in the siege of 256. After blocking the Roman countermine, the Persians went on to prop and fire not only the western part of Tower 19 but also a section of the curtain north of it. This caused part of the tower to slump more than eight feet (2.43 meters), although everything remained upright. Apparently, the mines at Tower 14 and Tower 19 were ineffectual, and historians have speculated that the Persians broke through the Roman defenses by using the siege ramp and the third, larger transit mine between Tower 14 and Tower 15 instead.[26]

The physical evidence of underground warfare at Dura-Europos is unique and important. While there are remnants of galleries and countermines at a handful of other siege sites from the classical era, none of them provide the remains of soldiers and civilian mineworkers. But one must keep in mind that Dura-Europos was, after all, one episode in a long history of military mining. As the Roman Empire waned and was assaulted by waves of European and Asiatic raiders, the history of military mining also waned. Barbarian military forces were incapable of conducting sophisticated siege operations due to lack of numbers, fragile logistical and supply systems, and lack of technical expertise. They usually tried blockade rather than siege approaches to reduce a fortified Roman city. Treachery proved more effective than mining.[27]

But when Roman forces conducted siege operations during the last few centuries of the empire, they continued to employ mining on occasion. In the power struggle between Julian and Emperor Constantius II, Roman forces loyal to Julian besieged a garrison loyal to the emperor at Aquileia, in modern-day northeastern Italy, in 361 CE. "A suitable place could nowhere be found for moving up rams, for bringing engines to bear, or for digging mines," writes Ammianus Marcellinus. The garrison gave up on receiving news of the death of Constantius, but this account certainly indicates that the besieger was seeking to use not only mining but also other types of siege approaches. After coming to power, Julian launched a campaign against Sasanian Persia. During the Roman siege of Ctesiphon, the capital of the Sasanid dynasty, in 363, mines proved effective. The Romans cleverly conducted sapping operations against the wall near the mining gallery so that "the clink of the iron tools" chipping away at the stones above ground

Crack in Tower 19, Dura-Europos. Here, during the 1932–1933 season, Robert du Mesnil du Buisson records information about the Roman countermine at Tower 19. Du Buisson was head of the joint French-American team that uncovered evidence of offensive and defensive mines in the siege of Dura-Europos in 256 CE. This evidence was soon covered up again as the team devoted more time to nonmilitary aspects at the site. Sasanian Persian besiegers managed to damage Tower 19—the evidence seen clearly in this photograph—after stopping Roman countermining efforts to protect the tower. Yale University Art Gallery, Dura-Europos Collection, neg. F-XIII-43.

would hide the much softer noise of digging underground. According to Ammianus Marcellinus, this underground gallery was not designed to bring down the wall but to gain entry into the enclosure. It emerged into the room of a house, probably a guard house just inside the wall, and allowed the Romans to capture the city. Zosimus writes of a similar use of a mine in the reduction of a citadel near a city he calls Besuchis, apparently conducted at the same time as the siege of Ctesiphon. There, too, the Romans dug a gallery to gain access to the interior of the fortified enclosure, emerging "inside a house," to capture the citadel.[28]

Doctrinally, the techniques of mining continued to be recorded along with all the other elements of siege craft. Vegetius, who wrote his prominent military manual *Epitome* sometime between 383 and 450, recommended wide and deep ditches around the outer edge of city walls to deter offensive mining. He also urged countermining as the best way to deal with a siege ramp or mound. Vegetius included nothing new on the techniques of sapping and mining, but he briefly discussed both methods of breaching masonry walls.[29]

As long as relatively advanced cultures waged war during this late Roman era, one saw mining and countermining appear as part of siege craft in action. The Byzantines continued a series of conflicts with Sasanid Persia well into the sixth century, and both sides employed these techniques. In the Persian siege of Dara in 503, the problem was in penetrating two wall circuits, not just one. The Persians started a mine gallery to approach the east side because only here was there no rock layer just under the surface. As recorded by Procopius, the attackers began their gallery "from their trench" and dug it "very deep" to avoid detection. They had progressed beyond the foundation of the first wall and were in the fifty-foot (15.24 meters) space between it and the second wall when a man among the Persian force surreptitiously warned the defenders of the mine. While pretending to collect missiles shot earlier by the Byzantines, he taunted the defenders and, through his words, managed to convey the treacherous information without arousing the suspicion of his comrades.[30]

The Byzantines immediately began to countermine. Procopius asserts that a man among them named Theodorus, who was "learned in the science called mechanics," suggested they dig the gallery "in a cross-wise direction" in relation to the Persian offensive gallery. They did so, making it quite deep while traversing the area between the two walls. The tactic worked. The Persians unwittingly intersected the countermine gallery and were attacked by defensive forces, who killed some and drove the rest back to the entrance of the mine; the Byzantines "decided by no means to pursue them in the dark." Soon after, the Persian commander opened negotiations and accepted a payment in silver for breaking off the siege.[31]

The defenders scored an impressive success at undermining a siege mound during the Persian siege of Amida, located on the border between Mesopotamia and Armenia, in 502–503. The city was not garrisoned by Byzantine soldiers, but its population was well organized and dug a countermine under the Persian ramp, excavating a large cavity without disturbing the surface appearance of the mound. When attacking soldiers crowded onto it, much of the area over the mine collapsed, killing many who fell into the hole. This success, however, failed to save Amida. A few days later a Persian discovered a breach in the wall that was not properly blocked up, and this allowed the attackers to gain entry into the enclosure to capture the city.[32]

A somewhat similar course of events transpired at the Persian siege of Edessa in 544, when the Byzantines dug a countermine under a siege mound, aiming at the middle of it. The Persians heard the digging as the countermine gallery neared its target and dug two shafts down to intercept it. When the Byzantines discovered this tactic, they stopped digging, filled up the outer end of their countermine, and redirected the gallery toward the near edge of the mound. They excavated a chamber just under the surface of the mound and filled it with dried timber soaked in "oil of cedar and added quantities of sulphur and bitumen." But after lighting their combustible, it went out before the flames spread to the mass of timber the Persians had used as the foundation of their mound. The Byzantines shoved more of their own wood in and restarted the fire. This time it spread, and smoke began to break through holes in the surface of the mound. To cover these telltale signs, the defenders threw small pots filled with coal fire and shot flaming arrows into the area to confuse the Persians. When the attackers realized what was going on below ground, they poured water through the smoke holes, but that failed to extinguish the smoldering flames in the heart of the mound. When they put dirt on the holes, it forced the smoke to break through at other places on the surface. Finally, the Persians abandoned the siege mound and instead conducted several attempts at escalade along the circuit, all of which failed. After this, they accepted a cash payment and lifted the siege.[33]

While the Byzantines conducted their countermining with determination and success, the siege of Petra in 549 demonstrated their adeptness at offensive mining techniques as well. The attackers brought down a section of a curtain between towers, but a substantial building happened to be located just inside the enclosure at this point; it served the purpose of a new wall. In their second attempt the Byzantines excavated a large chamber beneath a different section of curtain until that portion of the wall "stood for the most part over empty space," as Procopius puts it. They delayed firing the combustibles in this large cavity, which they abandoned when forced to lift the siege by a Persian relief force.[34]

But the Byzantines returned to Petra in 550 and mounted another siege. In the intervening months the Persians had filled in the cavity beneath their wall with gravel. But before that, they also had installed an innovative twist to reinforce the wall's foundation. Using heavy timbers and binding them together, they positioned this wood framework directly under the stone foundation. When the Byzantines dug another mine at this spot, which was the area that had the least bedrock, they were surprised to find the gravel and the timber framework. Working through these difficulties, their miners managed to excavate enough of the old cavity to cause the wall to fall. But when it did, the timber framework allowed a large section of the wall to sink intact. The result was that no breach of the circuit occurred, and the offensive mine proved a failure. Meanwhile, the Byzantines had developed three portable rams light enough to be moved about by forty men and brought them to an area of the circuit where the ground was too rugged for large battering rams. Through pounding, they managed to open a hole in the wall. After much hard fighting, the Byzantines captured Petra in 551.[35]

The long period of Roman history provides more detailed information about military mining than all previous eras. It is not surprising that the Romans, renowned for their engineering and practical application of technical expertise to the problems of siege craft, should excel at mining. But it is also interesting to see that mining techniques were shared by other cultures with sophisticated military systems. In this group we need to include the Gauls. Even though Julius Caesar tended to portray them as culturally inferior to the Romans, his detailed book on the Gallic Wars clearly demonstrates that they also knew how to mobilize large military forces and were conversant with operational theory and practice. Like everyone else, the Gauls saw the link between civilian mining and military mining and understood how to use tunneling both offensively and defensively. The Sasanian Persians also fully understood how to use mining to gain a fortified objective, as archaeological evidence of the siege of Dura-Europos vividly shows.

Our understanding of military mining in the classical era is crucially based on the accounts by historians who lived at the time and on archaeological evidence. But it also is severely limited by those sources. Most of the ancient historians paid relatively little attention to mining. Even Procopius fails to give us the kind of detail to be found in modern sources on the subject. While a handful of sites, most spectacularly Dura-Europos, have yielded important archaeological evidence, they collectively represent only a tiny proportion of the many sieges that took place during the long era of Rome's rise and fall.

The sporadic nature of mining—its employment in some sieges and not in others—is in part a reflection of the sporadic evidence to be found in available sources. But that cannot adequately explain why mining appeared in some sieges and not in others. To start mining, the besieging force needed many things, including secure logistics to feed its own men for the often lengthy time it took to dig galleries, access to technical experts from the civilian mining industry to guide and manage the digging, a hope that investing the time and effort returned a good chance of success, the right soil conditions and tactical circumstances, and a commander who had faith in mining's prospects.

While these conditions did not exist fully in all siege situations, they did often enough to make it clear that military mining was a significant component of siege craft throughout the classical era. Whether the knowledge of that technique would be carried on to successor powers in the Mediterranean region or into the heart of Europe or the Near East remained to be seen.

3

Medieval Mining

During the long medieval period, the trajectory of both commercial mining and military mining followed roughly similar courses. Both witnessed a sharp decline in activity during the early centuries, as European and Mediterranean civilizations dealt with the demise of the old Roman order and the slow, painful formation of whatever was to come in the way of ordered and sophisticated polities. Both strains then witnessed real progress by the thirteenth century, when commercial and military mining rebounded. Siege mining then assumed something like its former level of prominence in operations for the rest of the medieval era.

Another feature of this era was the shifting of siege mining into the heart of Medieval Europe as the many wars of the period encouraged growing operational sophistication. Mining experienced resurgence in the Near East with the onset of the Crusades, and it expanded into the British Isles as well. Chinese military history provides evidence that siege mining was a common practice in the Far East; at least there is ample evidence that technical manuals described its procedures in terms very similar to how it had been done in the Near East and the Mediterranean world (and now adopted in Europe as well). There is no evidence of cultural transference of siege techniques from the West to East Asia, so concurrent development must explain why Chinese besiegers operated along mostly the same lines as their counterparts on the other side of the globe. We have too few English-language sources to gauge the level at which tunneling played a role in siege operations and how often it was successful in China. It is not even clear how digging, propping, and firing a mine brought down earth-tamped Chinese walls as opposed to European masonry walls, but there can be no doubt that it was done.

Civilian tunneling declined during the early medieval period, with only "very modest works" being recorded, mainly for drainage or to gain access to water sources. But by the middle of the medieval period, what has been described as a "metal-mining industry" was developing from modern southern Germany through areas of southeastern Europe. Technical advances to improve the extraction of minerals and ore followed, but perhaps the most important new development was the creation of a specialized labor force to work those mines. In contrast to the Greeks and Romans, who largely used slaves and prisoners in their galleries, medieval mine owners employed a free labor force with special skills. Those miners were respected and well paid and thus posed as a readymade labor force that military commanders could hire for a specific siege operation. In fact, the development of a skilled labor force in the civilian mining industry was the key to the resurgence of military mining. Common soldiers and poorly read commanders were incapable of understanding or executing the digging of shafts and galleries. Hiring people who knew how to dig offensive galleries quickly became the best way to add the tactic of underground approaches to blockade, escalade, and treachery in the conduct of sieges.[1]

A new tool appeared in medieval commercial mining called the hammer and gad. It was a handheld pick that had a point on one end of the head and a hammer on the other end. One could use another tool, a sledgehammer, to drive the point of the hammer and gad into rock or soil. The size and length of commercial galleries were slightly different than in previous eras. Medieval galleries tended to be narrower than those of the Romans and were often coffin shaped to accommodate the loads that miners carried on their upper bodies. Galleries tended to be longer in the medieval era than in previous time periods, although exact measurements are scarce. Side tunnels for drainage or ventilation were added at sites that held the promise of extensive deposits and that were worked over long periods of time. In the silver mines located at Bere-Ferrers in Devon, England, which were worked at least from 1292 to 1348, many shafts and layers of galleries appeared in arrangements similar to the largest mine operations of the Roman era.[2]

The decline of military mining in the early medieval era has been well documented. Sieges were frequent during the early Carolingian period of the eighth century, but historians find little if any mention of tunneling.

Besiegers preferred to rely on escalade or blockade to achieve their ends. Subsequent time periods and other cultures yield some brief mention of sapping operations. For example, the Frankish siege of Bergamo in 894 witnessed the sapping of a wall, which led to its collapse and the capture of the city, while the Viking siege of Chester in 918 included a failed attempt to sap that city's wall. But no gallery mining is known to have taken place at these or other sieges during this era.[3]

But military mining experienced a resurgence by the twelfth century— the High Middle Ages—along with agriculture and other aspects of European culture. Charles Oman has argued that mining reached its prime in the medieval period by the thirteenth century. But he too often conflated sapping operations with tunnel mining, and thus we have to conclude that both siege techniques saw an increase in their occurrence within medieval siege craft by this time. As Michael Prestwich has reminded us, sapping and tunneling were intimately paired during this resurgence. He found no examples of long gallery mining, only the digging of short tunnels from positions very close to the defending wall. Mines also could be started in the moat fronting a wall, as happened at the siege of Dover in 1216.[4]

The increased use of sapping and mining contributed to changes in the design of masonry walls. Sharp angles in the wall proved vulnerable to both operations, as at the siege of Rochester in 1215. The result was that round towers became more popular. Thickening of the lower portion of a wall, especially its foundation, occurred also as did deeper and wider ditches in front of it. In some rare cases a series of buttresses supporting the interior of a wall greatly strengthened it against sagging or collapsing in case of undermining. As archaeologists have discovered, ditches that accompanied city walls in England during this period varied in dimension, but all were sizeable excavations. Bristol's defensive ditch was dug in the 1230s at 15.8 meters (51.83 feet) wide, while that protecting Coventry was 10.5 meters (34.44 feet) wide and 2 meters (6.55 feet) deep. Coventry's ditch was flat on the bottom but had steep sides. London's ditch, dating from the early thirteenth century, measured 22 meters (72.17 feet) wide and was located from 12 to 18 meters (39.37 to 59.05 feet) in front of the wall. A second ditch was dug between the early fourteenth and late fifteenth century that was 5 meters (16.40 feet) from the wall and 12–18 meters (39.37–59.05 feet) wide. Ditches such as these not only impeded mining but served as obstacles to escalade as well.[5]

Defensive countermining resurfaced along with the resurgence of offensive mining. Counterminers knew it was better to approach the siege gallery from beneath rather than from above, if for no other reason than they could insert smoke into it more easily. In some cases defenders used old tactics such as tunneling out to burn siege engines. This was attempted in the siege of Crema (1159–1160) in Lombardy. As was true of the classical era, historians of the medieval period often assume that hand-to-hand battles took place when countermine met offensive mine, but they usually provide few details. There is a report of chivalric combat between the Duke of Lancaster, who superintended the digging of offensive galleries at the siege of Limoges in 1370, with a defending knight named Jean de Vinemeur. In another instance Princess Anna Comnena recorded the use of chemicals by the Byzantine defenders of Dyrrachium against Norman besiegers in 1081–1082. They mixed pine rosin with sulfur and blew it through reedpipes into the offensive gallery to drive out the miners. Although archaeologists have uncovered the physical remains of countermines and offensive mines at a handful of medieval siege sites, none of them have yielded any material evidence of underground combat.[6]

Modern historians are somewhat at odds when evaluating the role of mining throughout the middle and late medieval periods. Some argue that it was not the first choice when deciding how to conduct sieges, and that conclusion rings true. Blockade, escalade, and bribery seem more common than sapping and gallery mining. Historians differ markedly, however, when evaluating how successful mining was when employed. John France argues that it "was the most consistently successful tactic used against fortifications" during the period 1000–1300. Still, he admits that there were failures and that local conditions affected mining operations. France believes that the rock present at Dover actually made it easier for miners because they did not need to prop up the gallery they excavated there. Charles Oman has assumed that if any mine was started close to a wall, the defenders would easily see signs of it and countermine the effort. But if the attacking commander started the mine at a distance and hid it from view, "he had a very fair chance of success."[7]

France's assertion that mining tended to be more successful than other methods of attack is not convincing. The techniques of sapping and mining were no different in the medieval era than they were in the classical era, and we have already seen that during the earlier period, neither tactic was consistently successful. Oman's assertion that starting a mine at a distance and concealing it would increase the chances of success is logical, but he provides no details about the length of gallery mining. The fact that mining was not the default choice of the besieger follows from the time it added to a siege, which was one of the more costly modes of operation for armies. If mining was widely viewed as having a high probability of success, commanders might have been willing to invest the time and resources needed to conduct it as long as they could afford to feed their army and did not mind giving up opportunities for other operations in the meanwhile.

Conclusions by historian Eric McGeer about Byzantine siege practices during the early medieval era also fail to convince that sapping and mining were high on the list of preferred siege techniques. McGeer argues that "tunneling to undermine the foundations was the tactic favoured by the Byzantines," citing the use of mining to capture Candia in 961. But a single instance does not support a general conclusion. McGeer also argues that the military manual written by Hero of Byzantium places emphasis on tunneling, so this must imply that Byzantine artillery was too weak to be relied on in siege operations. But based on McGeer's discussion, it seems that Hero actually was talking about sapping rather than gallery mining. Moreover, the purpose of manuals, even in the early medieval era, was to offer the reader all options.[8]

Throughout the crusading period in the Near East, sieges were a common occurrence, but the opponents conducted them in markedly different ways. Christian sieges of Muslim strongholds usually took longer and relied more heavily on blockade and starvation than on attacks because of limited manpower. In contrast, Muslim sieges of Christian strongholds were shorter, often lasting only a few weeks, and saw more aggressive measures such as large scale attacks because of greater manpower resources. Both sides employed mining and countermining, but tunneling was not the first choice for either.[9]

Roger II, a Norman who established a kingdom in southern Europe, employed sapping and mining perhaps more than any other crusader. The castles and city walls in southern Italy were not as strong as in the Levant, so his operations tended to be more successful than most. But actually exploiting a breach was another matter, and his forces often were unable to break through the circuit even after a successful collapse. Moreover, defenders often gave up before his miners had a chance to set fire to the combustibles. Roger II seems to have employed Sicilian Muslims as troops and laborers, and perhaps they provided his miners as well.[10]

Crusaders drew also on European miners to conduct tunneling operations during this era. Saxon silver miners dug the galleries in the siege of Toron in 1197, and Lombards constructed tunnels during the siege of Beirut in 1231–1232. Muslim besiegers called on civilian miners from Khorāsān and Aleppo to collapse the wall protecting Edessa and capture the city in 1144.[11]

The frequency with which commanders hired civilian miners rose dramatically during the middle medieval period, and this tends to be well documented. French kings, beginning with Philip Augustus, made regular use of them as did other monarchs, but the most extensively documented case is associated with the Forest of Dean in England. Miners there lived and worked on royal lands to extract iron ore and were often called on for military service. They worked under their own masters and were paid travel expenses as well as a salary. These commercial miners often labored in groups of thirty, forty, or fifty men while participating in a particular military operation. During the siege of Berwick in 1355–1356, the contingent reportedly numbered 120 men, although miners from some other sites joined those from the Forest of Dean for that unusually large gathering. For about two hundred years, the Forest of Dean miners were the first choice of military commanders who wished to conduct underground approaches. But by the fifteenth century, those men faced stiff competition from other communities of mineworkers. Coal miners from Liege and from coal pits in England became more noticeable in the military contingents assembled for sieges by that time.[12]

But while professional miners dominated the work of digging offensive galleries by the middle medieval period, they do not seem to have been frequently employed for countermining. The difficulty of assembling such men before a besieging force arrived probably accounts for this. Nevertheless, countermining increased by necessity in tandem with offensive mining, and it often was successful. Countermining galleries tended to be shorter and perhaps needed less specialized skill to create. Byzantine defenders at the siege of Durazzo (1107–1108) intercepted a Norman mine just as it reached the foundation of the city wall and inserted Greek fire into the gallery. After driving out the enemy miners, they collapsed the tunnel, and the Normans later raised the siege. Muslim countermining blocked French offensive galleries at the siege of Acre in 1189–1191. Countermining by the defending Templars during the siege of Saphet in 1266 broke into the top of a Muslim gallery and resulted in hand-to-hand fighting underground. It was a temporary setback for the besiegers, who later captured the city through treachery.[13]

The success rate of mining during the crusading period was checkered, reflecting the general success rate throughout the medieval and classical eras. But some historians argue that the Muslims came to excel the Westerners in siege operations conducted during the latter phases of the crusading period, when Western strongholds were consistently reduced. Yet they offer little in the way of hard evidence to support that contention. One of the few detailed descriptions of a gallery mine by anyone on either side of the warring divide appears in the memoirs of Usāmah Ibn-Munqidh. During the siege of Kafartāb in 1115, troops from Khorāsān began a tunnel from an advanced trench, and Ibn-Munqidh explored the workings just when they were in the process of setting up props under the wall's foundation.

The tunnel was dug from the trench to the barbican. On the sides of the tunnel were set up two pillars, across which stretched a plank to prevent the earth above it from falling down. The whole tunnel had such a framework of wood that extended as far as the foundation of the barbican. Then the assailants dug under the wall of the barbican, supported it in its place, and went as far as the foundations of the tower. The tunnel was narrow. It was nothing but a means to provide access to the tower. As soon as they got to the tower, they enlarged the tunnel in the wall of the tower, supported it on timbers and began to carry out, a little at a time, the splinters of stone produced by boring. The floor of the tunnel, on account of the dusts caused by the digging, was converted into mud.

But the gallery mine that Ibn-Munqidh describes proved to be only a partial success. When the miners set fire to the dry wood in the cavity, it slowly caused subsidence. "The layers of mortar between the stones of the wall began to fall. Then a crack was made. The crack became wider and wider and the tower fell. But only the outer face of the wall fell, while the inner wall remained intact." Because of this, the planned assault was canceled. But later that day a few soldiers sneaked into the partial breach when the defending Franks were off their guard and fought their way into the enclosure. The result was the fall of the city on September 5, 1115.[14]

Ibn-Munqidh's description of the offensive mine at Kafartāb provides a rare glimpse into the details of gallery mining by Muslim forces during the crusading period. Still, it does not prove that their mining practice was superior to that of the West; in fact, the miners had failed to achieve a clear breach of the tower wall. Their technique was exactly like that of everyone who practiced good gallery mining in any region. Muslim commanders regularly hired professional miners from the civilian industry, another common practice, especially favoring men from Aleppo, Diyarbakir, Mosul, and Khorāsān. Some have suggested that the miners from Khorāsān, the eastern province of Sasanian Persia, must have inherited the Sasanian expertise in mining techniques and continued them into the medieval period. If so, this represents an interesting continuity of military mining knowledge from one era to another.[15]

Western crusaders also inherited mining techniques from their Roman predecessors and continued practicing them with varying levels of success. Among the few detailed accounts of an offensive mine constructed by Crusaders appears in the account of an anonymous author, probably a priest from southern England, who accompanied the expedition against Lisbon in 1147. It was the only successful campaign of the Second Crusade (1147–1149).

At least two galleries were started. One aimed toward the west wall of the Moorish city between the Porta de Ferro and a flanking tower. A second was started by men from Cologne and Flanders against the east wall at a steep slope, digging straight into the sharp profile of the terrain to avoid the need to construct a shaft. According to the author, it "had five entrances and extended inside to a depth of forty cubits [sixty feet] from the front; and they completed it within a month." At sixty feet (18.28 meters) in length, the miners had dug the gallery about as far as they could without some sort of artificial ventilation.[16]

The defending Moors discovered both galleries but were unable to stop the one aiming at the east wall. Reaching the foundation, the Westerners dug out a chamber, filled it with flammable material, and fired it. On the night of October 16, it led to the collapse of thirty feet (9.14 meters) of the wall. But the defenders were prepared, having erected beams to create a secondary barrier to entry, so that follow-up attacks failed to break through. The next tactic was to move a siege tower eighty-three feet (25.29 meters) tall against the southwest corner of the city circuit near the Porta de Ferro. This approach led the Moors to negotiate, and the city surrendered on October 24 after a four-month siege.[17]

A survey of sieges during the crusading era leads one to assume that mining was by no means a sure tactic and that neither side dominated the other in its technique. When Frederick Barbarossa besieged Tortona in 1155, the defenders dug a countermine that intercepted and collapsed an offensive gallery, suffocating some of the miners; it obviously approached the Crusaders' tunnel from above. During the siege of Palma de Majorca in the Balearic Islands by James I of Aragon in 1229, miners collapsed three towers by firing props under them. Beneath another tower, they used a cable to pull the props rather than burn them. But these successes did not end the siege; when the besieger began to fill the dry moat with layers of timber and dirt, the Saracen defenders dug a countermine and tried to burn the timber layers from below. James instructed his men to extinguish the fire by letting water into the moat from a nearby source. He eventually captured the city by assault. Elsewhere, the defenders in the siege of Brescia in 1238 conducted a sortie and killed a good many of Frederick II's German miners. Sultan Kelaun employed an unusual method while besieging Markab in 1285. His miners collapsed a section of the curtain and part of a tower, but the follow-up attack was repulsed by the Knights of St. John. Muslim workers managed to extend a second gallery under a different tower, but Kelaun stopped their efforts for a while as he called on the knights to send their engineers to examine the mine. After looking it over carefully, they came to the conclusion that it was best to surrender on terms.[18]

While the written record brings us brief views of mining at many sieges of the crusading era, physical evidence is extremely rare. The Mamluk siege of Frankish Montfort in 1271 has yielded archaeological evidence of both sapping against the masonry wall and of an offensive tunnel. At what is left of the southwest corner of the outer wall, "about five stones from each of the three lowest courses have been removed." Sultan Baybars, who led the siege, reportedly offered one thousand dinham to the sappers for each stone they chipped out. Also, the remnants of a second and unfinished mine gallery at the west end of the south wall, upper circuit, has been found. Speculation is that the defenders gave up the city on terms when they discovered the sapping and mining operations.[19]

The Albigensian Crusade of 1209–1218 resulted in many sieges with fairly good documentation of operations. Historians have identified at least forty-five sieges and four field battles from this time. Most of the former ended in a negotiated surrender by Cathar forces to Papal troops; fewer than one-third of them were decided by escalade. At the siege of Carcassone in 1209, the besiegers constructed a cover for their sappers, who managed to dig a short tunnel under the wall foundation. The subsequent firing led to the collapse of a section, and the follow-up attack captured the southern suburb of the city. Counterattacks reclaimed this suburb, however, and the siege continued until a water shortage forced the defenders to surrender.[20]

At Simon de Montfort's siege of Lavaur in 1211, the Papal besiegers managed to reach and fill in part of the defensive ditch. But during the night, the Cathar defenders utilized a preexisting tunnel linking the city wall with that ditch to clear out what the besiegers had placed in it. In order to prevent them from doing this again, Papal troops constructed a pile of flammable material at the exit of the tunnel, where it entered the ditch. First they placed a layer of green wood and branches, then a layer of "dry wood, grease, [and] oakum." For a third layer, they spread green wheat and grass. When they set fire to the lowest layer, they created a slow burn that produced terrible smoke. The top layer of unripe wheat and grass prevented the smoke from rising through the pile and forced it into the tunnel entrance. This created such a poor atmosphere that defenders could not enter the tunnel and allowed the besiegers to fill in a section of the ditch and move a cover for the sappers forward to the wall. Papal forces breached the circuit, and an attack captured the place on May 3.[21]

The siege of Carcassone from September 17 to October 11, 1240, saw extensive use of mines and countermines. The city had been a stronghold of the Cathars under Viscount Raymond-Roger Trencavel, but it had fallen to a Papal crusading force in 1209. The son of Trencavel led a Cathar force in an attempt to take it back in 1240. When his men dug a gallery toward the barbican of a gate, the defenders heard the noise and countermined, also constructing a second wall at the targeted area. When the besiegers fired their mine, it collapsed part of the wall, but they failed to break through the secondary barrier. A second gallery aimed at a turret along a line of palisading, but the defenders countermined this one as well, intercepting the gallery and stymying the effort. A third mine collapsed part of the palisade line, but again a secondary line of works contained the damage. A fourth mine began some distance away from the angle of the city wall; defenders detected it in time to countermine and make a backup wall at that point. Trencavel's miners managed to fire their combustibles and reportedly collapsed sixty feet (18.28 meters) of the original circuit. The secondary wall, however, contained the attack. The fifth and last gallery also began some distance away and aimed at the barbican of another gate. Papal forces also discovered this one, began a countermine, and constructed another secondary wall. This time, the countermine intercepted and stopped the offensive gallery. Stymied in all five mining attempts, which must have been unusual in the long sweep of medieval mining, the Cathars mounted a desperate attempt at escalade that failed. A week later they lifted the siege at news of the approach of a relieving force.[22]

There can be no doubt that military mining was a familiar feature of sieges during the long and complex crusading era, both in the Levant and in Europe, but a good deal of murkiness exists as to its level of effectiveness. From a technical standpoint, it tended to be more successful than not. Most attempts to run an offensive gallery under defending walls succeeded if the defenders did not intercept it with a countergallery, and they often managed to bring down at least part of the targeted wall. But the follow-up attack very often failed because, by that time, defensive forces had become aware of the danger, and there were more ways to counter it than the tricky course of running a countermine gallery to intercept the besiegers. Constructing a second wall at the danger point was easy to do and often effective at saving the beleaguered city, at least for the time being. Another angle on the theme of effectiveness is that, in many sieges, mining began but was never brought to completion because the siege ended for other reasons.

Mining for crusading purposes took place not only in the Levant but also in southern Europe and parts of France because of the widespread nature of religious conflict during this era. Other wars in France, largely pitting English forces against the French, also involved mining on probably what was an equal level. The French king Philip Augustus besieged an English garrison at Château Gaillard from September 1203 to March 1204. His men filled sections of the ditch in order to cross it and gain access to the wall. They undermined part of the curtain to create a breach through which the infantry attacked and captured the outer ward. The middle ward was taken by escalade, while another mining project weakened the wall and, in combination with a catapult, brought part of it down. This led to the garrison's surrender. "The real work in this siege . . . was done by the miners," concludes Charles Oman. But he does not differentiate between sapping and tunneling, so it is not clear which technique was used at Château Gaillard.[23]

Similar vagueness in describing siege operations hamper our understanding of mining versus sapping in other English-French sieges of the era. When the Earl of Derby besieged the castle of La Réole, the operation was decisive in compelling the French garrison to surrender. According to the Saint Omer chronicle, the earl "sent his miners to the wall, and in a short time they had undermined it, so that they brought down a large piece of it." This apparently was an example of sapping, although the miners could have dug a short gallery under the foundation to achieve their goal. After capturing the work, the earl "rebuilt the wall" and installed an English garrison. A reliable source indicates that a genuine tunnel mine was dug by the French while besieging Pont-Audemer in Normandy during 1356. The head of the gallery reached a point only four feet (1.21 meters) from the outer wall when a relief column caused the lifting of the siege. The English then filled in the unfinished gallery. During the Reims campaign of 1359–1360, Edward III had brought along some English miners who played a key role in capturing the keep of Cormicy, a town north of Rheims.[24]

A bit more information is available for the English siege of Limoges in 1370. The Duke of Lancaster, son of Edward III, had charge of the mining and personally supervised the work. The French garrison countermined, but the besiegers "kept changing the line of direction of their own mine" to avoid intersecting with them. According to Sir John Froissart, the duke decided to fire the chamber at 6 A.M. right after it was ready. The result was the collapse of a section of wall into the ditch. The follow-up attack succeeded in capturing Limoges.[25]

In one of the last conflicts between the English and the French during this era, Henry V hired 120 professional miners and conducted two sieges of French coastal cities. Harfleur was located at the mouth of the Seine, with a wall circuit of two and a half miles (4.02 kilometers), twenty-six towers, and a wet ditch for protection. Henry's army reached the city on August 17, 1415, and started excavating tunnels, but the French effectively countered these efforts either through a defensive gallery or with a deep ditch in front of the target. Reports mention "a most determined underground conflict" without further details, but the English stopped mining afterward. Instead, they captured the barbican by assault on September 22, which led to the garrison's surrender. Caen was a larger city with thirty-two towers, twelve gates, and a wet ditch. When Henry's miners began their work, the French defenders discovered what they were doing and effectively countered them. Once again the English captured a fortified city by assault, not by collapsing walls, on September 4, 1417. In dealing with holdouts in the castle, however, Henry hesitated to mine. He found the castle beautiful, according to a medieval source, and did not wish to further anger the citizens by its destruction. Moreover, he would need to rebuild any mine damage in order to use the citadel himself. Therefore Henry caused the small garrison to surrender by threatening its members with dire consequences if they continued to resist.[26]

Henry's campaign gave rise to a literary reference to military mining. William Shakespeare, when he wrote *The Life of King Henry the Fifth* around 1599, was aware that one chronicler of the siege of Harfluer had criticized the English mining effort as contrary to doctrine. In his play the character Gower asks Captain Fluellen to go to the mines so that the Duke of Gloucester, the brother of King Henry V, can discuss their progress with him. But Fluellen replies: "it is not so good to come to the mines; for look you, the mines is not according to the disciplines of the war. The concavities of it is not sufficient; for look you," the French had dug "himself four yard under the countermines," and the captain feared disaster "if there is not better directions." The Bard thus provides the only exposure to military mining to be found in the world corpus of literature.[27]

Off the Continent, military mining had appeared in Britain only in the middle of the medieval period. According to Charles Oman, the earliest-known example of it dates to William the Norman's siege of Exeter in 1068. No details have come to us of this early example, but in a subsequent operation in the area, when King Stephen besieged Rougemont Castle near Exeter in 1136, a royal summons went out to civilian miners. Stephen ordered them, according to the author of *Gesta Stephani*, "to search into the bowels of the earth with a view to demolishing the wall." This mining effort failed to play a decisive role in the three-month-long siege, which ended with a negotiated surrender on generous terms.[28]

The earliest mine in England that has yielded physical evidence through archaeology dates to 1174 at Bungay Castle. Baron Hugh Bigod held the castle against the forces of Henry II, whose miners began a gallery under the southwest angle of the keep. What now remains of the gallery is twenty-six feet (7.92 meters) long. The miners started digging the tunnel in the northwest angle of an entrance tower. They also started two branches extending from close to the center of the still-extant gallery, but the branches remain unfinished.[29]

King John's siege of Rochester Castle in the fall of 1215 has become an icon of military mining. This is primarily because the miners employed the unusual tactic of stuffing the rendered fat of forty pigs in the chamber they excavated under the southeast corner of the massive keep, the tallest surviving example of its kind in England. Burning those carcasses produced an intense fire that consumed the timber props and collapsed the tower corner. But it was more common to use less costly items as combustibles. For example, at the Crusader siege of Nicaea in 1097, men gathered "twigs, stalks and sticks and dry reeds, pieces of tow and all sorts of kindling" for that purpose. The unusual operation at Rochester was undertaken by civilian mining masters Ernulf and William, whose eleven miners were paid for 159 days of work. Ironically, that success failed to win the siege for John's forces. Sometime later the rebels in the keep gave up due to food shortages.[30]

When Henry III besieged the rebellious Falkes de Breauté at Bedford Castle in 1224, he recruited thirty civilian miners from Saint Briavels in the Forest of Dean. They were commanded by John of Standon, who had directed miners during Henry's siege of Castle Bytham in 1221. Henry paid these men four and a half to five pence per day for their work, compared to two pence per day for infantrymen, and offered some of the miners land grants after the siege. Henry also supplied them with at least four hundred pickaxes during the course of the eight-week operation. Carpenters constructed a cover for the sappers and provided tunnel props to hold up the gallery. This small but experienced corps of professional miners produced a pretty quick victory for the king. They collapsed a segment of the curtain wall, which led to a follow-up attack that secured the inner bailey. Then they undermined the keep and on August 14 fired the combustibles. The result created "a visible crack in the wall" and filled the interior of the stronghold with acrid smoke, leading the garrison to surrender.[31]

The siege of Bedford proved the value of investing in a corps of professional miners and offering them substantial rewards for their work. This became a prominent theme in subsequent English mining history. Edward I often hired civilian miners to undertake the construction of galleries on several occasions, but they did not always succeed. When reducing Dryslwyn Castle to suppress a Welsh uprising against him, Edward's men employed above-ground sapping where limestone underlay the south and east side of the enclosure but used gallery mining on the north and west sides, where it was easier to tunnel through the earth. His miners had nearly completed a gallery when a group of knights entered to inspect the work. It collapsed and killed the warriors and apparently a dozen miners as well. The castle later fell after a twenty-eight-day siege on September 8, 1287, but the sources are vague as to exactly why that happened. Edward also included a dozen professional miners in the force he led to besiege Caerlaverock Castle in

1300. Although they failed to breach its massive defenses, the castle fell to royal forces after a lengthy siege. Edward's campaigns in Scotland immediately afterward also employed a master miner, who controlled twenty-one professional workers. A later monarch, Edward III, continued the policy of relying on professionals for such technical operations. At the siege of Berwick in 1356, Edward recruited 120 miners who drove a gallery nearly under the wall foundation before the garrison, consisting of only 130 men, negotiated its surrender.[32]

In the Mediterranean region mining continued to play a significant role in siege warfare after the Crusades, mainly because of expansionist moves by a military culture that had long nurtured the technique. When Ottoman forces under Sultan Mehmed II besieged Constantinople beginning April 9, 1453, they faced one of the most heavily fortified cities in Europe. It was protected by four miles (6.43 kilometers) of walls on the landside, nine miles (14.48 kilometers) along the sea, and a total of one hundred towers. The Ottomans dug fourteen mines, all reportedly beginning "a considerable distance" from the wall requiring lengthy galleries. One of the fourteen is said to have begun half a mile (0.80 kilometers) from the target, but that surely is an exaggeration—a gallery that long is not reliably documented until the Western Front in World War I. The problems of ventilation and the time necessary to dig such a long tunnel do not fit the technology of the era or the short duration (only seven weeks) of this siege. Both sides employed mine specialists; the Ottomans hired Europeans from Novo Brod, while the Byzantines had the services of a German named Johannes Grant. The defenders successfully neutralized the offensive mines through interception, but Constantinople instead fell to a massive assault.[33]

While the techniques of sapping and mining continued to flourish in the Euro-Mediterranean world and the Near East, they also were known in the rest of the world. An early reference to military mining in China dates to the middle to late fourth century BCE in the Book of Lord Shang (or Shang-Chiin Shu), attributed to Shang Yang, a minister in the state of Ch'in (359–338 BCE). This book suggests that commanders should work miners in eighteen-man teams to undermine a wall, using wooden props to hold it up until they could be fired to bring down the structure. The details indicate that the technique of sapping and mining had been worked out well before Shang Yang's time. In fact, historian Ralph Sawyer contends that the techniques "had already become highly systemized by the mid Warring States [403–221 BCE], if not a century or two earlier, and only minimal changes occurred thereafter prior to the perfection and application of gunpowder explosives in the Yuan [1279–1368 CE] and Ming [1368–1644 CE]."[34]

Chinese documents preserve some details about mining practices. They indicate that during the Warring States period (ca. 475–221 BCE), rather than stuffing the excavated cavity under the wall foundation with combustibles. Chinese miners wrapped rags soaked in lard or oil around the wooden props. Countermeasures were recorded in books issued during this time as well. These included the placing of "listening wells," dug at least 15 feet (4.57 meters) deep, every five paces (around 3.81 meters) along a wall. Big earthen jars with leather coverings stretched tautly over their openings were placed at the bottom of these wells. Their placement at regular intervals allowed the defender to use triangulation methods of precisely locating the position of the offensive gallery. Reportedly, this method could detect the sound of enemy mining up to five hundred paces (around 1,250 feet, or 381 meters) away. On intercepting the gallery, these books recommended the use of "poison smoke" deployed with pipes stuck through holes in a shield and protected by the use of thrusting spears to prevent enemy miners from dismantling them. The smoke was produced by building a kiln-like stove at the bottom of the listening well and blowing the vapors along the pipe to the enemy gallery with the use of a bellows placed on the surface nearby.[35]

The appearance of mining techniques in a variety of books on warfare is a reasonable indication that they had been used at least in a minimal way in the field. The siege of Ch'en-ts'ang in 228 CE during the Three Kingdoms period (220–280 CE) lasted only twenty days but included attempted escalade, the construction of siege ramps, and mining. The defenders stymied every attempt, digging "several transverse trenches out to intercept the approaching miners," and the siege was lifted. A unique approach to countermining occurred during the siege of Hu-lao in 423, when Northern Wei forces attacked Southern Wei cities. Defending troops reportedly dug six countermines, passing under the attacker's position and emerging in their rear. Four hundred soldiers passed through the galleries and came out to burn siege engines. This tactic succeeded in a temporary way; the attackers eventually captured the place. The defenders at the siege of Chien-K'ang, occurring sometime between 547 and 552, used countermines to neutralize siege mounds. They collapsed the forward part of these works and managed to set fire to siege towers on them. While the attackers, who were part of the emperor's forces combatting a revolt by the Hou Ching, gave up further such attempts, they managed to collapse a section of the city wall only to find the defenders had made a secondary wall to contain the breach. Imperial forces, however, captured the city by encouraging the betrayal of its garrison—someone opened the city gates at night.[36]

The siege of Yü-Pi in 546 is more thoroughly documented than most Chinese sieges of the premodern era. It pitted the forces of the Eastern Wei against defending troops of the Western Wei. The attackers employed many methods during this operation, all of which were thwarted until the siege was lifted after fifty days. The Eastern Wei dug ten offensive galleries toward the south wall, but the defenders dug a long trench along the inside of the wall. One of the offensive mines intercepted this trench, and the Western Wei troops managed to start a fire at its entrance, using bellows to force the smoke into the gallery. Although the texts describing this siege tend to be a bit confusing, it seems that other galleries approached the other three walls of the city as well. One is recorded as undermining the east wall, with a lateral digging to extend the attempted breach. The miners placed oil-soaked wood as props and fired them, but by that time the defenders had constructed a palisaded line at this spot to deny the Eastern Wei an opportunity to attack through the breach.[37]

An interesting feature of this siege, and presumably of many others in Chinese history, is that the walls being undermined consisted not of masonry but of tamped earth.[38] This traditional form of Chinese construction obviously was susceptible to mining. The method of building earthen walls was to lay down a layer of dirt and pound it with wooden mallets before adding the next layer. A major city wall could therefore have many layers of compacted earth, creating a very solid mass of dirt that nevertheless could be collapsed. The method of collapsing an earthen wall apparently was no different than that of collapsing a masonry wall.

By the time of the T'ang Dynasty (617–907), military manuals recommended stuffing combustibles into the excavation underneath the targeted wall rather than relying on oil-soaked rags wrapped around wooden props. There are examples of the besieger digging galleries fully underneath the defending wall and emerging within the fortified enclosure to help their comrades in scaling the wall. Fire became an element in the tactics of countering siege approaches. At the siege of Fengtian in 783, defenders stopped a siege tower by digging a gallery out to a point in the path of its approach, creating a cavity near the surface, and filling it with horse dung. When one of its wheels broke through and immobilized the tower, defenders then set fire to the dung and burned it up. The availability of gunpowder increased the possibility of using fire to stop a siege approach. During the second siege of Pien-ching in 1642, the defenders freely used blocks of it mixed with bundles of firewood, dropping them into the intercepted galleries of the besiegers. Once set ablaze, these bundles burned for hours, with periodic explosions of the gunpowder, and completely neutralized the offensive galleries.[39]

Military manuals of all periods in Chinese history offer advice for conducting underground warfare. The manuals of the Song era (960–1279) refer to "thunderclap incendiary spheres" designed to release noxious gases in a tunnel. They recommend using fans made of bamboo to encourage the gases forward toward the enemy while the fan wavers held "licorice water in their mouths as an antidote." The manuals also discuss using a shield made from rawhide to block the flow of gases coming from the enemy and, with large hand-cranked fans, forcing it back upon them.[40]

These brief glimpses of mining and countermining techniques throughout Chinese military history indicate a sophistication paralleling that of the West. Much depended on the level of military preparation. Cultures that developed large armies well administered and with access to resources and technical expertise were likely to employ many different siege approaches, including mining. Wherever mining took place, countermining was bound to follow. How far a particular culture's military force took these developments depended on decisions by the political and military leaders and how often their forces engaged in siege events.

In West Africa, while warfare and sieges were not uncommon, mining seldom took place. There were many fortified towns in this region, but siege techniques tended to be primitive. Armies were small and technical expertise not easily available. Escalade and trickery seem to have been the most common ways to end a siege. In his study of warfare in West Africa from 1400 to 1900, Robert S. Smith identifies only two documented instances of mining. A Muslim force under the Emir of Hadeija captured Marmar in the mid-1800s by mining its walls. Smith found another "possible example" in the siege of Abeokuta by Dahomean forces in 1864. Cannon were used in the early 1600s in at least one siege, and scaling ladders were common, but there is no evidence of the employment of siege engines.[41]

Much had taken place during the long medieval period concerning the trajectory of mining and countermining in siege warfare. During the early part of the era, the technique was barely used because the technical expertise, the doctrine, and the rich heritage associated with it had been muted and almost smothered by the centuries of cultural atrophy that followed the end of the Roman Empire. It was resurrected in the middle part of the era when states grew in size and power, and siege warfare increased in importance. The Crusades played a huge role in the increased number of mining and countermining examples in the field, as did the Hundred Years' War (1337–1457). In fact, every major conflict of the middle and later medieval period tended to be largely a war of position, with the acquisition of territory and fortified places high on the list of military objectives. Naturally, siege warfare and with it mining and countermining became ever more important components of military operations.

A survey of operations during the Hundred Years' War by a team headed by historian Clifford J. Rogers was based on the three contemporary histories that cover the conflict more thoroughly than any others. It reveals that sieges heavily dominated the military action. Using a broad definition of a siege, there were nearly 900 of them compared to only fifty battles and eighty-two skirmishes. Using a more narrow definition of a siege, one that comports with contemporary use of the term, the number was 350, which was still far more than open-field battles. For every siege that failed, three others were successful. This success rate increased during the latter decades of the war. It was impossible to calculate how often siege mining was used because the sources do not comprehensively mention its occurrence. But enough is known to conclude that tunneling supplemented the many other methods of winning a siege.[42]

We do not have enough information to know whether Chinese military mining equaled that to be found in Great Britain, on the European continent, or in the Near East. In areas such as Africa, mining hardly had a show because of the lesser development of powerful states and armies on the same level as those in the Euro–Near East region.

Still, the techniques of mine warfare were pretty common wherever practiced. Whether European, Islamic, or Chinese, the principles of gallery mining and countermining altered little from culture to culture, era to era. Attempts to undermine walls overwhelmingly were the common feature of offensive mining. While sparse information about the length of tunnels survives, we know that any that were more than sixty feet long had to be ventilated. Contemporaries too often conflated sapping a wall with undermining it, although in many cases the two went hand in hand. But there was a very real difference between trying to pry out some stones from a wall, on the one hand, and trying to dig a tunnel underneath it, on the other. In both cases the attempt was to collapse the wall, but the two approaches involved the need to overcome greatly different challenges. Tunnel mining was far more complicated, sophisticated, and risky than sapping, and it was less often implemented for those reasons.

Continuity rather than change dominated the techniques of military mining from its earliest origins in the ancient Near East to the introduction of gunpowder, which spurred the advent of the explosive mine in the fifteenth and sixteenth centuries. Innovation became a major feature of mining history for the first time when the primary method of bringing down a wall switched from fire to explosion.

4

Gunpowder

The most important watershed in the long history of military mining took place slowly and uncertainly during the fifteenth and sixteenth centuries as gunpowder made its way into the system of undermining enemy defenses. For the previous two thousand years, miners had dug, propped, and burned down walls, or they had used tunnels to gain access to the enclosure of a besieged city. Now they experimented with ways of planting large charges of gunpowder in the excavated chamber to bring down walls with a spectacular noise and crash. The change in means to accomplish this ultimate goal of mining also led to changes in the construction of the gallery, to the realization that one had to close off a good part of the tunnel to prevent back blast, and to figuring out ways to avoid the immense danger posed by gunpowder to its users; setting off the charge became a ticklish technical problem. Military mining entered a new age when gunpowder entered the picture.

But gunpowder was largely a technical change that did not alter military operations associated with a siege. The old problem of trying to capitalize on a successful mining to create a battle-winning victory remained. The noise and destruction created by gunpowder did not necessarily increase the chances of a successful follow-up attack. Moreover, counterminers quickly incorporated gunpowder into their craft and even constructed permanent countermine systems of stone at key cities to be prepared for a siege even before a besieger arrived.

Gunpowder was not even an immediate success, entering the picture furtively and with minimal documentation. The earliest mention of its use for military mining dates to the Florentine siege of Pisa in 1403, when Domenico di Matteo was hired to collapse a wall with gunpowder. For some reason the plan folded and was never implemented. There are indications that John Vrano used gunpowder in a countermine constructed against the Islamic besiegers of Belgrade in 1433. Another scant indication places a gunpowder-charged offensive mine at the siege of Orense, Spain, in 1468. In this case it was designed to bring down the tower of a cathedral, but historians have doubts about the reliability of the sources. In the 1487 Genoese siege of Sarzanello fortress, held by the Florentines, the besiegers failed to take the head of their gallery all the way under the wall's foundation. As a result, the explosion only produced "minor damage instead of opening a breach."[1]

The siege of Castel Nuovo at Naples in 1495 served as a watershed in the development of gunpowder-charged mining. Local forces rebelled against a French garrison and hired Francesco di Giorgio Martini of Sienna to superintend their mining efforts. Martini, an architect who also began writing a book entitled *Trattati* that discussed explosive mines in about 1475, managed what historians call the first fully successful and reliably documented instance of a gunpowder-charged mine in history. When he fired the mine on November 27, it brought down part of the west wall of the barbican, and the follow-up attack secured a portion of the structure. Another attack on December 8 captured the castle. There also is evidence that Martini dug the gallery not along a straight line, as was traditional, but in a winding configuration to reduce the escape of gases along the tunnel when the charge went off. He had already referred to such winding galleries in the final version of his *Trattati*, written in 1491–1492.[2]

Pedro Navarro, a mercenary in the series of conflicts raging in Italy at the time, also played a prominent role in making gunpowder-charged mines. His earliest application in the field took place at the siege of the Turkish fortress of San Giorgio on the island of Cephalonia in 1500. Historian Christopher Duffy, however, suggests that in this case Navarro only used gunpowder to set fire to combustibles in the cavity rather than to blow up a section of the wall. Moreover, there is evidence that superior artillery fire rather than the mine was the key to the besieger's success. Three years later Navarro superintended yet another siege of Castel Nuovo at Naples for Spanish forces. He undermined the same barbican that Martini had targeted in 1495 except that he dug toward the south rather than the west wall, springing the mine on June 12, 1503. The following infantry attack captured the castle. Not long after this success, Navarro also undermined Castle dell'Ovo near Naples, which led to the fall of the position. Having worked for the Spanish, he sold his services to the French in 1515 and undermined a Spanish position at Milan that year.[3]

As the gunpowder mine matured, the familiar pattern settled in—it sometimes proved decisive in determining the outcome of a siege but often made little difference in the overall picture. Sultan Suleyman the Magnificent attacked Rhodes with a large and determined force against the Knights Hospitallers who held the place in 1522. The Turks began siege approaches on July 29, which included offensive galleries, blowing two of them on September 4. The follow-up attacks, however, failed. Two more chambers were exploded on September 9 with the same failure of the infantry assault that followed. Three more attacks following three more mine explosions on September 17 also failed. Two additional mines that blew on September 23 also led to failed infantry assaults. By this time, the Suleyman had learned his lesson. On September 24 the largest Turkish attack thus far in the operation went in with no preparatory mines, hitting four sectors of the wall, but all were repelled with great difficulty. The Turks then switched their approach, crossing the moat and sapping the wall to open a breach at several places by late November. Follow-up attacks gained partial control of the breaches as the will to resist eroded among the citizens of Rhodes. Residents pressured the Knights Hospitallers to negotiate the surrender of the place on December 20. Mining failed to play a decisive role in this large operation; sapping combined with artillery fire barely provided the edge the besiegers achieved over the defenders.[4]

Despite the slow but steady development of explosive mines, the old fashioned way did not entirely disappear. The Spanish used a prop mine at the siege of Saint Pol in 1537, and sapping continued in other sieges. A good example of the latter occurred at the siege of Montalcino by a combined Spanish, German, and Italian force in 1553. The first effort, which began on April 3, was defeated because the sappers had inadequate shelter; defenders dropped rocks, fired small arms, and lighted up the area with incendiaries at night. The second sapping attempt, beginning May 29, approached a blind spot of the same bastion. Covered with heavy timber, the sappers concentrated not on collapsing the wall, but on cutting steps up its outer surface. The defenders not only constructed a secondary line to isolate the bastion but also raised the parapet on top of the bastion's original wall to keep pace with the sappers' progress. For several days, both sides engaged in a bizarre race at arm's length to see if the defenders could build as fast as the attackers could cut. But the siege was lifted on June 15 after a Turkish fleet appeared and forced the Spanish to divert their resources to protecting their coastal possessions.[5]

It is not surprising that there would be some holdovers of previous techniques such as prop mining and sapping in this two-century period of transition to gunpowder. The new method demanded more technical expertise and was more dangerous to the besieger than the old methods. But gunpowder came to rule siege mining, with contemporary experts publishing manuals to guide military engineers about how to use it. The earliest-known manual to describe the technique was the Sienese engineer Mariano di Jacopo's *De Machinis Libri decem* in 1449. Battista Della Valle's *Vallo* of 1524, a short and general treatise on the attack and defense of fortifications, offered a fuller description of mining and, because of its simple explanatory style, was widely read among military men, going through eleven editions in thirty-seven years. Vannoccio Biringuccio of Siena authored a multivolume treatise on the military art, the tenth book of which, *De la Pirotechnia*, appeared in 1540. It was widely read and went through further editions in 1550, 1558, and 1559. But the most sophisticated description of military mining yet published was G. B. Bellucci's *Nuovo Inventione* of 1598, although it had been written before 1554. Bellucci focuses on nuances, among these planting the mine so as to blow out a portion of the rampart in order to fill in a section of the defensive ditch and prepare the way for an attack or above-ground siege approaches.[6]

Manuals spread the basic ideas of gunpowder mining far and wide, and they also began to tackle technical problems associated with the new form of siege approach. One of the earliest realizations was that the gunpowder explosion created a dangerous backdraft, with much of its force channeled through the open gallery. This greatly lessened the force exerted upward against the targeted wall and endangered anyone in the gallery or even near its entrance. The first attempt at a solution to this problem was to dig the gallery in crooked fashion, often referred to as a zigzag configuration, or at least to add a sharp turn or two. Historian J. B. Bury believes that Francesco di Giorgio Martini invented this concept. As city engineer of Siena by 1485, Martini had realized that by inserting right-angle turns in the underground sewers, he could slow down the flow of water. He then applied the concept to military mining. An early appearance of a zigzag gallery occurred at the siege of Padua in 1513.[7]

But sharp turns could only slow down a backdraft, not eliminate it, and constructing a gallery in zigzag fashion created a new set of problems. Earlier tunnels had been short and straight, both characteristics contributing to their successful digging. Most importantly, it was far easier to keep the underground gallery on the correct course if it was dug on a straight line. But

Undermining the Wall of a Fortified City. This sixteenth-century illustration shows miners working under the wall of a fortified city. It depicts the working cavity as spacious in order to fit in the human forms. From Fronsperger, *Kriegsbuch*, 66.

adding sharp turns greatly complicated what historian Simon Pepper calls "underground navigation." He points out several examples of bad navigation in practice. At the French siege of Cuneo in 1557, the miners accidentally hit an old well shaft and had to veer around it. This caused them to lose their sense of direction, and they wound up exploding a charge under the ditch of the work rather than under the wall.[8]

To compensate for the increased difficulties, scientific methods of underground navigation were applied when digging a tunnel. The magnetic compass began to be used by the 1550s. More importantly, Gabriello Busca's *Della Espugnatione et Difesa della Fortezze*, published in 1585, argues persuasively for abandoning zigzag and sinuous configurations for the trace of the gallery. He advocated right-angle turns only and "described the use of astronomical instruments for computing the distance, line and inclination" of a gallery toward the target. As Pepper describes it, Busca urged "the simplest and safest right-angled geometry, with plumb-line verticals and ninety-degree changes in direction." Eventually, the straight gallery came back because it was far easier to plan and dig than one that incorporated any turns. But even with a straight tunnel, scientific means of planning and executing it greatly increased its chances of hitting the target as intended.[9]

Serpentine Gallery Design. The snaky configuration was one of
the early experiments in changing the design of attack galleries
to lessen the backdraft of the gunpowder-charge explosion. The
blast would less forcefully escape down the gallery, and more
of the force would hopefully go straight up. The Italian legend
says, "The Foundation of the Mine Exposed," and "Place of the
Effect." From Vannoccio Biringuccio, *De la Pirotechnia*, vol. 10
(Venice: P. Gironimo Giglio, 1540), 658.

The characteristics of the gunpowder explosion had a serious effect on
methods of constructing the gallery. G. B. Bellucci's *Nuovo Inventione* of 1598
(but written before 1554) was probably the first published discussion of these
characteristics. He noted that the force exerted by the explosion of the gunpowder charge expanded in all directions, not just up, and it naturally funneled down the gallery because that area offered the least resistance to its
expansion. Bellucci urged the blocking of the tunnel for some distance from
the powder charge before detonation to direct as much of the force upward
as possible. He was the first to discuss what a later generation would call
the line of least resistance—the direction in which the force of the powder
explosion found it easiest to travel. To create an effective mine, one thus had
to make sure that the line of least resistance was straight up to the surface.
Miners, in other words, had to block the gallery so well that the area above
the powder charge would be the easiest for the gasses to penetrate and thus
most of the force would go up rather than dissipate in other directions.
Bellucci continued to urge at least one crook in the approach gallery itself
to help in this process, at the same time advocating double fuzing to serve
as insurance for the successful detonation of the powder. Double fuzing
also would be helpful in setting off a large mass of powder because, in the

detonation of most siege mines, a certain amount of powder tended not to go off at all. Bellucci also advised the digging of two or three galleries besides the main one as a way to confuse the defender.[10]

Nuovo Inventione represented a milestone in the doctrine of mine warfare, establishing most of the basic problems and concepts as well as many of the solutions to the new form of gunpowder-charged mine. It also began a discussion of how to use this new weapon within the context of siege warfare, recognizing that countermeasures were already being developed to challenge its prospects for success in offensive mining. For centuries to come, theorists and practitioners would continue to work out these problems and improve basic techniques of offensive and defensive mining.

Counterminers managed to maintain the traditional equilibrium between aggressive and protective methods in underground warfare. Manuals continued to describe old methods of listening, including the use of water vessels and of pebbles resting on the taut heads of drums; in practice, defenders at the siege of Boulogne placed bells and rattles on the drumhead rather than pebbles. Authors estimated that one could detect an approaching gallery as much as fifty yards (45.72 meters) away under the right circumstances and recommended the use of augurs over picks and shovels to minimize the noise of offensive digging.[11]

But the big change in countermining was to employ gunpowder in defensive efforts. The first recorded instance of this dates to the siege of Belgrade in 1433, when John Vrano superintended a powder-charged countermine that stopped a Turkish offensive gallery. Venetian defenders at the siege of Padua in 1509 dug a chamber under a bastion in their own defensive line and sprung it just after troops of the besieging League of Cambrai captured the bastion. The explosion destroyed the work and killed many of the enemy soldiers; the siege was lifted three weeks later. A similar example dates to the French siege of the Spanish border fortress of Salses in 1503. The defenders rigged a mine under a redoubt and sprung it when the French captured the place.[12]

Counterminers added a new twist to their craft when they discovered that excavating a cavity near the gunpowder charge in the offensive mine could serve as a vent for the explosion. In other words, they could create a more attractive line of least resistance and divert the main force of the blast into the countermine gallery rather than upward against the defending wall. The first recorded instance of this tactic dates to the massive siege of Rhodes in 1522. One Turkish mine was sprung on September 22 but did not damage the wall because the Knights Hospitallers had extended a countermine close enough to the charge to divert its force. This practice occurred as well at the siege of Montalcino in 1553.[13]

Venting an offensive mine had many advantages. It avoided the confrontation that resulted when counterminers intercepted the offensive gallery. After all, no one could predict what would happen if underground hand-to-hand combat ensued. Venting neutralized the offensive mine after the attacker had invested much time and energy in the project. But this tactic had its limitation. It must have been more difficult to estimate exactly when to stop extending the countermine in order to get close to the charge without breaking into the offensive gallery, and one would always wonder if it had been better to continue until fully intercepting the attackers. Perhaps these problems explain why there are relatively few examples of venting an offensive mine in the available literature.

Far more often, counterminers relied on the standard method of meeting the offensive gallery and risking the results. One of the most thoroughly documented cases of this stems from archaeological evidence regarding the siege of Saint Andrews Castle in Fife, Scotland. In 1879 diggers found the remains of an offensive gallery and the countering efforts that intercepted it. The siege was conducted by the Earl of Arran against a group of men connected to the murder of Cardinal David Beaton, who was an opponent of Henry VIII. The castle rested on a rock formation, and in 1546 besiegers dug a gallery through the rock toward a tower. They began at a point 130 feet (39.62 meters) from the target and took the effort to make the gallery spacious. The remains are 1.8 meters (5.90 feet) wide and 2.1 meters (6.88 feet) high. The miners dug about halfway to the tower and were starting branches, probably for ventilation, before the defenders found them. But the counterminers had some difficulty starting their defensive work. There is evidence that they began two shafts west of the tower before deciding on a spot east of it. From this latter place they precisely hit the head of the offensive gallery from above. The attackers thereafter abandoned the mine. Months later, in 1547, a French naval force bombarded the castle into submission. One of the more popular tourist spots in Scotland, the mine and countermine at Saint Andrews Castle are well preserved in the rock substrata through which they were dug.[14]

In short, countermining kept pace with offensive mining during the period of transition from prop mines to explosive charges. It was a relatively simple task to incorporate gunpowder into countermining. The second-most-important change in defensive mining during this era was to add permanent countermines to a fortification. This advancement went hand in hand with a dramatic alteration in fortress construction that took place because of the increased use of artillery. Medieval fortress architecture, with its squared towers and single walls, gave way during the latter fifteenth and throughout the sixteenth centuries to the Italian Trace style, with its angled bastions.

Curtains were lowered and made much thicker usually by building a retaining wall several feet behind the older structure and then filling in the space between with dirt and rubble. Towers were rounded and ditches deepened and widened. It was possible to incorporate a system of permanent countermines into new and renovated fortresses, especially those high on the target list of potential enemies.[15]

Antonio de Sangallo the Younger pioneered the construction of a permanent countermine system at the Porta Ardeatina in Rome's defensive circuit in the 1530s. His prototype became the generally accepted way to construct such a system. It consisted of an underground gallery along the length of a vulnerable wall, shored with masonry to last many decades. Thinner sheathing filled in portals along the side facing the enemy so that miners could quickly break through them and begin digging galleries toward suspected mining activity during a siege. Listening posts were necessary in this system, placed at key locations and spaced apart from each other to cover the widest possible area. In this way anywhere from a third to a half of the countermine system necessary for the protection of a defensive wall was already in place on a permanent basis before any siege began.[16]

Historically, military mining had always been closely linked to civilian mining practices, but in this era it temporarily went forward faster than its commercial counterpart in one area. Judging by the available evidence, besiegers employed gunpowder in their work much earlier than civilian miners used it in their peaceful pursuits. The first recorded use of gunpowder in commercial mining dates to 1575 in northern Italy, with the next occurring only in 1617 in the Vosges Mountains of France. After this, its use spread pretty steadily to Saxony in 1621, Hungary in 1627, Austria the next year, Norway in 1635, and England in 1665. Powder-charged commercial mining spread to the rest of Europe and North America during the first half of the eighteenth century and to Latin America during the latter half of that century.[17]

The construction of tunnels to serve as permanent transit routes for civilian traffic entered a new era with the advent of the tunnel under Monte Visio in northern Italy during the latter fifteenth century. The Marquis of Saluzzo wished to connect his state to a neighboring one to the north for trade purposes, but the alpine Monte Visio stood in the way. He financed the construction of a tunnel, working out an agreement with French authorities in 1475 and hiring Italian engineers to plan and superintend the project. Work began in 1480 from both sides of the mountain, resulting in a slight bend when the two galleries met in the middle. The completed work was 2,400 meters (1.49 miles) above sea level and stretched for 72 meters (236.22 feet). At 2.47 meters (8.10 feet) wide and 2.05 meters (6.72 feet) high, it was "large enough for mules to pass through." The miners completed their work in a few months and opened the tunnel to traffic in 1480. As historian Bertrand Gille puts it, the tunnel under Monte Visio "denotes an act of will and material means which were both out of the ordinary."[18]

Civilian mining matured a great deal during this period to become a profession, and the design and dimensions of mining projects advanced as well. This era also witnessed the publication of the most important mining manual to appear, *De Re Metallica*, a book published in Latin in 1556. It was immediately followed by a German-language edition the next year and an Italian-language edition in 1563. The German Georg Bauer, who signed himself Georgius Agricola for this book, produced the most thorough discussion of commercial mining and metallurgy of the premodern era. It is graced with hundreds of detailed illustrations and thus serves as the major source of information on civilian mining practices during the Renaissance. *De Re Metallica* remained the standard text on the subject for the next two hundred years.[19]

Agricola details the dimensions of the shaft and gallery arrangement that was typical of commercial mining. The shafts generally were vertical or at a slight incline and usually no more than about eighty feet (24.38 meters) deep. A shed often stood over the top to prevent the elements from precipitating into it and to protect the men who worked the windlass and ventilation devices. The galleries were typically twice as high as they were wide, with enough room for two men and their loads to pass each other. Two men typically worked at the face to continue the drive; one dug the upper part and the other the lower part of the face, one usually a bit more advanced than the other: "Each sits upon small boards fixed securely from the footwall to the hanging wall, or if the vein is a soft one, sometimes on a wedge-shaped plank fixed on to the vein itself." Agricola provides detailed instructions on how to shore up a shaft and a gallery, advocating a wooden floor if wheelbarrows were used to carry out spoil and ore. He discusses the means of surveying for underground navigation, all based on the measurement of triangles, and details the use of fans, bellows, and wind sails for artificial ventilation if a natural draft of fresh air could not be encouraged through the mining system.[20]

The techniques of military mining mimicked those of commercial mining that Agricola describes, but there were some important differences as well. Military mines typically involved shorter galleries constructed as quickly as possible for temporary (and destructive) purposes. Those works were shored but usually kept at quite small dimensions. Civilian galleries were meant to last for longer periods of time and had to be spacious enough for the workers. But interestingly military miners saw the potential of gunpowder earlier than their civilian counterparts, mainly because of its role in blasting an enemy wall, and worked out the problems of this process before civilians.

In one way, however, the old relationship between military mining and civilian mining continued with the borrowing of labor. During the siege of Haarlem in 1572–1573, the Spanish used coal miners from the area around Liege in the Spanish Netherlands to construct offensive galleries. The Dutch defenders neutralized their efforts through countermining but eventually lost the city when a relief force was defeated and their supply lines were cut off.[21]

Powder-charged mines were seen outside the heartland of military mining during the sixteenth century, though not in a widespread or intense way. Military forces of the Mughal Empire of north India had not employed mining of any kind very much until the late twelfth and early thirteenth centuries. But when gunpowder became available, they utilized it in at least one major operation. Akbar besieged the massive fortress of Chittor in an attempt to reduce the last holdout against Mughal rule, the Kingdom of Mewar, beginning on October 20, 1567. Mughal soldiers began digging three saps (approach trenches on the surface) from about small-arms range of the wall and, when close to the target, began digging a tunnel through solid rock. They managed to excavate two cavities under the wall and charged them with 4,800 pounds (2,177.24 kilograms) of gunpowder in one and 3,200 pounds (1,451.49 kilograms) in the other. Miners rigged one powder train with a branch to each cavity, but only one charge exploded as planned when the train was lit on December 17. The infantry attack began as planned, and in the middle of it the other charge went off, killing nearly two hundred attackers and forty defenders and shutting down the assault. The Rajputs of Mewar constructed a secondary wall to seal off the partial breach. Mughal miners dug a third chamber at another place along the wall, "but it did not take fire properly," according to a contemporary account. This third blast killed thirty men of the garrison but failed to open a breach suitable for an attack.[22]

While mining failed the Mughals at Chittor, sapping worked. Akbar ordered the construction of a large approach trench protected by side walls and covered with a roof. This work inexorably neared another section of the defending wall. After closing in on the target, sappers began chipping away at the masonry wall on February 23, 1568. More importantly around this time, the garrison commander was killed, reportedly by a short fired by Akbar himself, after which morale plummeted within the fortress. Two days later the Rajputs sallied out to die in vain attacks on Mughal positions, and the siege came to an end.[23]

Elsewhere in Asia, the first gunpowder mine in Japan was constructed under Kameyama Castle during the siege there in 1583. Located in Mie Prefecture and a stronghold of the Ishikawa clan, the castle surrendered to an opposing clan after the mine explosion. Closer to the core area of mining activity but on the eastern fringe of Europe, gunpowder mining was introduced to Russia in 1535 by the Polish siege of Starodub, where an underground explosion breached a defensive wall. Polish forces mounted a major siege of Smolensk in September 1609. They employed gunpowder mines in an effort to breach the city wall, but Russian countermines foiled the effort. The final mining project, however, opened a passage for assaulting troops who captured the city after bloody fighting on June 13, 1611.[24]

After surveying the employment of powder-charged mines during the fifteenth and sixteenth centuries, we should consider whether gunpowder really represented a revolution. In a technical sense the answer certainly is yes. Powder explosions held the potential, if everything went right, to bring down a section of masonry wall in spectacular and possibly devastating fashion. But the key lay in proper placement of the cavity, proper charging of it, and using a reliable means of detonation, all of which was more complicated technically than prop mining. As the first mine effort at the Chittor fortress shows, powder trains were capricious instruments for setting off charges. Still, they continued to be used for hundreds of years, even in the famous Crater battle of the American Civil War in 1864, and they were anything but foolproof. Even with proper explosions, powder mines sometimes failed to bring down the targeted wall properly. Moreover, there is no reason to believe that besiegers were more successful than defenders during the period of gunpowder's introduction into mine warfare.

"Gunpowder could not alter the usual pattern of difficult and prolonged sieges," concludes historian Jos Gommans of the Mughals. The technology was not advanced enough to be fully reliable, and even if had been, sieges were complicated affairs. There were many other factors, including troop strength, the nature of a particular fortification and the surrounding terrain, morale factors, and logistics, that came into play to determine which side would succeed.[25]

Gunpowder certainly changed the technological aspects of military mining, but it did not change the operational factors that accompanied underground warfare. Technology was by no means a determinant of success in any consistent way. Mining remained one of many siege techniques; it was not elevated to a position of primacy because of gunpowder, yet it also was not reduced to a position of irrelevancy by it. One of the reasons for all this is that the defender also incorporated gunpowder into his range of skills and thereby enhanced his ability to counter the new threat. This was, in fact, one of the major consistencies in the long sweep of mining in warfare. Countermeasures kept pace with aggressive new techniques, and the result was consistent equilibrium between attack and defense in mine warfare. Gunpowder was a revolution of sorts but only in a limited, technical way.

5

The Maturation of Mine Warfare

The widespread use of gunpowder for military mining by the beginning of the seventeenth century presaged the growing professionalization of mine warfare. Reflecting the rationality of the growing age of science that century and the next, military men became interested in experiments with and the intense study of the technical aspects of setting off offensive mines and countering them with explosive charges. The process of using gunpowder was worked out to a fine degree during this period, compared to the early, rugged days of its introduction to siege mining. Moreover, by the end of the seventeenth century, advanced armies decisively shifted from hiring civilian miners to creating professional corps of officers and soldiers trained in the details of digging mine shafts and galleries. As commercial mining had become professionalized in the medieval era, so did military mining in the latter half of the seventeenth century.

Another important development of this era involved reorienting the age-old process of sapping into a new tactic with the old name to counter artillery fire from fortified positions. This reorientation also had an important effect on siege mining. During the previous long history of siege approaches, sapping had been defined as the process of chipping away at a masonry wall to create a breach through which attacking forces could enter the enclosure. In the seventeenth century sapping was redefined as the process of digging an approach trench on the surface of the ground, usually in zigzag patterns, to protect troops as they worked their way through a zone covered by artillery fire. Only when the sap, as the trench was called, neared the defending wall were the miners called in to dig short tunnels under its foundation to plant explosive charges. Sapping and mining had always been connected-but-separate tactics since the early development of siege approaches. In the late seventeenth century, that relationship became tighter, more mutually reliant, and more systematized than it ever had been. That is why when professional units of engineer troops were formed in the late seventeenth century, they were designated as Sappers and Miners.

Finally, the late seventeenth century came to be the age of Sébastien Le Prestre de Vauban, who masterminded the most important doctrine of siege warfare to date. While Vauban added little that was new to ideas about military mining, his contribution lay in codifying a comprehensive system of siege approaches that included the most advanced way to use saps and mines offensively. In addition, many military engineers conducted experiments and recorded technical data about the powder charge, its explosion, and the effect of the blast on surrounding earth to begin building a scientific and technical catalogue of the precise nature of subsurface gunpowder explosions. Never before had military mining been subjected to such scrutiny and its problems and potential so intensely studied by the men who relied on it in the field.

Those developments came to fruition during the later seventeenth and throughout the eighteenth centuries in the core areas of European mining history. But older siege-mining techniques persisted in some of the more peripheral areas of the core during the early and middle decades of the 1600s. The earliest gunpowder mine in England occurred at the siege of Lichfield Close in Staffordshire in 1643, more than a century after the first gunpowder mines were sprung in Italy. Medieval-style sapping operations, covered by penthouses or sows, continued to be used during the English Civil Wars (1642–1651), although they were not common. Sieges, however, often took place during this long, terrible conflict. Of 645 military actions, at least 198 were sieges, and mining operations occurred in many of them. In just one example, heavy rains collapsed offensive galleries during the first siege of the Royalist garrison of Lathom House in early 1644. Three miners were killed, and the siege later was lifted.[1]

Several English Civil War sieges left behind physical evidence of mine warfare that have been uncovered by archaeologists. A feature discovered 345 feet (105.15 meters) from the east gate of the defenses at Gloucester has been ascribed to an offensive mine the Royalists dug during the siege of the place in 1643. Covenanters sprang a gunpowder mine during their siege of

Newcastle upon Tyne that partially breached the defenses in 1644. Archaeologists much later found "an ovoid, funnel-shaped crater 5–6 m (16–20 ft.) across and at least 15 m (5 ft.) deep."[2] This may well be the earliest physical evidence of an explosive mine crater yet discovered.

The first Parliamentarian siege of Pontefract Castle, from December 24, 1644, to March 1, 1645, involved offensive mines at the keep and at the King's Tower. Royalist countermeasures included eleven or twelve pits dug by local colliers as beginnings of possible countermines. Archaeologists uncovered and excavated three of these pits, located very near the inner face of the curtain wall. Two of them were only shafts, while the third showed evidence of the start of a gallery but without any shoring apparent. This third shaft was circular at the surface and measured 2 meters (6.56 feet) in diameter. It descended 7.4 meters (24.27 feet), and the "lower 4 m [13.1 feet] of the shaft had been cut through solid rock," according to archaeologist Ian Roberts. The gallery extended east under the curtain wall for 2 meters, "creating a small rectangular rock-cut 'chamber'" 1.2 meters (3.93 feet) square. Chisel marks were still visible on the walls and floor, but there was no "evidence of sooting from candles or torches on the roof of the chamber." The other two countermine shafts were of similar size and depth to the third one, but they held no evidence of a gallery. All three were filled in sometime after the first siege and included many civilian and military artifacts.[3]

The English Civil Wars later reignited in Ireland, where mining and countermining had been quite uncommon. Sapping walls while covered by sows is evident at Enniskillen Castle (1594) and Sligo Castle (1595) during the Irish Nine Years' War (1594–1603). But the first reference to tunnel mining in Irish history occurred at the siege of Ballyshannon Castle in County Donegal in 1597. English miners were responsible for that event.[4]

Perhaps the first attempt to use gunpowder in a siege not only in Ireland but also in Great Britain took place at Listowel Castle, County Kerry, in 1600. Sir Charles Wilmot pushed the siege against the Fitzmaurices, a family that earlier had joined in the First Desmond Rebellion (1569–1573) against the English. Wilmot's troops dug a gallery with the intent to plant a gunpowder charge, but the tunnel hit an area of spring water and became flooded. They started a second tunnel to avoid that area and nearly reached a vault in the castle. At this point the garrison surrendered on November 5, and the mine was not sprung. In 1986, workers installing drainage ditches and sumps at the site uncovered features that have been identified as the second gallery. It was "dug into stony yellow clay" and measured 0.7 meters (2.29 feet) wide and 0.8 meters (2.62 feet) high. Archaeologist and historian Kenneth Wiggins has concluded that the complications involved in gallery mining and the sporadic availability of civilian miners at any given time or place accounts for the sparse appearance of tunneling in Irish history.[5]

The onset of civil war tensions in Ireland led to the siege of a Royalist garrison in Limerick Castle, today called King John's Castle, by a group of Irish rebels in 1642. This thirty-seven-day siege would be no more than a footnote in history but for the remarkable discovery of mines, countermines, and especially of timber shoring by archaeologists in 1990–1991. They found more than three hundred pieces of wood used in supporting the galleries, most of it astonishingly well preserved. This find represents the biggest accumulation of perishable components involved in siege mining and countermining discovered at a pre–World War I site. Built in the thirteenth century at a loop of the River Shannon, Limerick Castle was located within the walled English Town and near a separate walled Irish Town. Most of the siege activity took place at the east wall, where archaeologists have uncovered remnants of four offensive mines and four countermines. Mine No. 2 and Countermine No. 3 met under the north end of the eastern curtain but did not affect the masonry of that wall. No shoring timber was found there. But a great deal of shoring timber was found in Countermines No. 1 and No. 2 farther south as well as in Mines No. 1 and No. 3.[6]

This siege was part of the Confederate War in Ireland, which broke out in 1641 and lasted twelve years. The Old Irish joined with Catholic Old English against Protestant New English. Limerick Castle was besieged by 2,000–4,000 rebels who had no heavy artillery and could not isolate the stronghold, so they concentrated on tunneling toward its east curtain. The defenders managed to dig a ditch 8–10 feet (2.43–3.04 meters) deep and 12 feet (3.65 meters) wide covering part of the east curtain to inhibit offensive mining—archaeologists found the remnants of this ditch in 1995–1997. The defenders could do little about the fact that houses and a church with a walled enclosure were located quite close to the curtain wall. The besiegers took advantage of this to start the galleries inside the houses. The Royalists heard sounds of work in those buildings by May 26. But they had too little gunpowder to shoot at the buildings, and raids failed to burn them down, so they began countermining at the east side of the castle on May 30. Two countermine galleries began, with sloped entrances that started some distance back from the east wall, while others began with a shaft closer to the curtain.[7]

Excavation revealed that two other English countermines began in shafts. Countermine Gallery No. 1 was 1.55 meters (5.08 feet) wide and 1.69 meters (5.54 feet) high, extending from a shaft that was sunk 3.28 meters (10.72 feet) deep. Countermine Gallery No. 2 extended from a shaft that was

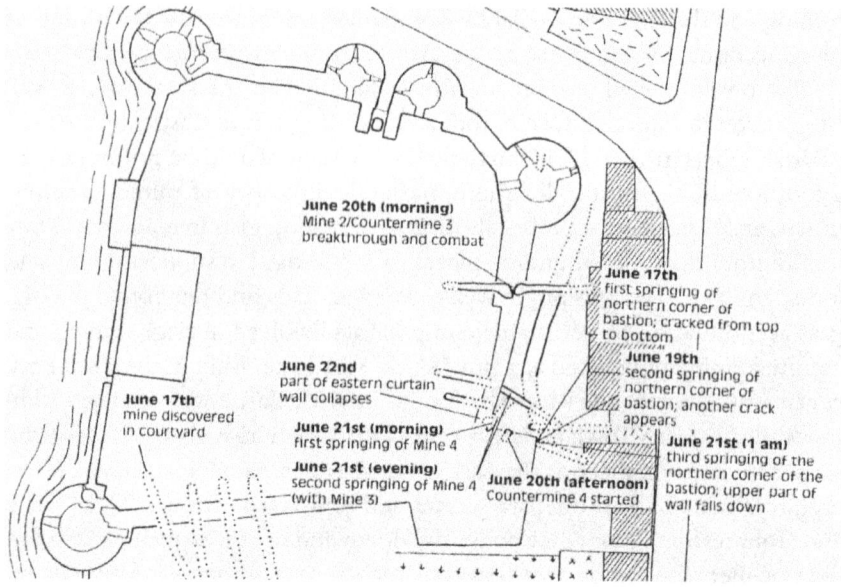

Siege of Limerick Castle (King John's Castle), 1642. From Wiggins, *Anatomy of a Siege*, 137.

2.91 meters (9.54 feet) deep. Measuring 1.29 meters (4.23 feet) wide and 1 me-ter (3.28 feet) high, it extended for 6 meters (19.68 feet). The Irish offensive galleries were longer than these. Mine No. 1 began at a house more than 12 meters (39.37 feet) from the target. Judging from the known dates of dig-ging and the distances involved in the mining and countermining at the east curtain, estimates show that both besieger and defender dug their galleries at the rate of 1.3 meters (4.26 feet) per day.[8]

Work also took place at the south wall of Limerick Castle by early June, with the offensive mines starting at Saint Nicholas's Church. According to documentary sources by eyewitnesses, countermines intercepted mines here at least three times, leading to underground confrontations, but ar-chaeologists uncovered no physical evidence of subterranean combat. Also according to the sources, in each of the three instances the Irish miners in-tercepted the Royalist countermines, and in at least two of them they tried to run water to flood the counterminers out. In the first instance the Roy-alists retained control of their gallery, and in the third instance they flooded their own gallery to keep the Irish from using it. Documentary sources also refer to the use of firearms at the junction of the mines and countermines, resulting in the killing of a friar who was part of the garrison and the deaths of up to eight Irishmen. One countermine on the south side reportedly was twenty-four feet (7.31 meters) deep, and there are accounts of more than one gallery collapsing due to water-saturated earth in this wet locality. The defenders also conducted a sally on the night of June 16 and entered one of the Irish galleries to temporarily drive out the miners. That gallery later flooded due to natural causes and was abandoned.[9]

On June 17 the Royalists noticed a crack developing in the southwest bastion of their castle and bore down with an augur to discover a mine gallery inside the enclosure. They poured water into the hole and assumed the gallery had come from Saint Nicholas's Church and extended under the south wall. A crack developed in the southeast bastion on June 19, resulting in the fifth and last underground contact, which took place at the east curtain on June 20. Members of the garrison found and entered Mine No. 2 by way of their Countermine No. 3, taking out shoring timber and setting a fire that smoked out the Irish miners after some shots were fired between the opposing sides. Archaeological evidence places this confrontation 1.2 meters (3.93 feet) from the outer edge of the east wall. The defenders also started another countermine, No. 4, at the east curtain that day.[10]

But the Royalist success on June 20 could not save the southeast bastion from further damage. Early on June 21 a major crack developed there, and later that day the Irish brought down part of the bastion by burning props, which also led to leakage of smoke through fissures in the ground; archaeologists found part of the rubble from this collapse. This was the work of Mine No. 4, which had caused enough subsidence since June 17 to develop the worrisome cracks in the bastion. This gallery ran only a short distance because it branched off from Mine No. 3 which in turn was a branch of Mine No. 1. Mine No. 3, which ran for a length of 9 meters (29.52 feet), probably had been started on June 12. Archaeologists found the shoring largely intact when they uncovered nearly half the length of this gallery. They also discovered that its floor rose and fell along its course, with a short section that was considerably higher than those forward and back from it. The Irish miners had shored this high middle section by setting up props on the side of the gallery and placing thin strips of light wood between them and the earthen walls. They also placed stones and packed clay along the floor, laying planks on top of that subsurface. This type of flooring seems more typical of long-term civilian mining than military mining.[11]

The excavated evidence also explains the complicated progression from Mine No. 3 to Mine No. 4. When Mine No. 3 neared the east wall, the miners realized that the foundation was much deeper than they could dig because of the water-logged levels below them, but the foundation was shallower to the south. So Mine No. 4 branched off from Mine No. 3 to reach that area. The unusual subsurface flooring described above was placed where Mine

Baseplates, Siege of Limerick Castle (King John's Castle), 1642. These items, recovered from Mine No. 3, have open as well as blind mortises for connection to upright pieces. Courtesy Kenneth Wiggins; from Wiggins, *Anatomy of a Siege*, plate 18.

No. 4 diverged from Mine No. 3 and was deemed necessary because of the wetness of the earth. The miners needed a hardened floor surface for firm footing and to better clear the spoil when they started to dig the new gallery. The entrance to Mine No. 4 was higher than Mine No. 3, and one needed a ladder to gain access to it. Mine No. 4 was, in short, quite shallow. When the defenders detected the sound of digging, they began Countermine No. 4 on June 20 but were too late. Mine No. 4 hit the foundation of the southeast bastion and then turned to undermine the end of the east curtain on June 21. The Royalist garrison surrendered two days later.[12]

Kenneth Wiggins has rightly pointed to the rare discovery of so many shoring timbers at the renamed King John's Castle in Limerick. Other than the timbers found by the French and American archaeologists at Dura-Europos in the early 1930s, there are no other significant finds of this kind outside of the Western Front of World War I. A total of 309 pieces of wood were taken from two Irish offensive mines and three English countermines at King John's Castle; 57 of them were framework elements with "mortise-and-tenon joinery which formed into rectangular frames placed at intervals," 19 were baseplates, 36 were upright props, and the remaining 2 were top plates. Many "lining timbers" had been placed between the upright

Wall Planks, Siege of Limerick Castle (King John's Castle), 1642. Made of oak, such planks were wedged between the earth and the upright supporting timbers to help prevent the walls of Mine No. 3 from slumping. Courtesy Kenneth Wiggins; from Wiggins, *Anatomy of a Siege*, 207.

props and the earthen walls of the galleries and in places between the top plates and the earthen roof. Most of the 176 timbers that could be identified by type of wood turned out to be oak, but birch, alder, Scots pine, and elm also were represented. The joinery exhibited either blind or open mortises at both ends of the baseplates. Two planks were so long, at 3.1 meters (10.17 feet) and 3.6 meters (11.81 feet), that they could not have been taken through the junction of a shaft and a gallery. The start of that mine had to have been in the cellar of one of the cage-work houses near the east wall, making a shaft unnecessary. The planks also must have come from the house structure itself.[13]

The archaeological finds at King John's Castle reveal that the timber framing must have been the work of professional carpenters. The uncovering of the galleries also reveals that the miners were of the commercial class as well. Miners in the region were displaced from their jobs and homes by the rebellion, and many of them wound up at Limerick. A handful of them are known by name in the written sources, offering their services to the English garrison, but those who worked for the Irish besiegers are lost to history.[14]

Limerick (King John's) Castle is a prime case study in underground siege warfare. It is one of a small handful of sites in the history of military mining that has been thoroughly documented by archaeological as well as documentary evidence. The most important finds are the more than three hundred pieces of shoring timber, an unusual resource that is supplemented by detailed physical remnants of several galleries. Combined with enough documentary evidence written by participants, the whole creates an unusually full picture of what would otherwise have been an obscure siege in a complicated war. The event also is unique in that virtually all activity was mining and countermining; little else was done by the besiegers, who managed to win the struggle against vigorous countermining. It is obvious that the Royalist garrison was behind the curve in countering their opponents; the Irish, although suffering setbacks, maintained the initiative in their offensive mining. The besiegers also had no apparent access to men with knowledge about using gunpowder in mines, for they relied entirely on old-fashioned prop mines. It is ironic that the siege of Limerick Castle was an old-fashioned affair occurring in the early stages of a burgeoning era of modernization. It took place after the introduction of gunpowder yet involved no powder explosions. But the siege demonstrated that earlier methods still could be effective.[15]

The last incidence of mining in Ireland occurred in the conflict between James II and William III in 1689–1691. While military mining occurred fairly often in Great Britain and Ireland during the English Civil War, it was last used there during the Jacobite uprising of 1745–1746. Infrequent use of mining techniques inhibited the adoption of new methods and reinforced the persistence of old ones. On the continent of Europe, in contrast, military mining innovated and flourished for a long time to come.[16]

Vauban sealed the importance of mining within siege warfare. He had such an influence on the history of military mining, as noted earlier, not because he developed new ideas about siege craft but because he succeeded at creating a system of siege approaches for current military conditions. Zigzag trenches, now called saps, had existed before his time. The concept of starting them from a deep trench called a parallel (because it was dug parallel to the enemy line) had also predated his career. The Turks had used a parallel in their long siege of Candia (now known as Heraklion) on the island of Crete from 1648 to 1669. Vauban's contribution was not in being the first to use saps or parallels but in bringing those elements together into a system for others to follow. As historian Janis Langins puts it, "his accomplishment lies not so much in the originality of his idea as in its rapid adoption, development, systematization, and application." Vauban used his system first at the siege of Maastricht in 1673. After that city's capture, everyone credited him for the concept of using a siege parallel even though he likely derived it from the Turkish example at Candia.[17]

The greatest contribution of Vauban's work on military mining lies in the setting for siege approaches. For centuries before his era, miners were forced to seek places as close as possible to the target before starting to undermine it. If they were lucky, they could utilize preexisting buildings as did the Irish at Limerick Castle or burial places as at Dura-Europos. Long galleries were rare in the classical, medieval, and Renaissance eras. Vauban's method now allowed miners to begin their galleries at the forward end of the sap; they did not have to choose distant locations from which to start digging and could dispense with sows and penthouses for cover. He wedded sapping and mining in a tight union that would endure. Only on the Western Front of World War I did the two tactics again separate, as heavy artillery fire made digging an above-ground approach trench too dangerous; galleries there typically started from friendly lines that often were long distances from enemy targets. But before 1914, Vauban created the best way to get miners as close as possible to defended positions without unduly exposing them to enemy fire. The many wars of Louis XIV of France helped spread the Vauban system to other European armies.

Warfare from the mid-1600s to the late 1700s tended to be positional in nature, focusing on the capture and defense of fortified cities and other strongpoints. Engagements in the open field were bloody, expensive, and uncertain in their outcome. They also were rarely decisive by producing war-winning victories. Louis XIV preferred sieges where the application of rationality, as embodied in Vauban's process, could offer better hope of success. The same was true of the king's opponents. The Duke of Marlborough fought four major field battles but conducted twenty-six sieges during the War of the Spanish Succession (1701–1714). Overall in this expansive conflict, sieges were 2.23 times more likely to occur than battles. They "provided the most common form of combat during this era," concludes historian John A. Lynn.[18]

Vauban's work would be the basis of mine warfare for the next two hundred years, at least up to the late nineteenth century. But a close examination of *A Manual of Siegecraft and Fortification* reveals that Vauban did not place a great deal of reliance on mining. He urges besiegers to wait until the sapping operation had taken them to the defender's ditch and their artillery had begun to crumble the masonry walls; only then were they to begin mining operations. He warns readers not to expect too much of the first blow. "It rarely happens that a mine is as successful as you hope, for there usually remain some escarpments that prevent you from carrying an assault to the top immediately." Be patient, he advises, and dig a second gallery to eliminate those remaining barriers before launching an assault.[19]

Vauban pays scant attention to countermining, and when he does comment, his vision is inadequate. If the opposing sides collided, his only advice is for the besieger to fire several musket rounds into the countermine gallery and then close the opening between the two mines to prevent the resulting powder smoke from escaping. The introduction of this smoke would be enough to neutralize the defending gallery. This cursory advice was no substitute for a thoughtful doctrine about how to deal with countermining.[20]

But Vauban is much better when detailing how to construct offensive galleries. He recognizes that they could vary in size and design depending on the need. The most commonly employed was a single design with one gallery and one powder chamber. A bit more complicated was a T-mine, with two branches and a triple powder charge. He advises digging the powder chambers one or two feet (0.30 or 0.60 meters) deeper than the gallery. Vauban writes about calculating the size of the powder charge by way of mathematical formulas. He estimates that fifteen pounds (6.80 kilograms) of powder would be needed to lift one cubic fathom (6.11 cubic meters) of earth but advises that amount to be increased by one-fifth to take into account many unforeseen factors. One should estimate the size and weight of the rampart to be attacked. Vauban provides in his manual a table spelling out how deep the chamber and how big the gunpowder charge should be to achieve a blow that would break the surface and properly damage the target.[21]

The Vauban manual deals with three different ways to arrange the powder when planting the charge. The older method was to place it in wooden barrels positioned in the chamber. Miners would uncover the barrels and break the sides to let powder flow out and fill spaces between them. The problem with this method was that the barrels took up so much of the limited space in the chamber and distributed the powder unevenly, leading to problems of detonation. Then someone had the idea to put the powder into bags and pile then up, slashing the fabric with knives to let it flow out. This solved the problem of space but not of poor detonation. The third method, endorsed fully by Vauban, was to lay down a wooden platform on the chamber floor and cover it with an inch-thick layer of straw. Then over this the miners placed coarse cloth, such as that used to make sandbags, pouring the powder loosely over the fabric to make a large uniform pile. This allowed the fuze to ignite the powder faster and more evenly.[22]

The fuze in general use during Vauban's era was relatively advanced. It consisted of a cloth tube filled with powder. One end was stuck in the center of the powder pile and the rest stretched out and secured in a wooden trough running along the gallery floor until reaching the mouth of the mine. In later years this would be called a powder hose. But Vauban's view of tamping is less than advanced. He writes of "closely joined beams, with dung used as cement and caulking to fill the gaps," and wooden wedges driven in with a sledgehammer to firm up the structure. He advises filling the entire length of the gallery with this tamping, taking care not to disarrange the fuze and trough. A priming device to light the end of the powder hose should be set up at the entrance of the gallery and covered until needed.[23]

Despite his lack of faith in mining as a reliable element in siege craft, Vauban provided what probably became the most widely read description, albeit brief, of mining techniques yet published. It was easy enough to write about the technical aspects of gallery design, powder placement, and detonation, but another matter to inform readers about how they could ensure that mining would win sieges for them. The tactical application of military mining, in other words, was little more than a technical service. But reaping operational benefits from it was a matter of coordination by the commander of his surface operations with the underground attack.

A significant development of Vauban's age was the growing reliance on a small corps of trained military men to conduct mining and countermining operations rather than obtaining civilian miners. Vauban convinced Louis XIV to form a corps of sappers and miners in 1673, and the Austrians followed suit ten years later. The concept spread less rapidly after that, reaching Prussia in 1742 and Britain in 1772, but the idea was sound and significant.[24] Military sappers and miners may not have had any experience or knowledge of civilian mining, but they were trained to meet the differing needs of siege mining. The fact that the corps combined training in surface trenching (saps) and underground tunneling (galleries) signifies the link between those two related operations.

Vauban represented crucial advancement in the maturation of mining technique during the latter part of the seventeenth century, which in turn centered this military craft firmly in Western Europe. Christopher Duffy believes the Turks also remained committed to mining during this period, especially after their use of it in the long and successful siege of Candia that ended in 1669. According to Duffy, the Turks measured the distance from their parallel to the target by stretching a cord along the surface of the ground under cover of darkness to know how long to dig the tunnel. They also kept the gallery straight by hanging a plum line from a peg driven into the surface above the mouth of the tunnel and aligned it with a candle placed at the head of the gallery. Turkish miners sat cross legged while they worked, making the gallery three to four feet high (10.91 to 1.21 meters), and removed spoil on trays that were dragged out with ropes. The Turks utilized a similar method praised by Vauban for distributing the gunpowder, piling it on cloth spread over the floor of their semicircular powder chamber. Their tamping consisted of filled sandbags and woolsacks, which were easier to carry and emplace than timbers.[25]

The Turkish methods described by Duffy represent good mining techniques, but they did not always ensure tactical success in a siege. Kara Mustafa's forces relied heavily on siege approaches and mining during the famous siege of Vienna in 1683. Their artillery was inadequate for the job of breaching the city's defenses, and they could not invest the place, so digging saps and starting mines as close as possible to three salient angles—the Burg-bastion, Burg-ravelin, and Löbel-bastion—became their main effort.[26]

Mustafa blew two mines on July 23 with limited effect on their targets, and the follow-up attack failed. Two days later a third mine exploded at Burg-ravelin and destroyed part of a palisade. Again, the infantry attack was repulsed. This pattern held true for the next week. The Austrians, effectively led by Ernest Rüdiger Stachemberg, countermined but seem to have had no real experts in mining technique. They blew a countermine at the Löbel-bastion on August 2 that failed to halt Turkish progress. Within nine days the besiegers had pushed their approaches until ready to cross the moat at Löbel-bastion. On August 10 the Turks sprang a mine, but "it misfired and they had to begin again," in the words of historian John Stoye. Two days later another Turkish mine created a rubble flow that made a rough causeway as high as the Burg-ravelin, and the follow-up infantry attack secured part of this position. A larger-than-usual Turkish mine exploded on September 2, which brought down part of the wall at the Burg-bastion. By this time, the Austrians evacuated the Burg-ravelin as a hopeless cause. Two days later, on September 4, another large mine blasted "a large hole in the wall to the left of the tip" of the Burg-bastion. It reportedly was thirty feet (9.14 meters) wide, and the Austrians barely managed to repel the Turkish infantry attack that tried to exploit the breach.[27]

It is clear that, despite their mining techniques, the Turks were having great difficulty combining their technical expertise with effective follow through by their infantry. While several of their mines were not well dug and their detonation process created problems, most of them achieved at least some degree of success in degrading the Austrian defenses. So far, however, every attack was repulsed. Stachemberg's forces relied on spirited action defending the breaches, but as time went on, the Turks were slowly gaining the advantage by wearing down the defenders' spirit and numbers.

On September 8 the Turks blew two mines that crumbled the tip of the Burg-bastion and a portion of the curtain. This left "only a small portion

of the masonry intact." Even with this most complete effect of all the mining efforts at Vienna, Kara Mustafa's infantry could not break through the
breach. His miners then concentrated on their most ambitious operation,
completing five mines along the curtain near the abandoned Burg-ravelin.
If technically successful, this effort likely would have led to the end of the
siege, as the defending force was weakening day by day. There were probably no more than 4,000 reliable defenders by this stage, and such a wide
breach would have been extremely difficult to defend. But Vienna was saved
by the approach of a large relief army of 60,000 men. Kara Mustafa could
use only 28,500 of his besieging force to meet this host. In a sharp battle
fought on September 12, the Turks were badly defeated, effectively raising
the siege.[28]

The Turkish siege of Vienna represented one of the more intensive uses
of mining in history. It also offered a cautionary lesson: no matter how good
the mining, tunneling was no guarantee of a successful siege. Only through
persistent application, combined with a willingness to absorb large losses
of manpower in failed assaults, were the Turks able to bring themselves to
the cusp of overwhelming the garrison with mining. Along the way they
worked out the bugs in their system and realized that expanding the plan by
planting several mines to be sprung at the same time against one section of
curtain was the key to success. It would be a lesson learned by the miners
of World War I as well.

While west and central Europe, along with the Ottomans, remained the
core area of military mining, not all armies within those regions embraced
it. Historian Pádraig Lenihan has argued that large-scale attacks rather than
siege approaches tended to be the reliance of the anti-French coalition involved in the Nine Years' War, conducted largely in central Europe from
1688 to 1697. He notes the storming of Namur Citadel on September 5, 1695,
as an example. Heavy artillery fire rather than mining opened a breach, and
a costly but successful assault settled the siege quickly. In short, even within
the core area of military mining, the technique was not uniformly used by
all armies.[29]

Nor did armies within the core area employ mining in a consistently
effective way. Every siege had its individual characteristics that mining engineers had to decipher. At the siege of Alicante, which lasted from December
1708 to April 1709 during the War of the Spanish Succession, French-Spanish
forces loyal to Philip V besieged an allied garrison that held a castle atop a
large promontory two hundred feet (60.96 meters) above the town. After
driving a gallery through the rock for three months, the French invited two
officers of the garrison to inspect their work. The ploy backfired; the officers believed that fissures in the rock would dissipate much of the charge

and decided to risk the results. When the charge was sprung on March 3, reportedly with an unbelievable 117,600 pounds (58.8 tons, or 53,342.46 kilograms) of gunpowder, it opened fissures on the surface of the parade that swallowed up fifty men and killed the garrison commander. The blast, however, failed to effect a breach in the defenses. Nevertheless, because a relief fleet could not reach the town, the garrison negotiated a surrender.[30]

In contrast, a French garrison at Tournai planted several countermines to blunt the efforts of an English besieging force led by the Duke of Marlborough and Prince Eugene of Savoy in July 1709. One countermine "blowed up and smothered severall of our workmen and likewise killd an officer & 28 men that was thereabouts," recorded Pvt. John Marshall Deane of Queen Anne's First Regiment of Foot Guards. The Royal Regiment of Foot of Ireland lost two officers and forty men in this blast. Later, on the night of August 18, French defenders set off three countermines that killed sixty of Marlborough's men and wounded others. Nevertheless, English forces captured the city by September 3.[31]

During the intense campaigns of the War of the Austrian Succession (1740–1748), about as many sieges as open-field battles took place. Most of the twenty-two sieges did not involve mining or countermining. The attackers won in seventeen cases by using a variety of methods that included artillery bombardment, breaching walls, and taking advantage of lapses in the defenders' diligence. When the defenders succeeded it usually was as a result of the arrival of a relief force.[32]

Either artillery fire or mining breached two places in the walls defending Brussels, the capital of the Austrian Netherlands, when Maurice of Saxony besieged it in January 1746. The garrison surrendered to avoid the destruction of the city. Mines may have been the operative mode for breaching two places in the walls of Asti, leading to the surrender of its French garrison on March 8, 1746.[33]

The sparse use of mining and countermining during this long war demonstrates that even in conflicts within the core area of mining, much depended on commander's decisions as to whether sappers and miners would be called on to do their work. Several other siege methods promised quicker and surer results than digging underground.

Nevertheless, by the middle of the eighteenth century, rational analysis of the process of mining and countermining rose to its most impressive height. Vauban began the concept of digging and springing test mines to study their effect as a prelude to their use in the field when he conducted experiments at Douai in 1686. He was the first to recognize the fracturing effect created by an underground powder explosion on the surrounding earth. In other words, he studied not only the crater, which was the surface effect, but also the unseen effect below the surface. Further experiments took place under the supervision of other French engineers at the artillery school at La Fère in 1725 and at D'Abouville in 1729. But those conducted by Bernard Forest de Belidor a few years later made the biggest impression on students of military mining. A professor at the French artillery school and later serving as the royal inspector of artillery, Belidor was influential in developing the sciences of hydraulics and ballistics but also turned his attention to the unseen forces unleashed by underground powder explosions. He rejected Vauban's assumption that most of the blast effect was funneled upward along the line of least resistance and theorized that it continued to be forced in all directions from the location of the charge, sending out shock waves through the ground. The crater and the material thrust out from it into the air was, in effect, just the tip of the iceberg, easily observed but not constituting the only or even the most important effect of the explosion. The difficulty lay in observing what was taking place underground. He also theorized that the gases produced by the powder explosion operated to exert pressure in all directions.[34]

To test his theories, Belidor organized an experiment at La Fère in 1739. He planned the placement of a powder charge with a line of least resistance amounting to ten feet (3.04 meters) and placed a series of galleries nearby. Gallery A was twenty-five feet (7.62 meters) from the charge, Gallery B thirty feet (9.14 meters), Gallery C thirty-five feet (10.66 meters), and Gallery D forty-two feet (12.80 meters) distant, all on the same plane as the charge. Then miners dug yet another gallery thirteen feet (3.96 meters) directly under the charge. Finally, they filled the chamber with 1,200 pounds (544.31 kilograms) of gunpowder and ignited it, the explosion creating a crater forty-five feet (13.71 meters) in diameter. All the galleries were severely damaged by the blast, including the one below the charge. The experiment proved that, even though much of the explosive force had gone upward, there was still plenty of energy to destroy tunnels within a certain radius in all directions.[35]

And yet some people remained unconvinced. Belidor thus agreed to a second demonstration, this time at Bizi in Normandy on the grounds of the Duke of Belle Isle, the minister of war. In this case the line of least resistance was twelve feet (3.65 meters), and four galleries were dug in the sandy soil and lined with masonry. Gallery A was located twenty-four feet (7.31 meters) from the charge, Gallery B was thirty feet (9.14 meters), Gallery C thirty-six feet (10.97 meters), and Gallery D, "lined with stout oak casing," forty-two feet (12.80 meters) distant. Another gallery was dug fourteen feet (4.26 meters) below the chamber. This time Belidor placed 3,000 (1,360.77 kilograms) pounds of gunpowder, which threw up dirt one hundred fifty feet

(45.72 meters) into the air. The blast created a "perfectly circular" crater sixty-six feet (20.11 meters) in diameter and seventeen feet (5.18 meters) deep. The results were devastating. Galleries B and C were completely broken except for twelve feet (3.65 meters) at each end. Forty-five feet (13.71 meters) of Gallery A was crushed, leaving intact only sixteen feet (4.87 meters) at one end and twelve feet (3.65 meters) at the other end. In Gallery D only twelve feet at the end farthest from the charge was left, while twelve feet of the gallery located under the charge was destroyed.[36]

After studying the extent of damage and the exact position of each gallery in relation to the chamber, Belidor concluded that his mine could have destroyed any tunnel at least fifty feet (15.24 meters) deep. That was the greatest depth at which countermines tended to be sunk during this era. He coined the term "globe of compression" to describe the impact of an underground explosion on the area around it and argued that a defender could destroy the offensive mines of a besieger without even breaking the surface of the ground with a crater. That was a significant fact because such a crater could be used by the besiegers for their own purposes. Even if visual confirmation could not be obtained through the creation of surface damage, there was no doubt that significant destruction was taking place underground.[37]

Belidor studied the effects of underground explosions more thoroughly than anyone before him and encouraged the deliberate overcharging of a mine to widen the zone of destruction. He also was the first to demonstrate how a crude trench could be instantly created by blowing the top off a length of gallery. Ironically, even after these two experiments, there were still some who doubted the existence of a globe of compression. Engineer officer Simon L. Le Febvre confirmed Belidor's conclusions through a similar demonstration held at Potsdam, Prussia, in 1754. With Frederick II in attendance, Le Febvre arranged for the placing of a chamber with 3,000 pounds of gunpowder and with a line of least resistance of fifteen feet (4.57 meters). He dug three galleries twenty-four feet (7.31 meters), thirty-two feet (9.75 meters), and forty-two feet (12.80 meters) from the charge, with a fourth tunnel sixteen feet (4.87 meters) under it. The galleries were three feet by five feet (0.91 by 1.52 meters) in the clear and lined with oak. Upon detonating the charge, the explosion created a crater sixty-six feet (20.11 meters) in diameter and eighteen feet (5.48 meters) deep, "free from rubbish, and perfectly smooth." The destruction of galleries mirrored that of Belidor's experiments.[38]

Finally, the doubts disappeared. There was no question that the globe of compression accurately described the effects of underground explosions, but some questioned the exact shape and extent of the zone of destruction. Henri Jean-Baptiste de Bousmard de Chantereine argued in his *Essai*

General de Fortification, published in Paris in 1797, that the globe of compression could destroy any galleries within a zone whose radius was "equal to four times the line of least resistance." In other words, if the line of least resistance was fifteen feet (4.57 meters), the globe of compression extended sixty feet (18.28 meters) in all directions from the powder charge.[39]

The first effort to consciously employ the globe of compression in mine warfare took place at the Prussian siege of Schweidnitz in Silesia during the Seven Years' War (1756–1763). Le Febvre, now holding the rank of major, conducted this mining operation, which began soon after the Prussians established their third parallel on the night of August 22–23, 1762. That trench was 150 paces (114.30 meters) from the ditch of Austrian Fort No. 11, also known as the Jauernich Fort. The miners sunk the shaft so that the roof of the resulting gallery was only four feet (1.21 meters) below the surface. Having extended the gallery fifty-two feet (15.84 meters) by August 26, that night Le Febvre sent a miner with one end of a rope crawling forward from one shell crater to another to measure the exact distance to the ditch. The man went as far as he felt it was safe and estimated the rest of the distance. By August 31, the gallery extended ninety-six feet (29.26 meters); Le Febvre decided to dig a chamber and pack it with 5,060 pounds (2,295.17 kilograms) of gunpowder. The Prussians exploded their mine at 9 P.M. on the night of September 1–2, then the next morning realized it had been planted far too short to fill the ditch with upturned earth. That day Le Febvre began a new gallery from the bottom of the crater, pushing it another twenty feet (6.09 meters) before the Austrians exploded a countermine on September 4. It failed, however, to halt Prussian digging.[40]

Water proved to be more of a hindrance than Austrian countermining at this stage of the siege. The Prussian gallery hit underground water on the night of September 5–6 as the gallery neared a total length of thirty-two feet (9.75 meters). The sodden earth collapsed part of the mine, forcing Le Febvre to begin a new one, his third so far, to the left of the old one. It reached a total length of fifty-four feet (16.45 meters) by the night of September 9, when a second Austrian countermine exploded, once again causing little damage to the offensive gallery. The Prussians continued digging until the night of September 10–11, when a third countermine exploded at midnight and closed part of their gallery. But Prussian miners cleared the damage during the day and pushed. Their digging continued as the Austrians sprung two countermines, one on each side of the Prussian gallery, on September 14. Neither explosion forced the besiegers to halt their operations. Le Febvre then exploded his second mine at 5 A.M. on September 16, still some distance short of the ditch. He started his fourth gallery from the bottom of this crater that night. but twenty-four hours later three Austrian countermines

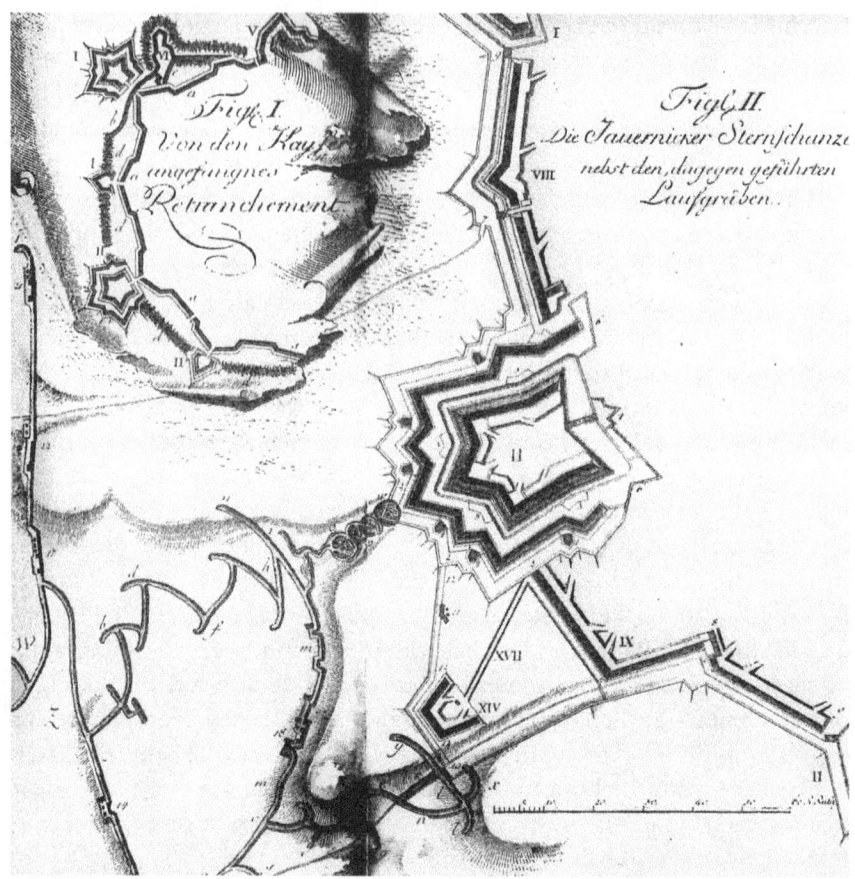

Prussian Siege Approach at Schweidnitz, 1762. The single approach sap and several mine explosions are indicated on this map of the successful Prussian operation. From J. G. Tielke, *Beyträge zur Kriegs-Kunst und Geschichte des Krieges von 1756 bis 1763*, vol. 4 (Freiberg: Barthel, 1778), plate 6.

destroyed it. The Prussians began their fifth gallery on the night of September 18 only to have another countermine completely destroy it that afternoon. Le Febvre began his sixth gallery between the two previously sealed ones, which extended more than six feet (1.82 meters) before another countermine destroyed it on September 19. The Austrians destroyed the seventh Prussian gallery, which also was less than ten feet (3.04 meters) long, at 6 A.M. on September 20. Once again Le Febvre began a new gallery, his eighth, but the Austrians destroyed it at 8 P.M. on September 22. Only by working a little faster than their opponents were the Prussians able to dig the twenty feet (6.09 meters) of their ninth gallery, plant a charge, and explode it at 10 P.M.

on the night of September 24. It made a crater a few feet from the palisade along the outer edge of the ditch at Fort No. 11.[41]

The siege operations at Schweidnitz had developed into an intense and deadly contest between besieger and defender. The Austrian counter efforts increased in efficiency the closer Le Febvre came to his goal, knocking out one Prussian gallery after another. Digging faster was at best only a slim edge over their opponents, so the Prussians could not guarantee success unless they found a better solution to the problem. They found that solution when Le Febvre decided to abandon his shallow approach and dig the next gallery deeper than the Austrian countermines.

Going back to the crater of his second mine, which had exploded on September 16, Le Febvre dug a shaft through the bottom to begin his tenth and eleventh galleries. He extended one fifteen feet (4.57 meters) and the other more than eleven feet (3.35 meters) before the Austrians exploded a countermine at 9 P.M. on September 26, destroying the shorter one. Le Febvre pushed forward with the other gallery, reaching a total length of sixty-five feet (19.81 meters) by October 4, when another Austrian countermine exploded but did no damage to it. Countermines that exploded at 3 P.M. on October 7, midnight that night, and 2 P.M. the next day failed to seriously damage the deep gallery, which stretched ninety-six feet (29.26 meters) before it reached the ditch of Fort No. 11. The Prussians moved 5,060 pounds (2,295.17 kilograms) of powder, carried by 170 men, into the chamber. The blast at midnight of October 8–9 was followed by a night assault of infantry that failed to break into the fort. But at dawn on October 9, Le Febvre saw that his mine had been close enough to push a great deal of dirt into the ditch, creating an effective ramp across it and up to the top of the fort's rampart. It was a breach inviting an attack in daylight, when the Prussian infantry could see their advantage. The Austrians realized this too and surrendered that day, October 9, after sixty-three days of vigorous defense.[42]

Schweidnitz was a watershed in the history of military mining. It was the first self-conscious use of Belidor's globes of compression and of using mines as a tool to progress the siege approach. At first Le Febvre did not deliberately explode mines well short of his target but later he came to do so in order to use the resulting craters as staging areas to begin new galleries. In this way he could make those new galleries shorter and avoid the need for artificial ventilation. He also was the first to explode mines to throw dirt into defending ditches to create assault ramps, which proved to be the final straw for the Austrian defenders. The Prussians relied on underground digging rather than surface sapping to cross the deadly ground between their third parallel and the ditch of Fort No. 11, violating Vauban's system in an important way. To do so, however, they had to engage in a race with the increasingly effective Austrian countermining efforts. The defenders had essentially stalemated the besiegers until Le Febvre tricked the Austrians by suddenly digging deeper than them. The Austrians failed to realize this in time to deepen their defensive galleries and paid the price for it. Schweidnitz became famous for the employment of the globes of compression, even though they were not used in the way that Belidor suggested as a countermeasure against offensive mining. But the siege validated the concept because Le Febvre, who had confirmed it through earlier experiments, now used mines in a new way.

Besides Le Febvre's innovation of using mine explosions to create craters as an offensive tool, the other salient feature of the siege at Schweidnitz was the persistency of the Prussian effort. This marked a new level of intensity in mine warfare. Previously, one or two underground encounters typically characterized a siege before other methods of approaching the target took precedence. At Schweidnitz the besiegers relied primarily on mining, which forced Le Febvre to press on every time the Austrians closed one of his galleries. Eleven Prussian galleries were started before the last one crowned the counterscarp of Fort No. 11, creating a fine breach that directly led to the fall of the city. Few other sieges could boast of a mining success equal to Schweidnitz.

The Austrians would have gained something on the besieging Prussians if they had already constructed the basics of a system of countermines long before the siege began. From its earliest origins in the sixteenth century, the concept of permanent countermines flowered throughout the eighteenth century in the core area of mine warfare. Beyond a few brief references to them found in general histories of fortresses and siege warfare of this era, no one has studied the art of permanent countermining, thus limiting our understanding of the concept and its practical application. But there are a handful of extensive permanent countermines still intact in the twenty-first century whose existence has been revealed primarily through the need for cities to draw tourists to their area.

At all such sites the permanent countermines tended to consist of only the foundation of an effective system. Designers dug a gallery just outside the defending ditch of the work and lined it with masonry. Often this gallery doubled as a counterscarp battery, allowing troops within it to fire back through portals into the ditch at enemy troops who entered it. The gallery also included underground portals facing outward toward the approach of an offensive gallery. These were like windows, arched and filled with thinner masonry so they could be quickly broken through and a defensive gallery dug forward when and where needed. In some cases these defensive galleries were constructed long before any siege, so they could serve as readymade listening posts as well. Although involving a considerable amount of time and expense, the construction of these permanent foundations for a countermine system cut down at least by half the time and money needed to build such a system from scratch. Because they appeared at dozens of cities, they obviously were considered worth the effort and resources.

Among the earliest examples of a permanent countermine can be found at Namur, at the junction of the Sambre and the Meuse Rivers in Belgium. Vauban planned permanent countermines to defend the citadel (the old medieval castle) and Fort William, constructed as a self-contained hornwork on the northwest sector of the city's defenses. The countermines were established between 1692 and 1695. Soon after that a similar system was constructed at the city of Huy, Belgium, in the Meuse valley. This multiple fortification complex included Fort Rouge and Fort Picard on the north side of the river and east of the town. The permanent countermine started from Fort Rouge, a strong work on the top of a ridge, and extended forward and under the smaller Fort Picard in front of it, perhaps to serve the double purpose of detecting offensive mines and to blow up Picard in case it was captured. This small countermine was completed by 1702. A photograph of the entrance to this countermine, taken in 1982, shows a rectangular-shaped gallery with no shoring.[43]

The most famous permanent countermine—at least as far as the modern tourist market is concerned—is located at Turin in northern Italy. It was tested during the French siege of the city from June to September 1706. The defenders used this system to crush two French offensive galleries and prevented French miners from entering the permanent countermine by exploding a charge that blocked their way with rubble. The siege was soon after raised by a relieving force. Only about 1.5 kilometers (0.93 miles) of the Turin countermine system survive. Elsewhere in southern Europe, Fort Manoel at Valletta, Malta, saw the construction of a major countermine system from 1723 to 1733. The Castillo de San Fernando at Figueres, Spain, also acquired a countermine system consisting of five galleries that were dug from 1753 to 1766.[44]

The Austrians invested more time, energy, and money in constructing large permanent countermines than anyone else. Engineers proposed a system for the fortress of Peterwaradin on the Danube (modern Petrovaradin in Novi Sad, Serbia) in 1764. It was constructed on four levels with a total of 16 kilometers (9.94 miles) by 1776. But the epitome of Austrian countermine construction occurred during the reign of Emperor Josef II (1780–1790). Forty kilometers (24.85 miles) were dug at Fortress Josefstadt (modern Josefov) by 1787. This probably represents the largest permanent countermine system ever constructed, although the one at Klodyo in Silesia (modern

Permanent Countermine Gallery, Maastricht. Completed by 1825, the permanent countermine system at Maastricht included 8.6 miles (13.84 kilometers) of galleries. Today, only a small portion of these defensive works is open to tourists. Earl J. Hess.

Kłodzko), constructed after the acquisition of the fortress by Frederick the Great in 1742, reportedly is the same size. Late in his reign Josef II commissioned the countermine system at Theresienstadt (modern Terezin), which amounted to several dozen kilometers of tunnels.[45]

While the eighteenth century represented the height of permanent countermine construction, the extensive system dug to defend Maastricht in the southern Netherlands was built over a period that extended to 1825. Vauban initially made plans for this system after he captured the city in his well-known first use of siege parallels in 1673, although construction began much later. When finished, the Maastricht kazematten, as they are locally known, included 14 kilometers (8.69 miles) of tunnels. Today they are among the most frequently visited of all permanent countermine systems.[46]

Taken as a whole, the period of the seventeenth through the middle of the eighteenth centuries was a watershed era in the history of military mining and countermining. This was largely due to the application of rationalism and professionalization as applied to the complicated process of utilizing mines in sieges. Vauban sparked this trend with his compilation of previous siege techniques mixed with his own theoretical twists (largely proven to be effective through his own application in the field). No one before him had done a better job of incorporating offensive mining into the context of siege approaches, cementing the important link between sapping operations and gallery mining. Almost everyone who studied Vauban's doctrine would be imbued with the need to begin mine galleries only after surface sapping had reached its closest point to the target in order to shorten the length of galleries and avoid the thorny problem of providing proper ventilation for them.

Equally important was Belidor's experiments to determine more information about gunpowder explosions and their effect on the underground strata as well as on the surface of the earth. Belidor's discovery of the globe of compression opened the possibility of a more sophisticated and nuanced use of gunpowder mines for defense as well as for offense. Le Febvre's aggressive conduct of offensive mining at the siege of Schweidnitz demonstrated what could be accomplished with persistent application of mining doctrine. All of the developments of this later period breathed new life into the process of military mining.

6

The New World and the Napoleonic Era

Mining appeared for the first time in the Western Hemisphere during the early eighteenth century, demonstrating both the spread of this military practice and its limitations on a global scale. British colonial culture was the medium for this cross-Atlantic transfer of siege technique, and the target was a fortified Native American village. The fact that this village was designed and built along the general lines of European fortification principles also demonstrates the universality of those principles. Nevertheless, the scale of mine warfare in the New World was miniscule compared to that of western and central Europe.

But even in the core area of mining history, underground approaches were utilized on a comparatively rare basis from the end of the Seven Years' War in 1763 to the onset of the Crimean War in 1854. Fewer commanders opted for it mostly because the pace of operations tended to accelerate, especially during the Napoleonic conflicts. Only in the Peninsular War (1807–1814) did the British resort to intensive mine warfare in several sieges. They also experimented with practice siege mines in preparation for actual use in the field more than any other belligerent. Despite a comparative lull in the trajectory of siege mining, the era kept the military craft of tunneling and countermining alive and well.

In 1711 the Tuscarora people rose against encroaching English settlers in the part of Carolina Colony that later became North

Carolina. This prompted the colonial government to send an expedition of white militia supplemented by a larger contingent of friendly Native people under the command of Col. John Barnwell. The combined forces conducted sapping operations against Catechna, a fortified town defended by members of the southern wing of the Tuscarora under Chief Hancock in April 1712. The village surrendered under terms.[1]

A second expedition, led by Col. James Moore, consisted of thirty-three whites allied with nine hundred Natives. In March 1713 it struck at the primary Tuscarora village of Neoheroka, which had been constructed not only with palisades typical of Native defensive architecture but crudely designed projecting bastions and a number of underground bunkers for sheltering women and children as well as warriors. Moore ordered the digging of an approach sap and erected a blockhouse and artillery battery. On nearing the palisade, the colonists and their Native allies dug a mine gallery under the work. They sprung the charge at 10 A.M. on March 20 but "with very little Success the Powder being damnified," in the words of the official report of the expedition. A follow-up attack, however, managed to secure part of the palisade line. The defenders constructed a secondary line, but the besiegers overran it, and the town fell on March 23.[2]

The explosion at Neoheroka was an inauspicious beginning for mine warfare in the Western Hemisphere, and it would not be followed for some time to come. But the concept of mining and countermining was part of the military cultural baggage brought to the New World by European forces and disseminated through texts to colonial forces as well. The first permanent countermine to appear in the Americas was at the French fortress of Louisbourg on the shore of Cape Breton Island, Nova Scotia, Canada. Constructed between 1720 and 1740, the fortress was a massive complex of masonry walls, bastions, and gun emplacements. Captured by British colonial forces in 1745 and returned as part of the settlement to end the War of the Austrian Succession, Louisbourg was again captured by the British in 1758 and largely demolished.[3]

Archaeological work to restore the grandeur of the fortress in the early 1960s uncovered a previously unknown permanent countermine. The French had opened a trench 4 feet (1.21 meters) deep at King's Bastion and lined its floor and sides with masonry, constructing a stone vault over the top. Then they covered everything with earth while shaping the glacis of the defensive ditch. Thus, this countermine was constructed in the open and then made into a tunnel. The gallery was 120 feet (39.57 meters) long with three branches, each one 40 feet (12.19 meters) long and ending in a powder chamber. When the British captured the fortress in 1758, they apparently knew nothing of this countermine; possibly, its entrance had been covered with debris. When they constructed a causeway in the area two years later, they inadvertently sealed the entrance.[4]

The second British capture of Louisbourg occurred during the Seven Years' War. In the latter stages of that global conflict, British engineers masterminded an effective mining effort to capture Havana. An expedition reached the shores of Cuba and established positions outside the city by June 8, 1762. The British commanders focused their efforts on Morro Castle, one of two masonry forts guarding the entrance to Havana harbor. Situated on a rocky promontory, the work was a formidable target. British troops began sapping their way toward it on the evening of July 17, aiming at the right bastion, and reached the ditch two days later. They then sapped along its outer edge until reaching the right face of the bastion, which faced the sea. Here the ditch descended deeply, much of it dug into solid rock. The miners managed to get into and across it by using a small projection left by the Spanish to serve as a traverse, which partially covered a length of the ditch for protection from fire by seagoing vessels. It now allowed the British to place their miners at the foot of the curtain at the loss of a handful of men to small-arms fire.[5]

The British planned two mines: one to blow a breach in the wall and the other to blow in the counterscarp and fill in a section of the deep ditch. If all went well a ramp and hole would be created to form an effective breach in Morro's defenses. Work began on the afternoon of July 19, much of it through solid rock. In ten days the miners were ready to plant the powder, with one charge established eighteen feet (5.48 meters) under the curtain. They sprung both mines at 2 P.M., July 30, with varied results. The charge planted at the bottom of a shaft dug into the glacis "had not a very considerable effect," according to the chief engineer's journal. But the mine planted eighteen feet under the curtain created an effective breach in the wall. British troops managed to cross the ditch over the imperfect ramp, attack through the breach, and capture the fortress.[6]

Morro Castle is a relatively rare example of a single mining effort leading to the capture of its object. And the fall of Morro was decisive in the fall of Havana. The British established batteries and pounded the fort on the opposite side of the harbor entrance, La Punta, until its guns were silenced. The garrison of Havana surrendered on August 13. Within six weeks after news of the city's capture reached London, a preliminary peace treaty was signed ending the Seven Years' War, and Havana was given back to the Spanish as part of its terms.[7]

In contrast to Havana, most mining efforts in North America during this era were at best partial successes. In fact, more were started than were finished. Even though the War of American Independence (1775–1783) lasted eight years and involved many operations that focused on positional warfare, very few mines were dug. Ironically, the first was attempted by Native Americans with the help of British military advisers. When the Shawnee besieged the American fortified settlement of Boonesborough in the Kentucky territory in September 1778, the advisers coaxed them into starting a mine gallery from the nearby Kentucky River's bank, which rose ten feet above the water level. For a week, they labored on the tunnel while attacking the stockaded fort at night. But a heavy rain on the night of September 17 collapsed the gallery, which so disheartened the Shawnee that they raised the siege the next day. The tunnel reportedly had been extended forty yards (36.57 meters) by then.[8]

The next mining effort was mounted by Americans loyal to the British who held Augusta, Georgia. When besieged by Revolutionaries in May 1781, the defenders dug a tunnel to a house between the lines and charged a powder chamber with the intention of destroying it, afraid their enemy would use the building as a sharpshooter's post. They succeeded in blowing it up at 3 A.M. on June 4 before the Revolutionaries could occupy the place. This minor success had no effect on the siege. The Revolutionaries mounted an attack on June 5 that captured Fort Cornwallis, the major work protecting the city, and Augusta fell into their hands.[9]

Another group of Loyalists holding the post of Ninety-Six in South Carolina became the target of a mining effort during the siege of that place in 1781. Beginning on May 22, Revolutionaries under Nathanael Greene managed to extend their earthworks to a point within seventy yards (64.00 meters) of Star Fort, the major work of the garrison. From there, they began a sap on June 3 that reached a point about forty yards (36.57 meters) from the work within a few days. Then the Revolutionaries started to dig a gallery on June 9. The Loyalists conducted an active defense, mounting several sallies that impeded the besiegers' work. They struck the entrance to the gallery on the night of June 9–10, wounding and coming close to capturing the Polish-born engineer officer Thaddeus Kosciuszko, who was in charge of the mining operation. Work continued, however, until the gallery was 90 feet (27.43 meters) long by the night of June 14. Within the next couple of days, it was extended another 35 feet (10.66 meters), to a total of 125 feet (38.10 meters), and only 4 feet (1.21 meters) from the ditch of Star Fort. Two lieutenants supervised each of two reliefs that pushed the gallery forward. By this time, however, a British relief force neared, so Greene mounted an attack against Star Fort on June 18, which failed. He then broke off the siege.[10]

A number of archaeological digs have revealed that more than 100 feet

(30.48 meters) of the gallery's 125 feet (38.10 meters) length remains intact. An examination using laser technology conducted in 2014 documented the exact dimensions and course of the underground work. It is not at all uniform, as most mining and countermining galleries tended to be throughout history. Dug by amateurs rather than professional commercial or military miners, the gallery floor rises and falls, sometimes sharply, along its course. The interior dimensions, which range from 2.5 to 5 feet (0.76 to 1.52 meters) high, also varies widely from one part of the gallery to the other. Laser scanning provided cross-sections of the gallery that indicate the interior assumes many shapes, from roughly rectangular to crudely oval. All of it betokens quick, rough digging by inexperienced men, with no finishing of floor, sides, or ceilings and no wooden shoring. There also is no shaft because the gallery had been started at the head of a sap, the miners instead digging an inclined entrance. The detailed results of the 2014 laser scanning illuminates an unusual rather than a typical example of military mining at Ninety-Six.[11]

Several factors influenced the relatively scant use of mining in the Western Hemisphere, including the underdeveloped military culture of the Americans compared to their British opponents. And the British Army preferred to rely on open-field battle; it did not have the resources in North America to engage in prolonged sieges that swallowed up time, men, and material. But even when a series of conflicts broke out in Europe, beginning with the French Revolutionary Wars (1792–1802), commanders often chose not to utilize mining in the resulting sieges. That trend continued during the Napoleonic Wars (1803–1815). Imparting a shift toward more highly mobile operations and shorter, more decisive campaigns, Napoleon Bonaparte rarely conducted a siege, although French troops reportedly dug an offensive gallery during his siege of Acre in 1799. During numerous operations in the Italian theater from 1805 to 1815, at least ten sieges took place, but mining and countermining were not reported as occurring at any of them. Nor did mining appear in Napoleon's main theater of operations in central and eastern Europe. Given the scope of campaigns in this main theater, the contrast in mining activity with the wars of the eighteenth century is stark.[12]

But when attention turns to the Peninsular War, mining and countermining were employed on something like the scale seen in the eighteenth-century conflicts. British efforts to tie down French resources in Portugal and Spain, with the aid of Portuguese and Spanish allies, resulted in many sieges during which tunneling played prominent roles. In the French siege of Saragossa from December 19, 1808, to February 20, 1809, the attackers broke through the defenses but fought house to house for many days within the city. They dug at least six short galleries to blow up various buildings, reportedly employing 3,000 pounds (1,360.77 kilograms) of powder in each one. These mine

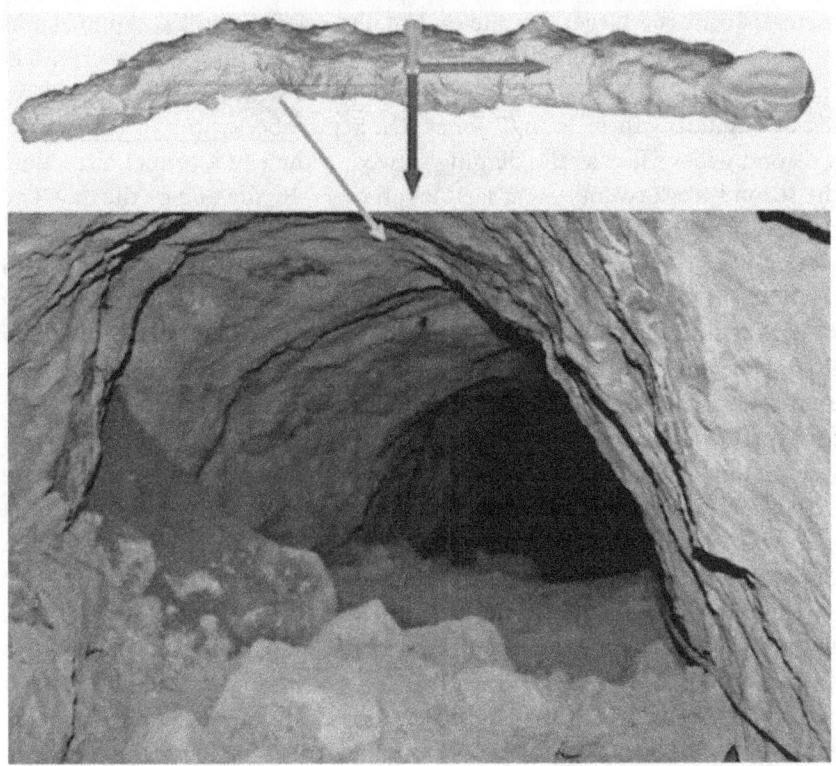

American Mine Gallery, Siege of Ninety-Six, 1781. Note the rough and uneven digging, with no shoring. The gallery was dug in haste by amateurs with no experience in underground navigation. It is one of the few historic examples of mining that remains intact, though not open to the public, with its dimensions and contours having been explored through laser technology. Courtesy of Ninety Six National Historic Site.

attacks were not decisive, for they merely destroyed structures; the Spanish defenders continued to fight doggedly until giving up after fifty-eight days of grueling combat.[13]

The Anglo-Portuguese siege of Burgos from September 19 to October 21, 1812, directed by Arthur Wellesley, involved heavy mining. Initial attacks failed, which led the small group of British engineers and miners on hand to begin sapping their way toward Burgos Castle from a parallel located fifty yards (45.72 meters) from the wall. The sap was six feet (1.82 meters) deep. When they got as close as possible without allowing the defenders to see into the sap, and upon finding that the soil was favorable for mining, they started a gallery from the saphead at 10 A.M. on September 25. Measuring the distance that night, the miners found they had started it sixty feet (18.28

meters) from the target. By the end of the twenty-sixth, the tunnel was eighteen feet (5.48 meters) long, three feet (0.91 meters) wide, and four feet (1.21 meters) high, "the soil stiff, and standing well without support," according to engineer officer John T. Jones. On September 27 the British started a second gallery just as the digging slowed in their first tunnel, extending the second work twenty-eight feet (8.53 meters) by the end of the day. The primary reason progress slowed in the first gallery was the inexperience of the infantrymen detailed to help the small corps of sappers. Although the detailed foot soldiers had been civilian miners before enlistment, they were not "accustomed to military mining" as were the sappers. Also, they used what Jones called "large English pickaxes" unsuitable for work in an enclosed space. But by the end of day on September 28, the first gallery was forty-two feet (12.80 meters) long, while the second one was thirty-two feet (9.75 meters) in length. By midday September 29, the miners had pushed the first gallery to sixty feet and reported they were under the foundation of the west wall of the castle. The engineer felt he could have reached this point thirty-six hours earlier if he had with him enough trained military miners.[14]

But when Jones examined the head of the tunnel, he found "large blocks of masonry obstructing the end of the gallery," so he ordered a chamber five feet (1.52 meters) long dug "under the wall." When finished about 10 P.M., it then received twelve barrels of gunpowder, 90 pounds (40.82 kilograms) in each, for a total of 1,080 pounds (489.87 kilograms). The men tamped fifteen feet (4.57 meters) of the gallery using earth-filled sandbags. Jones exploded the charge at midnight on September 29–30, bringing down the masonry wall but leaving most of the earthen rampart backing it intact. He believed the destruction should have been greater and concluded that the blocks of masonry observed in the gallery were "some old foundation, in front of the escarp wall," which deflected much of the blast. The follow-on infantry assault was mishandled, however, and failed. Many of the troops became disoriented in their approach and attacked an unbroken section of the wall to the right of the breach.[15]

Wellesley now placed hope on the second gallery, which was extended to seventy-five feet (22.86 meters) by the morning of October 3. At this point it hit "a large stone, which could not be removed, and the gallery had to pass over it." The air had become bad after forty-five feet (13.71 meters). Candles barely burned, and Jones often had to clear the tunnel of miners for half an hour "to admit air." On the night of October 3–4, two officers crept forward across the surface to gauge the distance to the target. Using cords, they measured it as seventy-three and seventy-four feet (22.25 and 22.55 meters) respectively. Measuring the length of his tunnel, Jones found it to be seventy-nine feet (24.07 meters) and felt certain that, "even allow-ing for the gallery not being in the shortest line," the head must have been under the scarp wall. He decided to dig another four feet (1.21 meters) to get behind the wall and make a sure thing of it. By early morning, the gal-lery was eighty-three feet (25.29 meters) long. Miners then dug a four-foot (1.21 meters) chamber in which Jones planted 1,080 pounds (489.87 kilo-grams) of powder in sandbags, leaving a one-foot (0.30 meters) empty space between the charge and the sides of the chamber. He tamped only twelve feet (3.65 meters) of gallery and estimated the line of least resistance was eight feet (2.43 meters) to the bottom of the scarp wall.[16]

When sprung at 5 P.M. on October 4, this second mine blew down a sec-tion of wall one hundred feet (30.48 meters) long, making "an excellent breach." But Jones's work was nullified when the infantry attack secured the breach but could not penetrate farther into the castle enclosure. Stymied once again, Wellesley authorized the digging of a third mine toward the Chapel of San Roman on the south wall of the castle complex beginning October 9. The gallery reached its target by October 16 and two days later, at 4:30 P.M., nine hundred pounds (408.23 kilograms) of powder created a breach. Once again the infantry secured the breach but could not penetrate farther. On October 21 Wellesley gave up and lifted the siege of Burgos.[17]

The mining operations at Burgos were fairly typical of a common theme in the history of sieges: engineers and miners were able to meet the tech-nical challenges, but cooperating infantry forces were unable (unusually so in this case) to take full advantage of their success. Mining relied heavily on cooperative effort among different branches of the besieging force. It had to be piggybacked onto the success of sapping operations, to begin with, and then had to rely on the troops to reap the benefits of a technically successful blast. These linkages often failed.

But the different branches eventually worked a bit better at the British siege of San Sebastián from July 7 to September 8, 1813. Wellesley's forces cap-tured a convent located south of a hornwork along their chosen approach to the seaport's land defenses. When sappers dug a parallel partway between the convent and the hornwork, they accidentally cut into an underground drain. An engineer officer explored it and found that the drain extended 240 yards (219.45 meters) to the hornwork, ending with a locked door. In other words, the drain served as a readymade mine gallery. The British planned to create a globe of compression at the end of the drain to blow in the counter-scarp and make a ramp across the ditch, which was reportedly twenty-four feet (7.31 meters) deep. The resulting charge was sprung at 5 A.M. of July 25 and created the desired ramp, but the follow-up infantry assault failed. Jones reported that smoke issuing from the crater was so heavy it led the attackers to hesitate too long. By the time they advanced, the French defenders had recovered from their shock and repelled the assault.[18]

Wellesley never relied solely on mining in his siege work, supplementing it with artillery fire. At San Sebastián his batteries managed to open two breaches in the defenses of the hornwork, but to reach them the assaulting infantry had to cross a seawall and advance over mudflats at low tide. The engineers paved the way by planning to breach the seawall, which was four feet (1.21 meters) thick and rose to a height of ten feet (3.04 meters) above high tide. They dug three shafts: the first was close to the wall, the second twenty-five feet (17.62 meters) from it, and the third forty feet (12.19 meters) from the second. All were sunk eight feet (2.43 meters) below the surface, with a chamber dug at the bottom of each for 540 pounds (244.94 kilograms) of powder. They were sprung at 2 A.M. on August 31 and worked perfectly, blowing down the seawall and creating craters thirty feet (9.14 meters) wide. Sappers then dug connecting trenches between the three to form a covered way by 10 A.M. When the assault went in an hour later, the French sprang two countermines that killed up to thirty British soldiers. Nevertheless, the attackers captured the hornwork in heavy fighting. All that was left at San Sebastián was the old castle at the end of the isthmus, and artillery battered it into submission on September 8.[19]

While artillery rather than mining had created the breaches that proved to be the turning point of the siege of San Sebastián, mining was essential in creating a path for the attacking infantry to take advantage of those breaches. It is noteworthy that during the long period of Napoleonic conflict, the British emerged as the most active siege-mining force of these conflicts. As early as 1792, upon the outbreak of the French Revolutionary Wars, British Army units incorporated mining into their training regimen. A large training camp of 7,000 troops held at Crowthorne Wood in Berkshire from July 23 to August 7 drew hundreds of spectators to witness a mock battle and a mine blowing up a square redoubt. The plume of debris rose forty feet (12.19 meters) high and left a crater forty feet wide and twenty feet (6.09 meters) deep, much to the delight of King George III, who witnessed the demonstration. Fighting the wars with a smaller force, the British sought ways to maximize field operations while minimizing losses in open-field battles.[20]

Engineer officer Jones, who was responsible for much of the mining that took place on the Iberian Peninsula, evaluated the experience within the context of doctrine for mine warfare. He stressed the important link between the sapper and the miner, the latter becoming indispensable as the saphead neared the enemy position. "The duty of a miner at a siege is to accompany the sapper, to listen for and discover the enemy's miner at work under ground, and prevent his blowing up the head of the rood [sap]." He could do this by "a thousand arts of chicanery, the knowledge of which he has acquired from experience." In addition, the miner's role in furthering the approach beyond the saphead by tunneling under the target took his importance beyond merely protecting the sap and could, as it did in blasting the seawall at San Sebastián, become vital to the success of the siege.[21]

The Napoleonic era witnessed the publication of an influential manual by Simon François Baron Gay de Vernon, *A Treatise on the Science of War and Fortification*, in two volumes in 1805. An English translation by John Michael O'Connor, an American officer in the War of 1812, appeared in 1817 and was used as a textbook at the U.S. Military Academy. The baron prefers using the terms "offensive mines" and "defensive mines" in his discussion and concludes that "perhaps the great effects of subterranean war are more in imagination, than reality." In that phrase he encapsulates an important strain of mine warfare: the difficulties in making it work as a decisive element in siege fighting and the emotional threat it posed to defending troops. Nevertheless, as in any good manual, Vernon provides a great deal of technical detail, including formulas and tables documenting the gunpowder explosion that lay at the heart of subterranean warfare.[22]

Vernon asserts that galleries could not extend more than forty yards (36.57 meters) before the need for artificial ventilation became imperative, which is farther than most examples in the history of mining would suggest: "Air becomes unfit for respiration, lights extinguish, and the men are soon struck senseless to the earth." Digging transverse galleries to create drafts of air was the solution. The powder hose, called in French a *saucisson* (sausage), had in the past been three inches in diameter, but recently it was found that half an inch in diameter worked well to lessen the output of gas from burning powder into the air of the gallery. It was still placed in a wooden trough, with one end in the powder charge and the other at the end of the tamping. A paper plug at the end could be set afire to begin the burning of the powder in the tube, although some smoke and gas would still enter the gallery. The alternative was called a mouse, consisting of a light that could be drawn through the trough to the charge by means of a cord running through a second trough. This eliminated the discharge of gas into the gallery.[23]

The defender needed only small, simple mines to counter offensive galleries. He needed only to deny ground and position to the attacker, and to cave in his tunnels without harming his own works too much. Larger, more complex galleries and powder charges were needed by the attacker. Vernon recognizes that an offensive plan could involve several mines placed in a line so that the craters would intersect and the holes could be incorporated to become a readymade sap. In this case, if one of the mines exploded a bit later than its neighbor, the line of least resistance would be toward that nearby crater rather than straight up. But Vernon prefers the use of isolated mines to reduce the complexity and the chances of faulty performance in multiple powder explosions. He recognizes, however, that complexity could be multilayered. Vernon writes of three tiers, or planes, on which offensive mines could be dug. The first would be established ten feet (3.04 meters) below the surface, while the two others could be placed at deeper depths to achieve maximum effect without interfering with each other.[24] This may well be the first discussion of a multilayered mine system, the kind that would become common on the Western Front of World War I.

Vernon devotes a good deal of discussion to permanent countermines, detailing four systems created by Charles Goulon of Vauban's era, Belidor, and Louis de Cormontaigne, all of the eighteenth century. He admits that none of those systems had been fully built at any one place but also finds faults with all of them. The system created at Verdun, which Vernon characterizes as a modified version of Goulon's, was probably the best. Argument over whether a permanent countermine system should be constructed in peacetime or after the outbreak of war centered on the fact that if built before hostilities, potential enemies could discover details of its design. But there was a consensus of opinion that at least the gallery, which was the base of any extension toward the enemy, should be built in peacetime or at the time of fortress construction. Listening posts should not be more than sixty-seven to eighty-three feet (20.42 to 25.29 meters) apart so listeners could hear all along the line of a work. Vernon also believes contact between countermine and offensive mine should be planned to take place as near the glacis as possible because that was where the attacker was most vulnerable and the defender strongest. Whatever the details, there was no doubt that countermines were needed. "Subterranean fortification is evidently an appendage of exterior fortification," Vernon concludes, "and is one of the chief accessary means to greatly increase the resistance of the latter."[25]

Discussion of permanent countermines and the techniques of offensive mining continued in the technical literature for decades after the end of the Napoleonic era. When Henry W. Halleck wrote the first truly comprehensive book on military affairs in the United States, published in 1846, he pointed out that the besieger could deal with a permanent countermine system by sinking a trench as close as possible to it and using overcharged mines to collapse it. English engineers advocated the placing of permanent countermine listening galleries at least twenty yards (18.28 meters) forward of the salient angles of the ditch. The thorny problem of ventilating a long gallery was addressed by Hector Straith, a British staff officer and professor of fortification and artillery at the East India Company College at Addiscombe. He discussed a ventilation system consisting of pipes 1.5 inches (3.81 centimeters) in diameter, joined together, with air forced through them by the use of bellows. This system needed to be constructed as soon as the tunnel was fifty feet (15.24 meters) long. Placing candles every fifty feet or at every turn in the tunnel would provide illumination without using up too much oxygen. Straith also endorsed the idea that the length of the tamping should be equal to 1.5 times the distance of the line of least resistance.[26]

New advances in detonating powder charges began to appear in the early nineteenth century. Working to address civilian mining needs, Englishman William Bickford invented the safety fuze in 1831. It consisted of a tube of gunpowder surrounded by a covering of waterproofed jute and could burn at the rate of about a foot (0.30 meters) every thirty seconds. In contrast, the old powder hose burned much more rapidly. The slower, more regular rate of burning made the Bickford fuze much safer to use in mining galleries. Its jute covering protected the powder against the elements and any casual interference that could set it off accidentally.[27]

From intensive use in the seventeenth and eighteenth century up to the end of the Seven Years' War, followed by sparse use from then to the middle decade of the nineteenth century, mine warfare had continued to mature in the area of doctrinal development. Mining and countermining always were included in the technical manuals on military affairs that were written during the nineteenth century. Those authors not only reinforced previous doctrine but also added new twists of their own to the technical knowledge and to operational recommendations. But actual use of mining and countermining depended on many factors beyond the control of engineers and technicians. The nature of operations in any given war and decisions by high-ranking commanders played key roles in whether a conflict witnessed the use of tunneling, charging, and exploding. While less mining took place from 1762 to 1853, that trend changed with the onset of the Sebastopol Campaign of the Crimean War in 1854–1855 and the fighting in India in 1857.

7

Sebastopol, India, and Electrical Detonation

By the mid-nineteenth century, no major mining project had drawn international attention since the Prussian siege of Schweidnitz in 1762. The Crimean War (1853–1856) garnered an enormous amount of international attention, mostly because of the siege of Sebastopol (1854–1855), which involved important mining operations. The French and Russians engaged in a vigorous and extensive underground war that was more thoroughly documented than any military mining effort yet seen in global history. For a time, the war beneath the surface was overshadowed by the surface operations conducted during the siege. But soon a new generation paid attention to the mine warfare at Sebastopol, bringing a noticeable increase of mining and countermining activity around the turn of the twentieth century.

Contemporaries paid much more attention to mining during the Indian Mutiny (1857–1858) than to the underground war at Sebastopol because of the drama inherent in the mutineers' siege of the Lucknow Residency in 1857. Indian mining at this siege tended to be crude and ill documented compared to French and Russian operations at Sebastopol, but the story of the outnumbered garrison's countermining captured world attention. In its level of underground activity, however, the siege of Lucknow was on a somewhat similar level as Sebastopol.

In the background during the era of Sebastopol and Lucknow, the second-most-important technological advancement in the history of military mining—only the introduction of gunpowder was more significant—related to slow and uncertain efforts to develop an electrical device to detonate gunpowder underground. This was the most important spur to increased mining and countermining in field operations around the turn of the century as those devices finally became practical. The modernization of military mining as a technical craft, and its increased use in the field, helped prepare the way for its full flourishing as an element of operations in World War I.

When the French and British campaign in the Crimea was directed at the Russian Black Sea port of Sebastopol (modern Sevastopol) in late September 1854, the place had strong masonry forts guarding the entrance to its harbor, but only about one-fourth of the projected land defenses had been constructed. The Russian garrison hastily threw up a line of earthworks to connect the few small works already existing. Thus, the siege that followed was one utilizing earthen fortifications rather than permanent stone construction. The allies could not isolate the garrison, and Sebastopol remained open to Russian forces within the theater of operations. Denied starvation as a viable weapon, the French and British tried periodic assaults, which usually failed or at best achieved limited results. In the end, the key to their success was a concentration of artillery fire that over time weakened the Russian defenders to the point of near exhaustion, leading to an evacuation after major assaults on September 8, 1855. It had taken a full year to reduce Sebastopol, during which mining and countermining had played a significant, although not decisive, role.[1]

French mining efforts against what the Russians denoted as Bastion No. 4 and the French called Flagstaff Bastion represent a dramatic illustration of underground attack and defense. Under the direction of their chief engineer, Brig. Gen. Adolphe Niel, the French sapped forward until placing their third parallel 150 yards (137.16 meters) from the bastion by November 2, 1854. Miners then sank three shafts just behind the parallel on November 20 and hit a level of clay 3–5 feet (0.91–1.52 meters) thick "between two layers of rock." From two of the shafts they started galleries nearly 100 yards (91.44 meters) apart, one toward the Russian salient and the other toward its right face. The miners, working in two teams over three daily shifts, dug on average 6 feet (1.82 meters) every twenty-four hours. The galleries were 5 feet high and 3.5 feet (1.06 meters) wide, barely large enough for working men. On reaching the extent of natural air flow, they ventilated the galleries by forcing fresh air through iron water pipes with a fan. The French used no shoring; they found the rock layer above served as an effective ceiling for the tunnel. While the combination of clay and rock favored French mining at Flagstaff Bastion, rock predominated in the British sector, explaining why no

English mining took place at Sebastopol even though the British Army had conducted practice mining exercises in 1844 and 1848. British engineers were keen on keeping up with the latest in siege mining but had no opportunity to practice it at Sebastopol.[2]

The Russians began countermining at Bastion No. 4 just as the French started their offensive efforts. Engineer officer Lt. Col. Eduard Ivanovich Totleben sank two shafts in the ditch of the work, one of them at the salient, and found the layer of clay used by the French. But at this point the clay was located 16 feet (4.87 meters) below the surface. Organizing three brigades, consisting of seventy-five miners and two hundred detailed infantrymen in each, Totleben worked them in eight-hour shifts to extend galleries from the two shafts and connect them with a transverse tunnel. From this he extended listening galleries toward the besiegers.[3]

As time progressed, the Russians dug nineteen more galleries to expand their countermine system. The tunnels were at least 3 feet (0.91 meters) high and 2.5 feet (0.76 meters) wide, but sections were enlarged in diameter to increase the flow of air. They ventilated the system by digging more transverse galleries. The Russian miners found that a candle could light up to forty yards (36.57 meters) of the tunnel and thus spaced them out accordingly to minimize the consumption of oxygen. They suspended work for two to three hours several times each day so listeners could try to hear sounds of French mining. As a result, the Russians managed to dig only about 2–3 feet (0.60–0.91 meters) per day. Totleben inserted training exercises into these work stoppages. He required some miners to resume digging in stages so others could hear and recognize the sound of their comrades' work. Totleben also required his men to estimate distance and the type of tools used by their comrades. The miners could detect a pick at thirty yards (27.43 meters), but the sound was dull and intermittent, while at fifteen yards (13.71 meters) it was continuous and louder.[4]

By late January 1855, the Russians had planted their most advanced listening post at the end of a gallery fifty yards (45.72 meters) long. Totleben worried that the French might be digging below his tunnels, so he sank two more shafts through a rock layer to see if he could find a lower level of clay. Then a French deserter confirmed that the two sides were working in the same shallow clay layer. On January 30, Russian listeners heard the first sounds of French offensive mining. Totleben thus directed the digging of a chamber and packed it with 435 pounds (197.31 kilograms) of powder. His men then used sandbags and timber to tamp 18 yards (16.45 meters) of the gallery and touched off the charge at 9 P.M. on February 3. The countermine did its work, collapsing 4.5 yards (4.11 meters) of the 120-yard-long (109.72 meters) French gallery, killing two French miners. It also created a

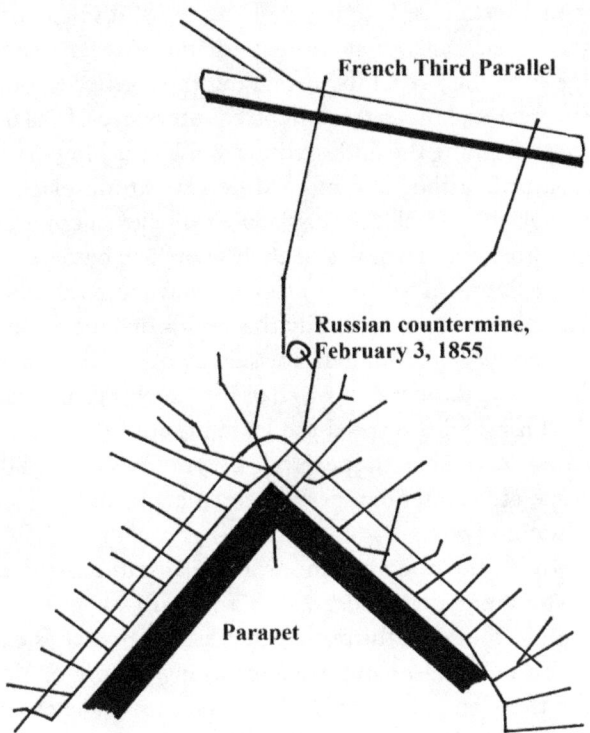

French Third Parallel

Russian countermine, February 3, 1855

Parapet

French Attack and Russian Defense Galleries, Siege of
Sebastopol, February 3, 1855. This map indicates progress
of both offensive and defensive efforts at what the Russians called Bastion No. 4—the French called it Flagstaff
Bastion—up to the first Russian countermine explosion
on February 3, 1855. Based on maps in Lloyd, "Mine
Warfare at Sebastopol."

crater 14 yards (12.80 meters) long, 9.5 yards (8.68 meters) wide, and 2.5 feet
(0.76 meters) deep that lay 49 yards (44.80 meters) from the counterscarp.
The French reacted by abandoning more than 30 yards (37.43 meters) of
their gallery, collapsing the part they evacuated with a blast of gunpowder
to deny Russian access to it.[5]

This was the first of many blows to come in the developing mine warfare
at Flagstaff Bastion. For the next two weeks after the February 3 explosion,
the Russians created an extensive network of countermines. Also, their slow
progress in digging two shafts through the rock layer finally paid off, allowing them to find a deeper layer of clay forty-eight feet (14.63 meters)
below the surface. They now had access to a lower level than the French

and established a system of listening galleries in it. At the same time, both sides expanded their shallow systems, with the Russians slowly gaining the advantage over the French by extending their galleries out to a point seventy-five yards (68.58 meters) from the counterscarp of Bastion No. 4. In effect, they were winning the underground war by pushing the French back from their point of farthest advance. More blasts took place. From early February through mid-April, the Russians set off eleven countermines; the French sprang three countermines of their own. The besiegers had sprung their first countermine on February 7 too far away from the Russian gallery and thus did no damage to it. Ironically, this explosion blocked up the French gallery and forced its abandonment. Russian troops took possession of the resulting February 7 crater and dug a shaft into the bottom, heavily shoring it because the blast had disrupted the ground, until the shaft was twelve feet (3.65 meters) deep. Then they exploded 325 pounds (147.41 kilograms) of powder to deepen the crater to nine feet (2.74 meters) and dug another shaft to serve as a listening post.[6]

By late February, the underground war at Flagstaff Bastion had reached a stalemate. The French tried to break it with a major mine attack. After weeks of digging, they had thirty-six galleries and branches extending toward the bastion by April 10 and began charging twenty-one of them the next day. Six of those charges contained 4,180 pounds (1,896.01 kilograms) of powder total, eleven had 2,508 pounds (1,137.61 kilograms), and the remaining four held 1,254 pounds (568.80 kilograms). That made for a total of 7,942 pounds (3,602.43 kilograms) of gunpowder carried into them in bags. The chambers were placed so that the resulting craters would nearly touch each other in order to create a line from which to start another parallel. In addition, the blasts might collapse Russian countermine galleries, which were ten to twenty yards (9.14 to 18.28 meters) away. The smaller charges were located on the flanks and intended to blow open the rudiments of communication trenches linking the third parallel with the expected new line of craters. This was deemed necessary because fire from the bastion's garrison made it dangerous for men to work on the surface. The French were in a hurry, given the proximity of Russian countermine galleries, so they only partially tamped their galleries.[7]

After preparing their big attack, the French set off the charges simultaneously at 8 P.M. on April 15, but only fifteen of them exploded. Three large craters appeared halfway between the third parallel and the bastion, while smaller craters were blown to both sides. Thirty Russians were killed at the listening post established in the February 7 crater and seventy were wounded in the bastion itself by flying stones. The French blasts of April 15 led to five days of combat on the surface for control of the new crater line.

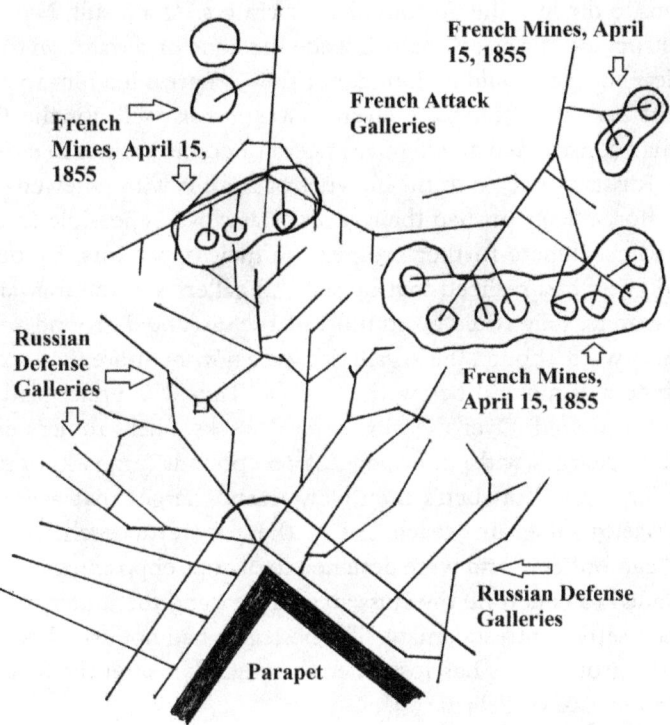

French Attack and Russian Defense Galleries. Siege of Sebastopol, April 15, 1855. The offensive and defensive efforts at Bastion No. 4 (Flagstaff Bastion) dramatically increased from February 3 to April 15, when the French blew several mines in a major effort to establish a fourth parallel. Based on maps in Lloyd, "Mine Warfare at Sebastopol."

By April 19, French artillery had silenced Russian guns in the bastion, and their infantry were able to secure the craters. The three largest ones were 128 feet (39.01 meters) long and 64 feet (19.50 meters) wide. French details dug connecting trenches to consolidate their hold on them. Then engineers fixed some of the failed mines and exploded six of them on April 21 and 23. Moreover, they dug two more shafts and exploded charges in them on May 3 to create more craters to use as shelters in no-man's-land. Most of the craters blasted at this shallow level tended to be 12–15 feet (3.65–4.57 meters) deep. By these means and after much trouble, the French completed their fourth parallel on June 18.[8]

But the French failed to capitalize on their hard-won success at establishing a fourth parallel. They found the ground so disrupted that it was dangerous to dig into the bottom of their craters. As a result, Niel ordered the construction of several shafts outside the zone of disruption to see if a deeper level of clay could be found, but slow progress led him to abandon that effort before reaching the layer; it was just as well, for the Russians already had constructed an effective system of countermines in it. For their part, the Russians reacted to the blowups of April 15 with renewed countermining efforts. They pushed their galleries as close as possible to the new parallel to checkmate further progress by their opponents. By this time, French diggers had realized that by starting galleries in the forward slope of their craters, they could avoid most of the disrupted ground and make progress forward, though these galleries were now no more than six to nine feet (1.82 to 2.74 meters) below the surface. Thus, the underground confrontation resumed. Over the next several weeks, Niel's miners exploded thirty-three charges with an average of 600 pounds (272.15 kilograms) of gunpowder, while Totleben's men blew thirty charges that averaged 300 pounds (136.07 kilograms) each. All of these were undercharged mines, termed "camouflets," and were designed to disrupt opposing galleries. The French failed to penetrate the Russian defenses, and the underground war once again settled into stalemate. The besiegers had not even been able to recover the ground they had lost when they retreated after the first Russian countermine blast of February 3.[9]

But Totleben was not satisfied with a stalemate. He planned a large attack to destroy the fourth parallel using the countermines he had established in the lower level of clay. By early September, the Russians had advanced a gallery so that its chamber was thirty-one feet (9.44 meters) below the floor of the parallel. Totleben's men packed 4,325 pounds (1,961.78 kilograms) of powder in it. But they never had a chance to touch it off because other methods of pushing the siege came to the fore to finally tip the balance in favor of the allies. Their concentrated and heavy use of artillery fire eventually paid off, degrading Russian defenses above ground to the point where infantry assaults were partially successful. Three days of unrelenting bombardment were followed by large-scale attacks on September 8. French troops secured a lodgment in the Kornilov Bastion on Malakov Hill, and the Russians evacuated Sebastopol that night, bringing the long campaign to a close. Because the powder was late in arriving, Russian miners had just begun placing the charge under the fourth parallel at the Flagstaff Bastion when they received the order to pull out.[10]

Sebastopol represented a watershed in the history of military mining and countermining that was unheralded by the generation of the 1850s but noticed by later generations. More digging and exploding of charges took place there than at any previous site in global history. Yet few people realized this because observers were fascinated by the use of extemporized earthen defenses on the surface and how they stood up to intense artillery bombardment. The extensive underground warfare was largely ignored, even though Niel authorized the publication of a long, detailed report of those operations. Totleben also wrote an extensive report. But neither book-length publication was translated from the original French or Russian into English. Totleben received a lot of international exposure, but that was mostly for his effective defense above ground rather than for his direction of the underground war.

The Russian countermine system at Flagstaff Bastion was huge, consisting of 3,639 yards (2.06 miles, or 3.32 kilometers) of defensive galleries in the shallow level of clay. The French approached this complex with 1,199 yards (0.68 miles, or 1.09 kilometers) of offensive galleries. While Niel's men constructed nothing in the lower level of clay, the Russians placed 138 feet (42.06 meters) of galleries there. Ventilation was accomplished not only by digging transverse galleries to create natural air currents but also by pumping in fresh air through tubes of sheet iron or india-rubber hosing. Totleben's men exploded a total of 83 charges, all of them on the upper level, using 231,000 pounds (104,779.83 kilograms) of gunpowder. In contrast, Niel's men exploded 107 charges using 131,733 pounds (59,753.98 kilograms) of gunpowder.[11]

The opposing tunnels at Flagstaff Bastion by themselves exceeded anything constructed in the history of military mining before 1854, but they were not the only ones created at Sebastopol. Both sides mined and countermined at three other locations along the extensive site. The French also dug saps aimed at Redoubt No. 1 and Bastion No. 5 but stopped them a little more than one hundred yards (91.44 meters) from the target due to heavy defensive fire. The Russians began digging countermines here in late May 1855, creating an upper and a lower level of tunnels. By early August, they had extended listening galleries on the upper level 48–96 feet (14.63–29.26 meters) out from the counterscarp and 48–54 feet (14.63–16.45 meters) apart from each other. The lower level started from the bottom of shafts that were sunk 57 feet (17.37 meters) deep, digging through 38 feet (11.58 meters) of rock in the process. At the bottom Russian miners found a layer of clay that was more than 6 feet (1.82 meters) thick and dug galleries 134–240 feet (40.84–73.15 meters) apart from each other, connecting them with a transverse gallery located 27 feet (8.22 meters) out from the counterscarp to increase air circulation.[12]

French Battery No. 53 happened to be near the end of these Russian galleries. To protect that emplacement, Niel's men dug two shafts and extended three listening galleries toward the enemy position. When the Russians detected this digging on August 14, they assumed it was an offensive mine. Totleben's men set a charge of 430 pounds (195.04 kilograms) of gunpowder and destroyed two of the listening galleries.[13]

The underground war at Redoubt No. 1 and Bastion No. 5 was on a smaller scale than at Flagstaff Bastion. The Russians dug thirteen shafts eighteen feet (5.48 meters) deep on the upper level, with thirty-five branches extending from the galleries dug outward from those shafts. The total length of Russian galleries on the upper level amounted to 2,521 yards (1.43 miles, or 2.30 kilometers). In contrast, the French dug only 130 yards (118.87 meters) of galleries on the upper level. The Russians also dug twelve branches extending from galleries on their lower level, where the tunnels amounted to a total length of 213 yards (194.76 meters). Both sides exploded eleven charges each at Redoubt No. 1 and Bastion No. 5. The Russians used a total of 4,325 pounds (1,961.78 kilograms) of gunpowder, while the French touched off 14,480 pounds (6,568.01 kilograms) of explosives.[14]

Mining took place at a third sector of the Sebastopol lines but on a much smaller scale than at Redoubt No. 1 and Bastion No. 5. Totleben's men constructed two levels of countermines to protect Kornilov Bastion at Malakov Hill. The upper level consisted of 1,782 feet (543.15 meters) of galleries, but the lower held only 144 feet (43.89 meters) of tunnels. The French barely engaged in any offensive action here, digging only 192 feet (58.52 meters) of galleries on the upper level and nothing lower. While Totleben's men blew no charges at Kornilov, Niel's men touched off three charges totaling 3,300 pounds (1,496.85 kilograms) of gunpowder. At the fourth and last sector at Sebastopol that witnessed countermining, the Russians sank eight shafts an average of 8 feet (2.43 meters) in front of Battery No. 34 by mid-April 1855. These shafts supported 273 feet (83.21 meters) of galleries.[15]

The extent of mining and countermining at all four locations at Sebastopol, plus the detailed documentation from both sides, provides an important case study. The Russians dug at a rate of five to ten feet (1.53 to 3.04 meters) every twenty-four hours in the upper level at Flagstaff Bastion, "according to the distance from the shaft," writes English engineer officer E. M. Lloyd, who studied the operations. On the lower level the rate was only two and a half feet (0.76 meters) for every twenty-four-hour period. In both instances the problem of hauling spoil out of the galleries was the key factor in this relatively slow rate of digging. In contrast to general principles of military mining, the extent of tamping was measured not by comparison with the line of least resistance, but with the size of the powder charge. One inch of tamping with sandbags was judged necessary for every pound of gunpowder. The Russians became the first in mining history to use electricity to touch off their charges. They utilized galvanic batteries, also known as chemical batteries or voltaic piles, rather than the more advanced electro-magnetic devices. Galvanic batteries produced electrical charges by chemical reaction rather than by turning cranks as with electromagnetic batteries. A lead consisting of three copper wires encased in gutta-percha communicated between the voltaic pile and the powder charge. The Russians also laid traditional powder hoses to back up the electrical system in case it failed. In contrast, the French used powder hoses touched off by a combination of Lariviére fuzes and Bickford safety fuzes. The former was a quick match enveloped by a waterproof sleeve. This represented the most advanced non-electrical means of detonating mine charges. French engineers set up the Bickford fuzes so that one end was stuck in a forty-pound (18.14 kilograms) bag of gunpowder in the chamber while the other end entered a small box outside the gallery. This box had holes in the side to accommodate all the fuzes used in one series and was filled with four pounds (1.81 kilograms) of gunpowder. On lighting that powder, the fire communicated to the Bickford fuzes so they would begin burning simultaneously.[16]

Although the number of mines and countermines blown at Sebastopol was larger than at any previous operation, a significant number of them were duds. The Russians tried to explode a total of ninety-four charges, one of which failed due to a mistake in setting it up while eight others failed because their electrical wires were disturbed by enemy blasts. The French attempted 107 detonations of which 20 failed, all because of poor connections between the Lariviére cord and the Bickford safety fuze. Thus, the Russians demonstrated the higher level of reliability of electrical detonation, although few observers paid much attention to it. The French also created bigger charges than their opponents. In part this was because several of them were offensive mines designed to create craters for use by troops as a fourth parallel. But most of their charges actually were countermines designed to confront enemy galleries. At Flagstaff Bastion, the scene of the heaviest mine warfare at Sebastopol, the Russians touched off seventy-five mines from February 3 to August 28, 1855. The charges in those mines ranged from 108 pounds (48.98 kilograms) to 757 pounds (343.36 kilograms) each, for a total of 21,655 pounds (9,822.54 kilograms). The French blew ninety-eight charges, ranging from 125 pounds (56.69 kilograms) to 4,180 pounds (1,896.01 kilograms) each for total of 119,750 pounds (54,317.68 kilograms), at Flagstaff Bastion. Personnel losses on both sides were relatively light. Fifty-four Russians were killed and 137 wounded for a total of 191 men. Of that number 24 were killed and 32 wounded inside the galleries; the rest were above ground. The total of 56 underground casualties represented 29 percent of Russian losses attributed to mine warfare. On the French side, Niel lost 9 killed and 94 wounded for a total of 103 men.[17]

The Russians generally dominated the French in mine warfare at Sebastopol. They forced their opponent to retreat with their first countermine blast on February 3 and kept them at a distance from Flagstaff Bastion for the rest of the siege. They were about to destroy the fourth parallel when other factors led to the city's evacuation. By the end of the confrontation at Sebastopol, the Russians had protected nearly four miles (6.43 kilometers) of their line by an effective system of countermines. Yet all of this occurred not because of any Russian superiority in theory or experience. The key was Totleben, a Baltic German from Latvia who brought his skills as a military engineer to the service of the tsar and created a legendary reputation in Russian history. During the course of the siege, he was elevated in rank from lieutenant colonel to lieutenant general. He drew several conclusions from the mine warfare at Sebastopol and published them as suggestions for future operations: It was necessary that the besieger be "energetic and persistent" in pushing forward not one but several galleries toward the target. They should not slacken in the face of heavy countermine work, for "too much prudence and circumspection for the sake of avoiding loss will almost always result in complete failure for the besieger." Countermines were best dealt with by approaching them from below or above and collapsing the galleries with camouflets.[18]

Totleben's advice for the aggressor reflected the persistence Le Febvre had already employed at Schweidnitz in 1762 and foreshadowed the persistence to be employed by the British, French, and Germans along the Western Front of World War I. But he reserved most of his advice for the defender. There were two objectives: to destroy enemy galleries and to do so without making craters on the surface that the besieger could use for their own purposes. Therefore it was best to dig countermine galleries deep enough so the defender could use smaller charges to cave in galleries without making craters. Such charges should be part of a countermine system boldly pushed forward as far as possible to meet the enemy well in front of the position defended. Totleben believed that transverse galleries were the best way to ventilate the system, but one could bore holes from the galleries up to the surface, too.[19]

Stopping work now and then was crucial for detecting sounds of enemy mining; when detected, the course of action was clear. Dig straight toward the sound and stop short of it, plant a charge, and wait for the aggressor to approach near enough. This last effort was the most difficult in Totleben's view. Most men underestimated distance underground, and thus many countermine charges were blown too soon and only damaged their own gallery. It was possible to teach miners how to judge distance, as Totleben did at Sebastopol, by having them practice with friendly mine activity, especially in the transverses. But in the end, one had to have the experience to judge distance accurately and the stamina to wait until the head of the offensive gallery was closer to the countermine charge than the line of least resistance (from the charge up to the surface) if one wanted to seriously damage the aggressor's work. It was better to collide with the offensive gallery and fight a hand-to-hand battle underground than to blow the countermine too soon.[20]

Totleben defended the digging of his countermines on the lower level at Bastion No. 4, noting that if the French had used that level for offensive purposes, they could have destroyed his extensive countermine system at the upper level. Moreover, he was convinced he could have destroyed the French fourth parallel by working from that lower system.[21]

The lessons learned at Sebastopol were consistent with the developing tempo of mine warfare into a more extensive operation than ever before in world history. In the sixty-four years from 1854 to 1918, mining and countermining increased its profile as a major element of warfare in the West. In part that was because, by the turn of the twentieth century, a new generation of military engineers would pay more attention to the example of Sebastopol, which had seen the most thoroughly documented mine warfare to date. Only the mining operations during the American Civil War (which were documented on a comparable basis) and the massive mine warfare of the Western Front (which was even more thoroughly documented) occupy the top tier of mining episodes in terms of available information. Prussian offensive mining at Schweidnitz was more thoroughly documented than any previous mining episode but not as well as at Sebastopol. Although observers tended to shortchange the significance of mining and countermining in the Crimean War, probably because it failed to play a decisive role in offensive operations, the siege of Sebastopol started a trend that culminated in hundreds of galleries and thousands of explosions from 1914 to 1918.

Another link between Sebastopol and the Great War lay in the fact that the French and Russians were the first in global mining history to conduct simultaneous mining and countermining efforts at several locations along a siege line. That came about because Sebastopol was not a typical or traditional siege. Prior efforts had been conducted against smaller, self-contained targets with masonry walls unsuitable for more than one or two siege approaches at a time. But Sebastopol was a semisiege conducted along a line thirty-five miles (56.32 kilometers) long against earthworks with many salients and angles, all of which potentially invited siege approaches. Several sieges and semisieges of the American Civil War, such as Vicksburg and Petersburg, invited multiple simultaneous underground approaches and countermeasures because they too involved long lines of earthworks. But the Western Front of World War I became the biggest, most complex se-misiege in history, spawning literally hundreds of mining operations at the same time.

The generation that fought World War I was more aware of the Sebas-topol legacy than those who had fought the Crimean War. A short book by Capt. Fernand Taillade, an engineer officer, titled *Sebastopol Guerre de Mines*, was published in 1906 and influenced Lt. Col. Alfred H. Brooks, an American engineer who served as the chief geologist for the American Ex-peditionary Force in France during World War I. Brooks was impressed by the Sebastopol experience: "The scale of these operations far exceeded that of any previous wars."[22]

Totleben's lessons resonated within the general history of mining and countermining; they were just as relevant for his day as for the years of the Great War. But they also were only possible if engineer officers had large re-sources of men and material as well as support from their commander, for Totleben's lessons were best suited for large-scale mine warfare. Through-out most of world history, the majority of commanders could not afford the kind of time and commitment of resources to conduct extensive mining or countermining efforts.

Moreover, Sebastopol was the first time that electrical detonation was used to touch off mines in the field. It proved its worth in the very low rate of Russian misfires compared to the French. Niel's men continued to use traditional methods (although the Bickford fuze was only about twenty years old) and experienced a much higher rate of misfires. After the fall of Sebastopol, French engineers planted mines, anywhere from six to thirty feet (1.82 to 9.14 meters) deep, to blow up harbor defenses and used chemi-cal batteries to detonate them with great success. At Fort Nicholas, for ex-ample, they placed 106,000 pounds (48,080.79 kilograms) of powder and managed to set if off with electricity. A report noted that the French exper-imented with mechanical boring equipment during the siege. They trans-ported three machines, each weighing 1,980 pounds (898.11 kilograms), to create camouflets, although no details are readily available on their use.[23]

Sebastopol influenced Russian permanent-fort construction as well. Im-mediately after the war, Russian engineers planned a new masonry work at Cape Ak-Burun (or White Cape), the narrowest point of the Kerch Strait, in eastern Crimea. Called Kerch Fortress or Fort Totleben, it included a counterscarp gallery with portals for digging countermine galleries toward the enemy.[24]

Interestingly, outside the context of Sebastopol, permanent counter-mines continued to be developed by the major military powers of Europe. The U.S. Army sent a delegation of three officers to observe the Crimean

War, and as part of their task, the officers also spent a great deal of time studying the various military systems of the Continent. Maj. Richard Delafield, an engineer, wrote about the permanent countermine systems he found. One of them was designed like the Russian system at Fort Totleben, with several galleries starting from the counterscarp. A transverse, or envelope, gallery was dug parallel to the counterscarp to connect the galleries. From this transverse gallery, listening posts were extended toward the potential enemy usually no farther than the outer edge of the glacis, which was a zone of earth surrounding the fort outside the ditch carefully sloped to prevent enemy troops from sheltering close to the walls. The other countermine system that Delafield noted began within the fort, starting from the ditch surrounding the masonry barracks, and extended under the rampart of the work. These interior galleries could be further extended to meet the enemy or could be used to blow up the ramparts and the besieging forces if the fort fell into enemy hands.[25]

Sebastopol represented a surge of sophisticated mining and countermining in a part of the Western world that had seen relatively little of that activity in the past. At about the same time, the Indian subcontinent witnessed an interesting resurgence of mining and countermining that grew into a clash of cultures centered on this old technique of siege warfare. As the British East India Company's military forces, strongly supported by British army officers and engineers, continued to war against indigenous powers and enlarge company control over the subcontinent, many sieges took place, a good number of which involved mining.

A technical discussion that included a cultural debate ensued about whether European concepts of siege approaches were appropriate for the "mud forts" typically seen in the Maratha Confederation, which became the chief opponent of company forces in the early nineteenth century. An English officer anonymously authored a book entitled *Observations on the Attack of Mud Forts*, in which he disparaged the effectiveness of mining them. The officer had seen defenders stymie offensive efforts through countermining and advocated instead the creation of breaches with artillery fire mixed with infantry attacks. Edward Lake, a lieutenant in the Madras Engineers, countered this idea when he published his *Journals of the Sieges of the Madras Army* in 1825. Based on his experience fighting the Marathas, he argues that if indigenous forces were good at countermining, then:

> it ought to be an incentive to us, not to allow any of the Natives of India to excel us in so important a branch of the art of War. However expert the Natives of Hindostan, where that author served, may have been in the practice of Mining, it is absolutely impossible, that their

Chiefs could have directed them with the same science as the Company's Engineers, to whom they were opposed. If the latter had been at the head of a body of well trained Miners, the result of their labours must therefore have undoubtedly been success instead of failure. In those parts of India, where I have served, the Natives have little or no knowledge of Mining.[26]

Lake goes on to argue that the nature of Indian fortification strongly invited mining. He notes that many Maratha forts deviated from European models to create weaknesses in their defensive capabilities. The "ill flanked outlines" allowed sappers to approach on the surface of the ground and lodge at the foot of the rampart "without the necessity of approaching it by subterraneous galleries." Moreover, the earth "of which the works are composed, is soft enough to be penetrated with ease, and yet of sufficient tenacity to stand without woodwork of any description." Only a few Maratha forts were meticulously planned and constructed along European lines; those works demanded attack galleries but were still vulnerable to effective mining procedures.[27]

Not only the experience of company forces during the Maratha wars but also previous European experience against other local powers in India tended to strongly support Lake's views. For example, in the siege of Caroor in 1760, Capt. Richard Smith supervised sapping for two hundred yards (182.88 meters) in seven days and nights until reaching the crown of the counterscarp of a Native fort. Then he mined so as to blow in the counterscarp, at which the garrison surrendered without risking the follow-up assault. This certainly does not mean that mining played a decisive role in every siege, but it does indicate that the practice was far from useless. In 1818, during the Third Anglo-Maratha War, the Native fort at Amulneir surrendered before a Captain Coventry of the Madras Engineers had an opportunity to dig mines. But the officer wanted to perform an experiment to see how well a charge could bring down one of the circular towers that typically adorned Maratha forts. He instructed his men to dig a gallery beneath one and planted 1,100 pounds (498.95 kilograms) of gunpowder (which "was of inferior quality, being made by the Natives"), with a line of least resistance measuring twenty-two feet (6.70 meters). The experiment amply fulfilled Coventry's expectations, bringing down the tower and destroying a part of the connecting curtain as well.[28]

The siege of a small Native fort at Nowa, twenty-four miles (38.62 kilometers) northeast of Nandair in West Bengal, showcased European methods of underground warfare. The work formed an "oblong square," with the longest face running forty-six yards (42.06 meters) and the shortest thirty-six yards (32.91 meters). It was protected by a fausse-braye, a first line of defense located twenty-nine feet (8.83 meters) in front of the main wall that consisted of a ditch thirteen feet (3.96 meters) deep and thirty-five feet (10.66 meters) wide, with a glacis leading up to it. The crown of the glacis rose twelve feet (3.65 meters) above the level of the land, making the distance between the ditch bottom and the top of the glacis twenty-five feet (7.62 meters). Nowa had been constructed "on the European system" and demanded a European approach. The company brought 4,400 men against it, mostly Native troops and a few British officers, to reduce this strong work held by 500 men.[29]

The British began sapping on January 8, 1819, and eight days later had finished sixty yards (54.86 meters) of the approach. They reached the crest of the glacis on the night of January 23 and began to cut away at it to provide a place for sinking a shaft on the evening of January 25. The "stiff clay" allowed for fast digging without shoring, and the miners sank the shaft twelve feet (3.65 meters) the first night. They continued digging the next day until the shaft was a total of twenty-six feet (7.92 meters) deep from the crest of the glacis. At that depth they began extending a gallery to the left, parallel to the ditch, aiming at a point opposite a tower. By January 28 the branch extended twenty-eight feet (8.53 meters). From there the miners dug another branch to the right for eight feet (2.43 meters) before hitting the ditch. This opening provided ventilation and allowed them to get their bearings; they could see that the main branch was on the same level as the bottom of the ditch.[30]

Engineers selected the spot for two chambers and loaded them on January 29, the left one with 900 pounds (408.23 kilograms) of powder and the right one with only 315 pounds (142.88 kilograms). They laid a powder hose for detonation and finished tamping the next day at 2 P.M. By this time, the garrison knew trouble was imminent and entered into negotiations for surrender. But after those talks fell through, the British touched off their mines at 1:40 A.M. on January 31. The blasts blew in the counterscarp and filled in the ditch opposite a breach that had been created by intense artillery fire. The way was now open for an infantry attack across the filled-in ditch and through the break in the fausse-braye that had every chance of success. Within ten minutes, according to Lake, company infantry had secured the breach and within an hour had captured the fort. The besiegers had lost only 4 men killed and 71 wounded in the assault and a total of 24 killed and 181 wounded during the twenty-four-day siege. This offensive-mining success was accomplished by one engineer officer—an ensign of the Madras Engineers—with three European assistants (one of whom was killed and another wounded). They had the services of seventy pioneers, only twenty of whom "had some little knowledge of mining." No wonder that Lake effused over this example of European siege craft applied to an Indian fort: "The siege of Nowa indeed deserves, in its general features, to be held forth as a model of universal practice."[31]

In contrast, what is known of Indian countermining in the early nineteenth century fails to display much in the way of professional knowledge or effective technique. British engineer Lt. Col. E. W. C. Sandes related a story of crude countermining during the Second Maratha War (1803–1806). The defenders dug a gallery out from the counterscarp of a fort and charged it by laying loose powder along the floor of the gallery. They then ignited it with live coals, which low-caste men hired to perform menial labor threw across the ditch onto the train.[32]

Lake's point was not only that British knowledge and technique were superior to that of indigenous forces but also that Vauban-style siege approaches were relevant in any part of the world. They were most relevant when confronting Native forts constructed along European lines, but he argued they were useful when confronting irregular fortifications as well. Lake placed his faith in Vaubanian principles, "those old, established, well known rules, which have prevailed in Europe for more than 120 years; namely, to work up to, and crown the crest of the glacis by sap, to blow in or pierce the counterscarp and to fill up, if necessary, or otherwise to provide for the effectual passage of the ditches, before the breaches, effected by the battering gun, or by the mine, be assaulted."[33]

British engineers often employed effective mining techniques in their continuing conflict with Native forces in India after the Maratha wars. An army under Lord Combermere besieged Bharatpore, a strongly fortified city in modern Rajasthan, in December 1825 and January 1826. The troops established their second parallel and started to mine forward in January. The Jat defenders countermined but failed to prevent the British miners from exploding a charge under the northeast angle of their perimeter on January 5. This blast produced a breach that was not wide enough for an effective infantry attack, and the second mine was stopped by Jat countermining. Another charge exploded beneath Long-Necked Bastion, also called Pathan Bastion, on January 16 that also failed to create a suitable breach. A charge planted under the northeast angle of the perimeter contained 10,000 pounds (4,535.92 kilograms) of powder. When it went off at 8 A.M. on January 18, it blew open a breach that allowed the British infantry to get through and capture the city.[34]

All of this tended to be a prelude to the high-profile mining and countermining at the siege of the Lucknow Residency during the massive revolt against British rule in India that broke out in 1857. The siege lasted from

"LYING IN WAIT."

Lying in Wait—Capt. G. W. W. Fulton, Bengal Engineers, Siege of Lucknow, 1857. Fulton, serving as an effective garrison engineer, is depicted here as prepared for personal combat underground when British countermines seemed about to intercept the besiegers' galleries. He later was killed on the surface by artillery fire rather than in underground battle. From Sandes, *Military Engineer in India*, 1:352–353.

June 30 to September 25 and pitted 6,000–10,000 mutineers against 1,720 defenders. The mutineers started an offensive gallery on July 14 and blew it six days later but had badly miscalculated the distance. After digging 160 feet (48.76 meters), they were still 140 feet (42.67 meters) short of the redan on the north side of the residency perimeter. Nevertheless, they launched an infantry attack, which was repulsed. Then the garrison began countermining. Chief engineer Maj. John Anderson chose eight miners from Cornwall serving in the Thirty-Second Foot to instruct others in the work. The countermining was superintended by garrison engineer Capt. G. W. W. Fulton of the Bengal Engineers. Soon listeners detected four more galleries approaching the residency works, three of them aiming at the southern sector and one toward the east. Diggers sank countermine shafts and established more listening posts. "Captain Fulton would sit, revolver in hand, at the end of a British countermine waiting for them to break through," notes historian E. W. C. Sandes.[35]

The repulse of another above-ground assault on August 10 spurred the mutineers to increase their efforts at mining until they were digging fourteen galleries by September 5. That day they sprang two mines, but the follow-up attack again was repelled. Fulton was killed by artillery fire on September 14, only eleven days before a relief column led by Maj. Gen. Henry Havelock and Maj. Gen. Sir James Outram fought its way through the besiegers to add more men to the garrison, but they could not raise the siege. British countermining increased during the second phase of the confrontation at Lucknow Residency. Outram reported that his men dug twenty-one shafts totaling 200 feet (60.96 meters) deep and constructed a total of 3,291 feet (1,003.09 meters) of galleries extending from them. The mutineers advanced a total of twenty mines and exploded three of them during this second phase, leading to loss of life among the garrison. Two more were exploded but with no effect on the defenders. Meanwhile, British countermining blew in seven Indian galleries. According to Outram, miner met counterminer in eight other instances. Mutineers broke into British galleries, but in every case the counterminers had detected sounds of digging beforehand and were ready for the encounter. British miners fired into the offensive gallery through the hole connecting mine and countermine and charged, capturing or destroying the Indian gallery without the use of explosives.[36]

A second relief column, under Sir Colin Campbell, finally raised the siege of the Lucknow Residency on November 14. On examining the Indian works, British engineers found that their shafts were on average 8 feet (2.43 meters) deep and the galleries were 3 feet (0.91 meters) tall and 2 feet (0.60 meters) wide. Each had an arched roof and usually no shoring. One Indian gallery was 200 feet (60.96 meters) long and another was 300 feet (91.44 meters) long, both with minimal ventilation; mention was made of "air tubes" with no details about them. The British rarely used shoring in their countermines, even though the soil was "light and sandy." One of their galleries extended 298 feet (90.83 meters) and another 192 feet (58.52 meters). According to Capt. C. B. Crommelin, neither of them needed ventilation, but he provided no explanation for this. Lights burned in the 192-foot-long (58.52 meters) gallery but would not burn in the longer structure. How miners were able to work in either of them without artificial ventilation was never explained.[37]

The underground war at Lucknow was quite extensive, similar in its multiple simultaneous approaches and countermines to what had taken place at Sebastopol only two years before. But it represented an uneven quality between the belligerents. Mutineer mining efforts completely failed, in part due to faulty calculations but mostly due to aggressive and effective countermining by the garrison. The British established a good system of listening posts and detected every near approach, managing to intercept eight attempts and neutralize the offensive gallery with little if any loss of life. The garrison also conducted a successful underground attack on the besiegers. The British sank a shaft on August 17 just inside their defensive wall and extended a gallery fifty feet (15.24 meters) into the contested zone between the two forces, digging two branches at its end under the Johannes House, held by the mutineers. At dawn on August 21, they sprang the two mines, which demolished the structure, and then conducted a sortie that routed the mutineers from the area.[38]

Events at Lucknow indicated that Lake had been right. The level of mining expertise among indigenous Indian forces was not terribly high, and as long as trained engineer officers could rely on civilian miners who happened to serve in the ranks, the British seemed to have a qualitative edge over their opponents. Defensive countermining proved to be one of the keys to the survival of the British and allied Indian soldiers as well as the civilians who took refuge in the residency compound. Equally important was the garrison's ability to repel several assaults and withstand artillery fire, but that takes nothing away from the significance of Lucknow in the history of military mining and countermining. The fact that the siege took place outside the core area of mining in Europe also supports the conclusion that Western forces had an advantage over regional powers around the world when it came to underground conflict.

While the mid-nineteenth century witnessed impressive mining in Europe and India, it also witnessed a new surge of technological innovation in underground warfare. After the introduction of gunpowder, the next important technical development lay in the application of electrical energy as an alternative to powder trains, powder hoses, and Bickford fuzes to detonate mines. The possibilities began in the late eighteenth century and proceeded in fits and starts as the complicated process of developing effective batteries played out in several Western countries. By the middle of the nineteenth century, it was obvious that electricity held the potential for improving the reliability of exploding underground charges. Totleben's use of chemical batteries at Sebastopol proved the point.

Electrical devices applicable for mining eventually divided into two major types: one was a battery that generated electrical current by combining different chemicals, and the other created current through a magnetic process. The former, which was the earliest form of a battery, was often called a galvanic, voltaic, or electrochemical device. When Luigi Galvani, a professor at the University of Bologna, pressed a metal object to the nerve of a partially dissected frog and observed the reaction, he theorized that animal electricity was generated in the brain and published his findings early in the 1790s. Alessandro Volta, a professor at the University of Pavia, disputed that theory and argued that the source was external to organic bodies. He created electrical energy by constructing a pile consisting of alternating disks of copper and zinc, each disk separated by a piece of cardboard soaked in salt water. Volta published his results in 1799. This voltaic pile, as it was popularly known, became the dominant form of generating electricity for many decades. It produced a steady, low-level current for a period of time until the materials had to be replaced. Toward the middle of the nineteenth century, a number of people had improved on this basic concept to increase the length of time the battery could generate usable power. William Grove of England developed the most popular model. With their constant flow of energy, electrochemical devices were widely used to power telegraph systems and in the electrotyping and electroplating industries.[39]

Electromagnetic devices were the second major way to generate power. Hans Christian Oersted at the University of Copenhagen conducted the earliest experiment to develop this device in the 1820s, and many others in different countries continued his work. By 1831, Michael Faraday of England had developed the first truly workable machine. It rotated a copper disk between the poles of a magnet and depended on external help—someone turning a crank. Electromagnetic devices could produce stronger current than electrochemical batteries, but the effect was not long lasting. A surge of improvements on the basic concept from 1840 to 1880 made the electromagnetic device a fierce competitor with electrochemical alternatives. Charles Wheatstone of England had developed the most popular electromagnetic machine by the 1840s.[40]

The utility of these devices in exploding gunpowder became evident by this time. In 1839 the Royal Navy used both chemical and magnetic machines to break up the wreck of a warship, the *Royal George*, which had sunk in 1782 at Spithead near Portsmouth, for salvage purposes. Both types were useful, but most people came to conclude by the late 1850s that electromagnetic devices were more suited for exploding gunpowder because of their greater power and event-centered use.[41]

Equally important as the power generator was the connection between the wires and the gunpowder. This device was technically termed a "deflagrater" but popularly called a "squib" and was developed simultaneously by different men in different countries during the mid-nineteenth century. An example of a squib, which "ignites an explosive by the heating effect of an electric current," was created by Prof. Robert Hare at the University of Pennsylvania in 1829. It consisted of a tube with small iron wires placed a short distance from each other and filled with gunpowder and fulminate of silver. The electric current traveled across the points of the wire through this mix and ignited it. The resulting small explosion set off the larger charge in which it was imbedded, transforming the electric current into heat that could set off the larger charge.[42]

It was not easy to convince practical men that electricity could be harnessed for the purpose of setting off explosives. The popularity of the Bickford fuze and the continued use of powder hoses inhibited the growth of electrical detonation in commercial mining.[43] At Sebastopol, the Russians were the first to use electricity in military mining. They employed chemical devices to set off all their countermines, based on Totleben's advice, but the French continued to use powder hoses to explode their offensive mines. French engineers, however, used chemical devices to set off explosive charges that brought down the Russian masonry defenses of Sebastopol harbor after the port's capture. Apparently, powder hoses were used at the Lucknow Residency—there is no indication that electrical detonation made a show at that siege. During the American Civil War in the 1860s, electricity was not used in mining or countermining.

In fact, despite the rate of successful firings achieved by Russian counterminers at Sebastopol, it took several decades for military men to fully accept electricity as the best way to spring underground charges. Writers of handbooks recognized the possibilities inherent in electrical detonation, but they often fully described the old methods as well to give their readers a choice. In his *Treatise on Field Fortification*, originally published in 1834 but reaching its fifth edition by 1860, Capt. John Shortall Macaulay of the Royal Engineers provides full information about the many ways one could touch off underground charges but promotes voltaic batteries as the best method, declaring that electricity is "instantaneous and certain." If the wires were properly protected, they could be set up and maintained for a long while before the time of detonation. Macaulay especially was impressed that, by 1860, it was possible to set off several charges simultaneously or in succession, a useful procedure for countermining. One could also place several wires in a large amount of gunpowder to more fully explode the mass, a possible solution to a problem that had dogged miners ever since the sixteenth century.[44]

Macaulay also points out that the next generation of explosives, beyond gunpowder, could be incorporated into mining. Guncotton, or nitrocellulose, had been developed in Europe by 1846 and proved to be six times more powerful than gunpowder while producing much less smoke. Cotton, a source of cellulose, was treated with concentrated sulfuric acid and nitric acid to produce cellulose trinitrate. It was further treated with alcohol or some other liquid to reduce the danger of exploding while in storage, and thus often was called "wet" guncotton. For twenty years after its development, guncotton was too unstable for safe use, but by 1865 English chemist Frederick Augustus Abel had developed the first workable process for its manufacture. Twenty years later it was being applied as the first smokeless powder for artillery and infantry arms.[45]

Macaulay theorizes that guncotton could be used along with gunpowder to increase the effect of underground explosions. The gunpowder would presumably expand and press on the guncotton to spread the effect and explode more of the charge. He also advises applying some guncotton to the end of the electrical wires because it would explode more readily than the squib. But, when Straith and Macaulay were writing their books, guncotton was still too unstable to be safely used in charging mines or setting up detonation devices. It and electrical detonation itself had to wait until the late nineteenth century before becoming important elements in military operations.[46]

The brief mention of a drilling device shipped by the French to Sebastopol reminds us that great strides had been made in developing mechanical drilling machines for the civilian mining and tunneling industries. From this point one can talk confidently about a separate branch of mining called tunneling. By the early and mid-nineteenth century, the practice of creating permanent transit galleries had greatly increased, mostly due to the rapid spread of railroads, which needed level tracks across the countryside and through hills and mountains. The two most famous tunnel projects were the Hoosac Tunnel in western Massachusetts (1855–1876) and the Mont Cenis Tunnel linking Italy and France (1857–1871). Before the second quarter of the nineteenth century, the typical way to cut a passage through rock was to drill small shot holes, fill them with powder, and then blast, cleaning out the rubble and starting again. Forty feet (12.19 meters) per month was typical progress. But drills powered by compressed air were available by about 1860, when nitroglycerin also became available. These revolutionized the process of tunneling. Both were used at the Hoosac Tunnel to increase the rate of digging this four-and-a-half-mile (7.24 kilometers) gallery up to 184 feet (56.08 meters) per month by 1873. At the Mont Cenis Tunnel, which was seven and a half miles (12.07 kilometers) long, the rate of progress soared from 23 to 231 feet (7.01 to 70.40 meters) per month because of mechanical drills and nitroglycerin.[47]

The Hoosac Tunnel and the Mont Cenis project were watersheds of new drilling and blasting technology on hard-rock tunneling. Soft-ground tunneling had earlier seen its own watershed in Marc Isambard Bunel's Thames Tunnel project in London back in 1825, which used the latest mechanical devices for burrowing through clay. Further advances that benefited both hard-rock and soft-ground tunneling included the use of compressed air to power devices for extracting water from galleries. This allowed the construction of permanent transit tunnels at and below the water level at the site of the Kattendyk Tunnel in Antwerp, Belgium, and at the Hudson River Tunnel in the 1870s and 1880s in the United States.[48]

Military engineers remained behind the curve in the development of new technology employed in civilian mining, whether one looks at transit tunneling or ore mining, throughout the mid-nineteenth century. Nevertheless, doctrinal and technological change was slowly growing in military mining and countermining with the publication of numerous handbooks, treatises, and engineering manuals that publicized some of those new developments and continued to give full credit to the significance of gallery mining for offensive and defensive purposes in army operations.

A procession of factors slowly came to modernize military mining. Gunpowder was the first in the fourteenth and fifteenth centuries. Vauban contributed to this process with his system of siege approaches that incorporated short mining galleries in the late seventeenth century. Experiments with test mines provided data on underground effects, leading to the concept of the globe of compression. The siege of Schweidnitz in 1762 employed all of those elements in one of the more successful sieges in history. Sebastopol continued this trend with its multiple simultaneous mining and countermine efforts. That siege also saw the debut of electrical detonation in 1855.

The next major conflict to significantly employ mining, the American Civil War of 1861–1865, was something of a holdover from the past. It did not employ electrical detonation but did see at least one example of multiple simultaneous underground approaches along a long line of earthworks at the siege of Vicksburg in 1863. Moreover, mining took place at relatively few sites, although several other campaigns provided possible venues for it. In many ways, the Civil War represented the continued marginality of regions lying outside the core of mining history, which was still centered on Europe.

8

The American Civil War

The Western Hemisphere had witnessed virtually no military mining except for a handful of brief applications in the British colonies and during the war that separated them from Great Britain. This part of the world lay outside the core area of mining activity, but the professional army of the United States retained its European-orientation throughout the early nineteenth century, meaning it also retained its grasp of mining and countermining doctrine. Although it never applied that theory in practice, the army could draw on it at any time.

The American Civil War (1861–1865), however, involved more military mining than occurred in all of Western Hemispheric history before the firing on Fort Sumter in 1861. In addition, American military forces would never again employ mining after Appomattox in 1865. The Civil War, therefore, stands out in stark relief within the context of American military history as far as mining and countermining is concerned. The conflict occurred soon after the intensive mining experience at Sebastopol, a high-profile siege that occupied the awareness of Civil War officers even if the mining there was muted by an awareness of the role played by earthen field fortifications in that siege. Within half a century after Appomattox occurred the greatest mining and countermining employment in world history on the Western Front of World War I. With its unique position as a major war positioned within the resurgence of military mining during the latter half of the nineteenth century, the Civil War oddly displayed a retrograde nature in the history of mining. No new devices such as electrical detonation were employed. But the mining operations during the Civil War are unusually well documented, not only in official reports but especially in personal accounts by the men involved in and affected by them. This offers us the first truly in-depth look into the experience of military mining in history.

Civil War commanders and engineers had no experience with military mining in the field at the start of their conflict, but it was general knowledge that mining and countermining had occurred in numerous sieges of the past. No special handbook on it had ever been published in the United States until northern engineer Maj. James C. Duane's *Manual for Engineer Troops* came out in 1862. Duane points out that underground siege approaches usually were dug at shallow depths compared to commercial mining, and thus galleries sliced through "the more recent formations of earths and sands," requiring much shoring. "It is in the adjustment and fittings of these linings that the chief art of the military miner consists."[1]

Duane follows the Vaubanian approach to mining doctrine. He states that the besieger should proceed with surface approaches, establishing parallels and digging saps toward the target until reaching the glacis of the enemy work. There, he had to go underground with digging shafts sixteen to twenty-one feet (4.87 to 6.40 meters) deep, from which he extended galleries. The shafts could also be dug from the floor of a parallel or even from behind a bank of earth or some natural cover. Duane does not recommend digging the gallery more than sixty feet (18.28 meters) before installing some sort of ventilating system. The options for this included drilling vent holes up to the surface, constructing transverse galleries, and laying tubes to bring in fresh air and extract bad air. These tubes could be made of wood or tin, but the best was "vulcanized india-rubber," with a fan or blower to force air into them.[2]

On determining the size of the charge, Duane spells out the terminology involved. If the radius of the crater to be formed at the surface of an explosion is equal to the line of least resistance, it is termed a "one-lined crater." If the radius is double the line of least resistance, it is a "two-lined crater," also called a "common crater." Anything less than a two-lined crater was termed an "undercharged mine," and anything larger than it was called "overcharged." According to Duane, the overcharged crater also is called the globe of compression. The term "camouflet" refers to any charge that fails to produce a crater on the surface and is used principally in attacking enemy galleries. It often is applied by drilling small holes horizontally from a countermine gallery toward the enemy gallery, six inches (15.24 centimeters) wide and up to eight feet (2.43 meters) long, and placing twelve to twenty pounds (5.40 to 9.07 kilograms) of powder into the hole.[3]

To detonate the charge, Duane prefers the old powder hose, "a tube of strong linen" filled with powder and laid in a wooden trough under the tamping. A length of portfire stuck in the end of the hose and packed all around with clay—this prevented sparks from reaching the powder hose before the portfire had properly burned—would enable the operator to light it and get out of the gallery before the charge went up. Duane also writes of the Bickford fuze as an effective means of touching off the charge and discusses voltaic batteries but expresses little faith in them: "The care and attention required to isolate the wires, and the difficulty of arranging securely so great a length of them, render this application of the battery, however desirable, hardly available for military purposes." While dismissing the most advanced method of detonation, Duane also describes utilizing rockets placed in a wooden trough. This was an old method of ignition, but its viability seems highly questionable. The rocket would certainly move forward through the trough, but if there were any bends, it would be necessary to place another rocket at the corner so the previous one could ignite the next one and send it on its way. While this method remained part of the long-term theory of mining, there is no evidence it ever was used successfully. He then prescribes that the length of gallery that needed to be tamped, if tightly packed, could be one and a half times the line of least resistance; if loosely packed, it had to be twice the line of least resistance.[4]

Duane effectively explains the basics of mining and countermining technique even though he fails to promote the latest trends in detonating charges. But he provides no help to commanders or engineers in terms of the principles of mine warfare. That is not surprising, for the Civil War occurred in the latter part of an era when military manuals were tightly focused on technical rather than theoretical issues. They told a practitioner *how* to do something, not *when* to do it. There was no theory about how to ensure that the successful explosion of a mine could be translated into the capture of a work because that depended heavily on a wide variety of factors out of the control of the miner. But armed with the technique, commanders and engineers at least had a chance of judging when and where it would be efficacious to employ it.

Halfway through the war, the Union siege of Vicksburg, Mississippi, from May 18 to July 4, 1863, involved far more mining and countermining than had taken place in all previous Western Hemispheric history. Along siege lines that stretched for eight miles (12.87 kilometers) north, east, and south of the town, Federal engineers advanced ten approaches toward selected Confederate works. Following Vauban's outline, they dug saps to get as close as possible and then began mining galleries. The approach that drew the greatest attention was in the center, where it snaked along a narrow ridge toward a work the Federals called Fort Hill—it was called the Third Louisiana Redan when Vicksburg National Military Park was created at the turn of the twentieth century. Dug in sections, the sap extended for 508 yards (464.51 meters) until it reached the foot of the exterior slope of the redan's parapet on June 22. The work had no ditch, so Capt. Andrew Hickenlooper, an artillery officer serving as an engineer, directed that a level place be cleared at the end of the sap where he intended to start mining operations. The sap was so close to the work that it was possible for Federals to lay a hand on the exterior slope. Hickenlooper procured some short-handled shovels and picks for his miners, who he had selected from the Forty-Fifth Illinois, a regiment that contained many civilian coal miners in its ranks.[5]

Thirty-five miners working in three reliefs and eight-hour shifts began digging the gallery at 9 A.M. on June 23. Hickenlooper used two men at a time, one with a pick and the other with a shovel. To move the spoil out of the gallery, he placed a man every ten feet (3.04 meters) along its length. Ground in the Vicksburg area consists of loess soil blown into the region thousands of years earlier. It is a fine soil that, after hundreds of years of compression, tends to remain in place when cut, so mining progress was swift. The Federals dug a gallery three by four feet (0.91 by 1.21 meters) in dimension for twelve feet (3.65 meters) the first day directly under the redan and, by June 24, had extended the gallery a total of forty feet (12.19 meters). Then they dug a branch to the left, where the Federal miners could hear Confederates digging a countermine above them; it was actually located to the left of their gallery. Hickenlooper's men dug without artificial ventilation, and the air in the tunnel was "almost stifling," as reported by an observer, the candles placed at intervals along the gallery burning dimly. Each swing of the pick, this man noted, "slices down six inches [15.24 centimeters] of the tough subsoil of Mississippi," but the Federals did not have to dig very far.[6]

When the Unionists first heard countermining above, they knew it would now be "a race to see which side would get in the first blow," as Hickenlooper put it, "whether we would be blown out or they blown up." The Confederates had begun countermining on June 23. Directed by engineer Maj. Samuel H. Lockett, they dug a shaft inside the redan that was thirty feet (9.14 meters) wide because Lockett wanted to try venting the effects of the Union charge with a big hole to minimize damage. The Confederates created benches to make it possible to physically throw up shovelfuls of dirt from the lower depths. They reached a point nine feet (2.74 meters) down before they heard the sounds of Union miners at work.[7]

But Hickenlooper won the deadly race. By June 25, his gallery extended a total of ninety-five feet (28.29 meters), a length that included two branches, each fifteen feet (4.57 meters) long, added to the gallery of sixty-five feet (19.81 meters). Navy vessels in the Mississippi River cooperating with the army provided 5,000 pounds (2,267.96 kilograms) of powder and three hundred feet (91.44 meters) of safety fuze. The powder arrived in barrels, which were carried through the sap to a safe distance from the redan, where they were broken open and the powder transferred into grain sacks. Each man then carried one sack weighing 25 pounds (11.33 kilograms) into the gallery, dodging shell explosions. Nearly one hundred sacks of powder were needed to make a total charge of 2,200 pounds (997.90 kilograms), with 700 pounds (317.51 kilograms) in each of the two branches and 800 pounds (362.87 kilograms) in the central gallery. Hickenlooper planned to touch off all three charges simultaneously. He laid two strands of safety fuze to each charge. Then he tamped the gallery with a combination of sandbags and timber, placing four feet (1.21 meters) of sandbags first, then timbers, then another four feet of sandbags, and so on until reaching a point only twenty-five feet (7.62 meters) from the mouth of the tunnel. In short, he tamped forty feet (12.19 meters) of the sixty-five-foot (19.81 meters) gallery.[8]

All was ready by 1 P.M. on June 25, and Hickenlooper received orders to touch off the charges at 3 P.M. Federal infantrymen waiting near the gallery were told to stand on tiptoe to avoid concussion as shockwaves rumbled through the earth. Hickenlooper watched closely as Fort Hill "commenced an upward movement, gradually breaking into fragments and growing less bulky in appearance, until it looked like an immense fountain of finely pulverized earth, mingled with flashes of fire and clouds of smoke, through which could occasionally be caught a glimpse of some dark objects,—men, gun-carriages, shelters, etc." The plume of dirt and debris ascended seventy to one hundred feet (21.33 to 30.48 meters) into the air in what Hickenlooper called a "perfect success." For the Confederates, the blast was a deadly surprise. Much of the redan was destroyed, and a suffocating atmosphere of dirt and debris engulfed everyone near it. At least six men working in the countermine were instantly killed, and a total of about fifty Confederates in and near the redan perished in the mine explosion.[9]

While the three charges had done their work, Hickenlooper estimated they destroyed only half of this small earthwork; the damage was limited by the wide countermine shaft, which dissipated much of the force of the explosion. The crater measured fifty feet (15.24 meters) in diameter and twenty feet (6.09 meters) deep, its floor sloped regularly "like a large bowl from the bottom up," as reported by a Union soldier who occupied it during the ensuing battle. "The up-heaved earth was soft, and our feet sank deep into the loose dirt."[10]

The infantry attack that followed the blow up soon developed into a vicious, static battle. Federal troops easily occupied the crater but could not advance beyond it because the Confederates had previously constructed a parapet fifteen to twenty-five feet (4.57 to 7.62 meters) west of the hole's location. Union officers fed in a total of nine regiments, each of which detailed two companies at a time to fight in the crater, rotating them in and out until another regiment was ready to replace the previous one. This went on for more than twenty-four hours. The Confederates' parapet served as the solid defense of the ruined fort. From it, Confederate troops moved up to the western lip of the crater to battle the Federals at very short range across the upheaved earth. The Federals brought in building timber (hewn logs) from a demolished cotton gin near the siege lines and placed it on the lip of the crater to protect their heads while they fired through portholes dug underneath it, but Confederate artillery rounds knocked off the timber. Both sides threw hand grenades across the western lip of the crater, and the Confederates also lit artillery shells and rolled them over the embankment. Many casualties resulted on both sides from this practice. Attackers and defenders alike also shoved their rifled-musket barrels over the lip to fire randomly at the enemy. "My God," said the Union brigade commander most immediately in charge of the fight, "they are killing my bravest men in that hole." As the fighting continued through the night of June 25–26, regimental officers guided their men toward the crater by looking at the flashes of grenade and artillery shells that lit up the awful scene, placing seventy men in the hole, relieving them in a couple of hours, and then counting the casualties.[11]

The battle sputtered out by the evening of June 26, as the Federals gave up further efforts to exploit the mine explosion. Both sides had fought to a standstill, with 243 Federal casualties and ninety-four Confederate losses. "That ended the farce of capturing Vicksburg by tunnel," wrote a bitter Union veteran after the war.[12]

What had happened at the Third Louisiana Redan was in many ways an old story. Technical success in digging the gallery, setting the charge, and exploding it was not followed up by success in the infantry assault. In this case it was because the defenders had ample warning to prepare. The large countermine shaft saved much of the redan, and the new parapet played a key role in containing the breach.

The Federals did not give up on mining. Hickenlooper sought a way to exploit the approach he had so carefully constructed. Even before the battle of June 25–26 ended, he supervised the construction of a casemate inside the crater to protect Federal soldiers from grenades and constructed a parapet to the right and left from this dugout. On the morning of June 28, he began digging a new gallery toward the northwest, aiming at what was left of the Third Louisiana Redan's left wing and the connecting infantry trench. This new gallery ran for twelve feet (3.65 meters), then veered north for fifty feet (15.24 meters). For at least fifteen feet (4.57 meters) of this distance, Hickenlooper passed through earth disturbed by the previous mine explosion. Here he had to shore the gallery by placing a four-sided frame to support boards holding up the earth. His miners sharpened the ends of some boards and then drove them forward into the earth on the sides and ceiling, only then digging out the area between these three sides for a few feet before driving more boards forward into yet undug ground.[13]

By June 30, the Confederates realized that a new gallery was being dug and became desperate to counter it. Lockett arranged for the construction of a barrel containing 125 pounds (56.69 kilograms) of powder and a fifteen-second time fuze to be rolled over the western lip of the June 25 crater It was aimed at the dugout Hickenlooper had constructed under the assumption that the structure housed the entrance to the new gallery. Lockett's men successfully rolled the powder barrel toward the target and it went off, sending timber into the air, but the casemate mostly held firm.[14]

Although the barrel failed to stop Hickenlooper, it did alert him to the danger of prolonging work on the new gallery. Therefore, the Federals decided to explode the mine as soon as possible rather than wait for a planned assault along the Vicksburg siege lines set for July 6. After digging a total of sixty-five feet (19.81 meters), Hickenlooper's men scooped out a chamber ten feet (3.04 meters) square for 2,000 pounds (907.18 kilograms) of powder and tamped the gallery as they had done on the twenty-fifth.[15]

Hickenlooper set off the mine at 3 P.M., July 1. This time it would be solely for the purpose of harassing the enemy; there would be no follow-up assault. Once again the Federals won the race with a spectacular explosion that raised "a tremendous mass of earth and cloud of dust . . . with great velocity high in the air, expanding as it rose until the top thereof appeared like an immense spherical dome supported by a huge pillar under its center." Confederate troops within a few feet of the explosion were stunned, their hats knocked off and many of them hurled backward down the slope of the narrow ridge occupied by Fort Hill. Several counterminers were killed underground while half a dozen live men, including at least one slave laborer, were blown toward Union lines. This explosion did more damage to the earthwork than that of June 25. What had been left of the Third Louisiana Redan was destroyed, leaving behind a gaping hole fifty feet (15.24 meters) long, thirty feet (9.14 meters) wide, and twenty feet (6.09 meters) deep. The Confederate parapet that had been so useful in holding the June 25 attack was mostly destroyed at its left end. The result was "almost a practicable breach for an assault," as Lockett put it. But there was no assault. The Confederates lost about one hundred men in the explosion and the intense firing that took place right after it. The Federals did not report any losses.[16]

While the two offensive galleries at the Third Louisiana Redan represented the heart of mining at Vicksburg, more galleries were started at other locations. Engineer Capt. William Kossak took over a Union approach at the northeast sector of the long Confederate line on June 19. This sap snaked along a narrow ridge toward Stockade Redan, a large earthwork that covered Graveyard Road as it headed toward Vicksburg. By that date, the head of the sap was twenty feet (6.09 meters) from the counterscarp of the redan, but Kossak concluded he could dig no farther because of numerous obstacles in the way. He reinforced the head of the sap with gabions and sandbags, converting it into an outpost, as his sappers dug branches of the approach to right and left parallel to the counterscarp of the ditch. They covered the open sap with several layers of fence rails to prevent the Confederates from tossing grenades into it. On June 22 the Federals could hear the sounds of countermining, so Kossak started two galleries of his own to right and left of the outpost at the head of the sap, both of them aiming directly at Stockade Redan, to protect the outpost and sap. He worked at a deeper level than the Confederates.[17]

Lockett's subordinates had some difficulty arranging their countermine effort at Stockade Redan. On June 24 they tried to spring two charges, but both of them failed. They had experimented by filling gas pipes with gunpowder but found that powder "confined in a long tube, when ignited, will burst the tube a few feet from the end and will not burn farther." That delayed their work for nearly two days. The Confederates removed both charges and dug new chambers to be closer to the target, as they estimated, and found quick matches to set off an old-fashioned powder train. Both chambers were eight feet (2.43 meters) deep and thirty-five feet (10.66 meters) from the counterscarp. One contained forty-five pounds (20.41 kilograms) of powder and the other eighty pounds (36.28 kilograms). When they were blown on the morning of June 26, they collapsed the roofs of both Union galleries. The charges were too small to form craters on the surface, yet they not only destroyed Kossak's galleries but also pulverized the ground for thirty feet (9.14 meters) around without doing much damage to the outpost.[18]

The two Confederate mine explosions had created a zone of rupture that Kossak could not dig through with speed or safety, so he planned a bold venture. To completely flank this area, he started a gallery from the approach some distance back from the outpost. First he began a decoy sap to fool the Confederates into thinking his main effort would be directly toward Stockade Redan. Then he constructed a traverse to hide the fact that he was starting to dig an underground gallery toward the right. The only problem was that a deep, steep-sided ravine lay to the right of this approach, but Kossak developed a plan to deal with this problem. His gallery descended one foot (0.30 meters) in every three feet (0.91 meters) so that at sixty feet (18.28 meters) long, it had gone down twenty feet (6.09 meters) in elevation. At this point, his miners' candles "being extinguished by the extreme heat and foulness of the air," Kossak dug a ventilation hole to the surface. Then he continued digging the gallery another sixteen feet (4.87 meters) to reach a level just under the floor of the ravine. At this point he was twenty-five feet (7.62 meters) lower than at the mouth of the gallery and had thus far carefully planned his digging to adjust to the shape of the terrain. He soon began to ascend as the men continued, going seventeen feet (5.18 meters) while lifting the gallery one foot (0.30 meters) in every three feet (0.91 meters). Reaching a large log, Kossak dug an exit from the gallery so that its mouth was hidden from Confederate view by the log. Here he dumped the spoil so that the miners did not have to carry it all the way back to the mouth of the gallery. "This new dump and air hole brought plenty of circulation of air into the mine," the captain reported.[19]

This gallery was a remarkable achievement due to Kossak's careful calculation of length, depth, and terrain features. It was four feet, six inches (1.40 meters) tall; three feet (0.91 meters) wide at the bottom; and two feet, six inches (0.79 meters) wide at the top. His men set shoring frames every four feet (1.21 meters) and placed sheeting only on the ceiling. Kossak believed that Confederate countermines would inevitably approach from above and "could only crush our tops." His men worked twenty-four hours a day in six-hour reliefs, "suffering much from the extreme heat and want of air," before opening the combination dump and air hole at the log. A map of the gallery indicates that a second dump and air hole also had been constructed at some point in time fifty feet (15.24 meters) short of the main one at the log, increasing the flow of fresh air through the gallery.[20]

Continuing from the log, Kossak's men soon neared Stockade Redan. They dug seventy feet (21.33 meters) farther than the dump area, going under the ditch and parapet. Kossak planned a major attack. He wanted to use 2,200 pounds (997.90 kilograms) of powder and rounded up 175 feet (53.34 meters) of safety fuze but never was able to put the plan into operation. The captain received verbal instructions to stop work at 10:30 A.M., July 4, because negotiations held the previous day had led to the Confederate surrender of Vicksburg that morning. His planned blow up undoubtedly would have been successful; there is no indication that the Confederates were countering this remarkable gallery.[21]

While the Third Louisiana Redan and Stockade Redan presented the venue for the most intense mining efforts at Vicksburg, lesser work took place at several other places along the siege line. Just south of Stockade Redan lay a Confederate earthwork called Green's Redan. Federal troopers dug a sap toward it and reached a point where they could annoy its garrison. In fact, Brig. Gen. Martin E. Green was killed by a Union sharpshooter on June 27, leading Major Lockett to begin digging a countermine to destroy the head of the sap. He sprung a charge on July 2, but it did no damage because the Confederates had not dug far enough and used only twenty-five pounds (11.33 kilograms) of powder. The explosion, however, led Kossak to begin digging two galleries of his own in this area. He called for sixteen men of the Fourth West Virginia whom he knew to be civilian coal miners and started work on the night of July 3. Early the next morning Kossak calculated that the Confederate gallery was on the same level as his own and but eight feet (2.43 meters) away, so he arranged for two hundred pounds (90.71 kilograms) of powder and a length of safety fuze to be sent to the new gallery. The material proved to be unnecessary because soon after word circulated that Vicksburg would be surrendered.[22]

A similar story developed at other points along the line. Near the center, the Federals advanced a sap toward the Railroad Redoubt and by June 28 detected signs of a Confederate countermine gallery. Engineer Lt. Peter C. Hains dug a listening gallery toward the assumed enemy effort and changed the direction of his sap a bit to put more distance between it and the countermine. The Confederates sprung a charge of 125 pounds (56.69 kilograms) on the night of July 2, but it was exploded too soon, failing to do much damage to the Union listening gallery. The explosion "made some fissures, in the north wall, through which smoke and dust came," recalled James H. Dean. "The roof did not fall as it was well supported with timbers, several of the working squad, of which I was a member, was seriously shocked by the concussion, three of whom were sent to a hospital."[23]

A bit north of Railroad Redoubt, another Union sap approached the Second Texas Lunette. On the night of June 19, the Confederates started a countermine gallery from the ditch of the lunette. By June 28, they were able to set off a charge at 2 P.M., but it failed to hurt the Federal trench. They misjudged the distance, exploding the charge about twenty feet (6.09 meters) from the head of the sap. "Falling clods of earth struck several of the men but no one was seriously injured," reported the Union officer in charge of the sap. The Federals began digging an offensive mine from the saphead on the night of July 2, but hard clay limited their progress. Meanwhile, the Confederates had dug a new countermine and prepared to spring it that same day, but developments overtook the efforts of both sides outside the Second Texas Lunette. The surrender also put a halt to Confederate plans for blowing up the head of a sap that approached the Twenty-Sixth Louisiana Redoubt west of Stockade Redan. The engineers had planned to use one hundred pounds (45.35 kilograms) of powder in that effort.[24]

After the bloodletting of June 25, Maj. Gen. Ulysses S. Grant gave up any attempt to explode offensive mines at only one point and instead relied on advancing all Union approaches until ready to spring several charges simultaneously. A follow-up attack would then have a real chance of success. By June 30, all ten of the Federal approaches were within 5–120 yards (4.57–109.72 meters) of the Confederate lines, and Grant planned to spring mines and launch a major assault on July 6.[25]

To oppose this effort, Lockett had established eleven countermines at six locations and planned to use 100–125 pounds (45.35–56.69 kilograms) of powder in each. The charges were placed from six to nine feet (1.82 to 2.74 meters) below the surface, and the major was confident all of them would have been ready by the time the Federal saps closed in. But whether this plan could have stopped the Unionists is an open question. As noted by Grant's engineers, the Confederates did not have a good track record in underground warfare at Vicksburg. Their countermines were charged with quite small amounts of powder, and they had a habit of underestimating distance. The charges rarely did any significant damage to the Union saps and galleries, merely pulverizing the ground nearby and making it dangerous for the Federals to dig quickly through the affected areas. Lockett defended his record after the war by noting that "it was very difficult to determine distances underground, where we could hear the enemy's sappers picking, picking, picking, so very distinctly that it hardly seemed possible for them to be more than a few feet distant, where in reality they were many yards away."[26]

Lockett could be forgiven for charging his countermines short because the problem of estimating distance underground had bedeviled military miners for centuries. But Totleben, who keenly recognized the significance of this problem, dealt with it effectively at Sebastopol by training his miners how to overcome it. The real problem was that no one in either the Union or the Confederate armies had any experience in military mining and countermining. Duane's engineering manual contained no information on how to estimate distance underground. By every measure, Union engineers at Vicksburg outperformed their Confederate opponents. Kossak did an impressive job in planning his gallery to go under the floor of a deep ravine and ascend once again to accurately strike at Stockade Redan. Nothing like this had been accomplished by the Confederates, who could not even accurately gauge how far to dig a straight-level gallery a few yards toward a Union sap. Even if they had been accurate in estimating distance, the small size of their charges would have inflicted minimal damage. It also bears mentioning that the majority of officers who supervised mining and countermining at Vicksburg were not even trained engineers. Like Hickenlooper, they were infantry or artillery officers from volunteer units (as opposed to regular officers) who were detached for engineer work. They had to read Duane's book in the field and try to implement the technical instructions as best they could. But the Federals did their homework better than the Confederates.

Could offensive mining have played a larger role in success at Vicksburg for the Union army? Two of the trained engineer officers with Grant thought not. "Mines could not make an easier way into the enemy's line than existed already," concluded Capt. Frederick E. Prime and Capt. Cyrus B. Comstock. "Their only use was to demoralize the enemy by their explosion at the moment of an assault." In their view only volunteers placed much store in the use of military mining. "More importance was attached to them by officers and men than they deserved."[27]

There is no doubt that mining and countermining had a strong emotional effect on the rank and file. "It upset us more than every thing else," wrote Confederate W. O. Dodd. "We did not think it fair, as it gave us no chance." On the Union side, fighting in the crater on June 25–26 seared itself on the consciousness of those who survived the bloody pit. "The men called it 'the slaughter pen,'" wrote the regimental veterans who authored a history of the Thirty-First Illinois. "It was a hell within a radius of five hundred feet [152.40 meters]." But Prime and Comstock underestimated the possibilities of a decisive breakthrough by mining. Confederate division commander Maj. Gen. John H. Forney saw the Third Louisiana Redan as a key point in the defensive line. Capture of that position would have enabled the Federals to enfilade a long stretch of connecting works to right and left and could have opened the Confederate line for a decisive rupture. The Confederates held on there by only a thin margin.[28]

The mining and countermining at Vicksburg left a physical legacy for some time to come. On July 3, just before the capitulation, a Confederate soldier visited the Third Louisiana Redan and was shocked at the devastation. "I saw a good many green flies on top of the works, indicating the spot where several poor Rebs had been buried" when the countermine caved in due to the second Union charge, which exploded two days before. Ten days after the surrender, a party of Union officers visited the same site and found the crater still just as wide and deep as when created. "The dirt [from the explosion] was thrown down the ravine and plastered all up the side of the trees for a hundred yards [91.44 meters]." Two days later Union details began leveling the Confederate siege line. While filling up the crater at the Third Louisiana Redan, they uncovered a number of bodies that had been buried by the two blasts. What they found was a "mass of legs, arms, and other decaying remains." For years afterward, farmers plowing the top of this ridge turned up human bones, which could have been remains of men destroyed by the two mine explosions.[29]

Concurrent with the siege of Vicksburg, another Confederate garrison on the east bank of the Mississippi River was besieged by Federal forces at Port Hudson, Louisiana. Beginning May 22 and lasting until July 9, 1863, it involved much smaller forces than had the operation to the north but represented the last Confederate resistance for control of the Mississippi. Federal engineers began mining about July 1 and aimed at two strongholds in the Confederate line of earthworks, the Citadel and the Priest Cap. Engineer Capt. John C. Palfrey, who was in charge of this effort, noted that the soil retained its shape and needed no shoring. His miners could work for sixty to seventy feet (18.28 to 21.33 meters) before the air became "so bad that candles would no longer burn." Palfrey began his offensive mine toward the Priest Cap at the end of a zigzag approach sap ten feet (3.94 meters) away from the ditch of the work. He decided to dig an inclined gallery from the start rather than a shaft in order to save time.[30]

The Confederates discovered the effort almost immediately and assigned Capt. Louis J. Girard to supervise the countermining. He dug quickly, placed a charge, and touched it off on the morning of July 4. It was premature and mostly blew up the ten feet of earth that lay between the ditch and the head of the sap, opening up the approach to Confederate fire down its length for some distance. Two Black men working on the Federal side were stunned by the explosion, but no one was injured. Palfrey's men quickly built a traverse in the sap to protect everyone from enfilading fire.[31]

That ended all Confederate efforts to counter Union mining at Port Hudson. Palfrey's men pushed forward vigorously until the gallery at the Priest Cap was 27 feet (8.22 meters) long by July 6. Maj. Gen. Nathaniel P. Banks had great faith that both mines would be successful, so he carefully planned a follow-up assault at both places. His breaching batteries opened a hole in the parapet of the Citadel as well. Here the gallery had to go under the ditch of the work, which was 10–12 feet (3.04–3.65 meters) deep and 12–14 feet (3.65–4.26 meters) wide. The gallery at the Citadel eventually reached a length of 160 feet (48.76 meters) and chambers were dug to right and left under the Confederate line. The Federals were almost ready to charge both mines when the garrison surrendered on July 9.[32]

Port Hudson was one of several locations where an offensive mine was started but not completed. During the Atlanta Campaign of 1864, a large Federal force penetrated northwest Georgia while dealing with one heavily entrenched field position after another. One of the strongest Confederate positions was anchored on Kennesaw Mountain, which stalled the Union drive for two weeks. Maj. Gen. William T. Sherman ordered a major assault on June 27 that failed to break open the defensive line. At one of the three points of attack, a small rise of ground later called Cheatham's Hill, Federal troops lodged within a few yards of the Confederate line and dug in at the end of the assault, partly protected by a slight rise of ground between their location and the enemy works. They began a gallery on June 28, taking advantage of the slope to dig straight forward without a shaft. Starting 105 feet (32.00 meters) from the enemy line, they estimated the gallery would be 16 feet (4.87 meters) below the surface once they reached the parapet. The Unionists had to improvise everything, using ordinary spades and shovels rather than specially made miners' tools, and moved the spoil out in sacks and baskets, dumping it at the end of a trench to make it appear that the dirt came from the digging of their field fortifications. Still, the Confederates suspected something was up and reportedly used the age-old technique of placing pebbles on drumheads that were laid on the ground to verify enemy digging. But they did not start any countermining. The gallery was not yet finished when the Confederate army evacuated the Kennesaw Line on the night of July 2–3. Reports of its length vary from 15 to 85 feet (4.57 to 25.90 meters) by that time. There is no report about shoring or ventilation. The red clay of northwestern Georgia tends to remain solid when cut, so timbering was probably unnecessary. The lack of information on ventilation may indicate that the gallery was quite short rather than 85 feet (25.90 meters) long.[33]

Another operation that produced the start of an offensive mine without much progress toward its completion was Cold Harbor in 1864 in Virginia. Occurring at the end of the Overland Campaign, a major, bloody offensive beginning in early May, Federal forces had arrived at Cold Harbor, only a few miles northeast of the Confederate capital of Richmond, a month later. Grant, now a lieutenant general, ordered a major assault there against Confederate field fortifications that failed on June 3, but some Union troops remained lodged close to the enemy line and dug in for protection. At a place called Edgar's Salient, they were only thirty yards (27.43 meters) from the base of a slope upon which the Confederates were positioned. The Federals dug an approach trench to the base on June 5 and started a gallery the next day, aiming it horizontally into the slope rather than starting with a shaft. By the evening of June 8, the gallery had reached a length of thirty or forty feet (9.14 or 12.19 meters) when the Federal high command decided to reverse its June 4 order to conduct siege approaches at Cold Harbor. Instead, Grant wanted to break contact and conduct a flank movement around the Confederate position. Work continued on the gallery at Edgar's Salient for another day before the order filtered down to the brigade level. The gallery would have had to be at least one hundred feet (30.48 meters) long in order to place a charge twenty feet (6.09 meters) below the apex of Edgar's Salient, and it was less than halfway to that point when all work stopped. The primary labor had been provided by Company A, Engineer Battalion with help from volunteer regiments in the nearest brigade. This was the only instance in American military history of mining conducted by trained engineer troops.[34]

The Overland Campaign ended when Grant broke away from the confrontation at Cold Harbor and crossed the James River to attack Petersburg, an important rail center thirty miles (48.28 kilometers) south of Richmond. The first round of fighting outside the town on June 15–18 failed to capture the city, but Federal troops lodged only 125 yards (114.30 meters) from Pegram's Salient in the Confederate line of earthworks. On June 21 Lt. Col. Henry Pleasants had the idea to dig a gallery under that salient from the Union line. Pleasants was a civil engineer experienced in coal mining, and his regiment, the Forty-Eighth Pennsylvania, contained more than one hundred men who were coal miners. He received the enthusiastic support of his superiors, especially Ninth Corps commander Maj. Gen. Ambrose E. Burnside. Pleasants began digging on June 25 from the floor of Poor Creek, located behind Union lines, which allowed him to dispense with a shaft. Managed by Sgt. Henry Reese, a Welsh-born miner, the regimental work force dug in shifts of two and a half hours, with a whiskey ration at the end of their shift. The men dug fifty feet (15.24 meters) the first day and managed forty feet (12.19 meters) on succeeding days, amounting to nearly two feet (0.60 meters) every hour. The miners improvised as they proceeded, filing down the flukes of big army picks to make them more easily managed within the confines of the gallery and hauling spoil out in hardtack boxes or in sacks. Unskilled men of the Forty-Eighth Pennsylvania and detailed troops from a Black regiment provided labor for the nonmining tasks associated with the gallery.[35]

The digging needed to be shored, and Pleasants scrounged timber from the track of a nearby railroad, also putting into use a sawmill six miles (9.65 kilometers) away to cut lumber. He placed candles and lanterns every 10–20 feet (3.04–6.09 meters) along the gallery. The work hit a snag when the miners encountered wet ground that nearly collapsed their shoring. After dealing with this problem, they hit a layer of marl with a "consistency like putty," as Pleasants put it, and was a perfect material for fashioning pipes,

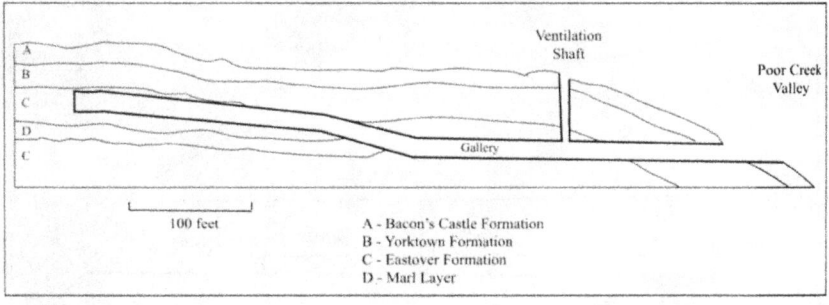

Ventilation
Shaft

Poor Creek
Valley

A
B
C
D
C

Gallery

100 feet

A - Bacon's Castle Formation
B - Yorktown Formation
C - Eastover Formation
D - Marl Layer

Geological layers

Geology of the Pleasants Mine Gallery, Petersburg, 1864. Although relatively shallow, the gallery went through four superficial geological layers because it began horizontally in the bank of Poor Creek. Federal miners also found a mastodon bone 10,000 years old thirty feet (9.14 meters) below the surface. Hess, *Into the Crater*, 14.

badges, and other souvenirs. Regardless, he had to incline the gallery 13.5 feet (4.11 meters) in the next 100 feet (30.48 meters) to climb above that layer. At an elevation of 144 feet (43.89 meters) above sea level, Petersburg lay at the intersection of a former bed of the Atlantic Ocean and the beginning of the Piedmont region. Pleasants's miners therefore dug through three superficial geological layers, the top two of which had been exposed by the creation of Poor Creek and the third layer, consisting of the marl, they encountered underground. Thirty feet (9.14 meters) under the surface, the Federals found a mastodon fossil that was at least 10,000 years old.[36]

Pleasants had initially planned to dig not one but two galleries so he could connect them with transverse galleries for ventilation, but that would have doubled the effort. Instead he fashioned a workable ventilation system for his single gallery that consisted of a small air hole twenty-two feet (6.70 meters) up to the surface. It opened just behind the Union trench and was therefore not visible to the Confederates. He then built a square wooden tube along the floor of the gallery and placed a partition made of sacks to serve as a door near the air hole. A fire was kept burning under the air hole and inside the partition to create a draft that forced bad air up through the hole and drew fresh air in from the mouth of the gallery through the wooden tube. Unlike the Federal miners at Vicksburg, Port Hudson, Kennesaw Mountain, and Cold Harbor, Pleasants had access to a theodolite, which was very helpful in judging distance. He took several readings on the surface so that by triangulation he could accurately gauge the distance between Poor Creek and Pegram's Salient.[37]

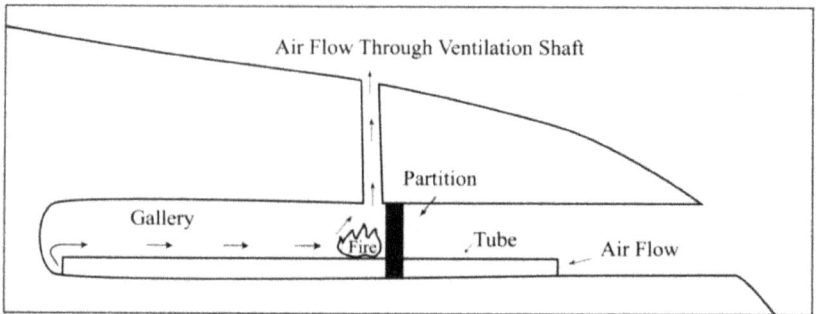

The ventilation scheme

Ventilation of the Pleasants Mine Gallery, Petersburg, 1864. Henry Pleasants drew on his prewar experience in coal mining to arrange this system of pushing old air out of the gallery while drawing new air into it. Hess, *Into the Crater*, 15.

The Confederates suspected they were in danger by late June and started countermining. Under the direction of Lt. Hugh Thomas Douglas of the First Confederate Engineers, they dug two shafts, one on either side of the salient, sinking one eighteen feet (5.48 meters) and the other fourteen feet (4.26 meters). Extending galleries forward, Douglas's men linked them with a transverse gallery. Even by July 26 there was no firm evidence that the Federals were mining at Pegram's Salient, but countermining continued. The Confederates worked two detachments of diggers, most of whom were detailed infantrymen, at both shafts, with fifteen men in each detachment digging in twelve-hour shifts. They made eight feet (2.43 meters) each day despite stopping every fifteen minutes to listen. Douglas also directed similar efforts at two other salients along the line as precautionary measures (one of them was being approached by a clearly visible sap). One of the countermine galleries at Pegram's Salient reached a length of eighty-two feet (24.99 meters) and the other seventy-two feet (21.94 meters) by July 26. Douglas slightly inclined both so that they were about ten feet (3.04 meters) deep at this point. In contrast, Pleasants gallery was twenty-five feet (7.62 meters) deep. Without knowing it, Douglas had already lost the race. Ironically, his left gallery intercepted the line of Pleasants's gallery but was several feet above it. Nevertheless, Confederate engineers constructed a new parapet just to the rear of Pegram's Salient to serve as a secondary line of defense.[38]

Pleasants brought his work to a close by late July. When his men finished the gallery on the morning of July 17, it measured exactly 510.8 feet (155.69 meters) long. They began digging branches to right and left, but later that

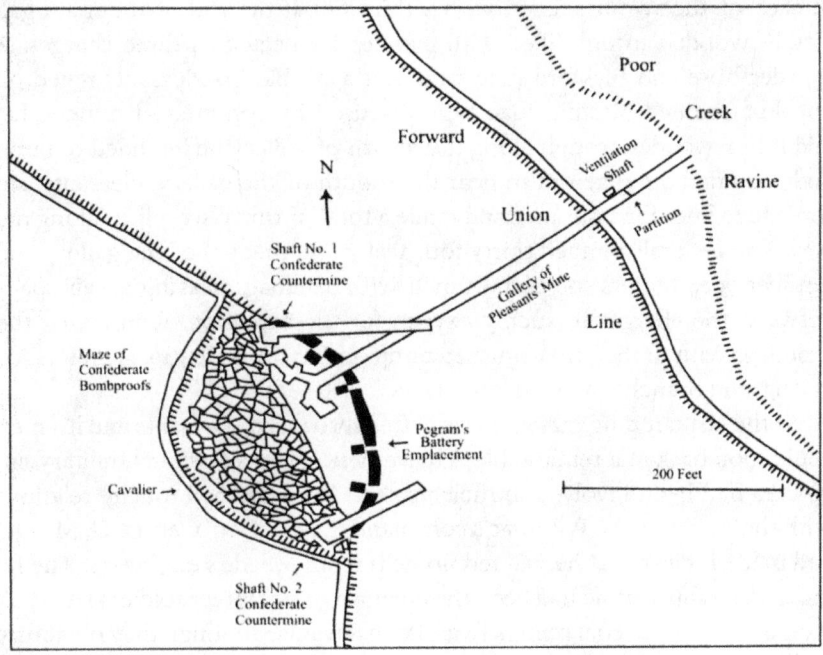

Pleasants's mine and Confederate countermines

Pleasants Mine Gallery and Two Confederate Countermines, Petersburg, 1864.
This diagram shows the shallow angle in the Confederate line, with numerous
individual bombproofs and a cavalier—a short secondary line of defense—
behind it, that was the target of the Pleasants mine. Confederate engineers
started two countermines, one of which crossed the path of but was several feet
above the Union gallery. Pleasants and his men won the underground race and
exploded the mine on July 30. Hess, *Into the Crater*, 23.

day two Confederate deserters told the Federals about Douglas's countermines. Now the Unionists worked faster, digging 25 feet (7.62 meters) of one
branch and 15 feet (4.57 meters) of the other in the next twelve hours. They
could hear digging and other noises of work in the salient, which spurred
them on. By the time the branches reached lengths of 38 and 37 feet (11.58
and 11.27 meters), the total length of digging by Pleasants' men equaled
585.8 feet (178.55 meters). They had moved 18,000 cubic feet (509.70 cubic
meters) of earth during the past month.[39]

Pleasants arranged for 8,000 pounds (3628.73 kilograms) of gunpowder,
which arrived in 320 kegs weighing 25 pounds (11.33 kilograms) each. His
men transferred it to bags and carried them one by one into the gallery
beginning on the night of July 27, placing them in wooden bins constructed

in each of the two branches, with a third bin at the end of the main gallery. A wooden trough filled with powder connected all three charges. A powder hose and Bickford fuze were not available, so Pleasants relied on an old-fashioned blasting fuze typically used by commercial miners. He laid it in a wooden trough along the length of gallery he intended to tamp and extended the fuze out to near the mouth of the gallery. Pleasants laid three fuze lines for assurance and made a total of thirty-six splices along the way. The Federals tamped thirty-four feet (10.36 meters) of the gallery and ten feet (3.04 meters) of each branch with sandbags, leaving an air space between the charges to supply oxygen for the explosion. Reinforcing the sandbags with timber, they finished tamping fifty-four feet (16.45 meters) of gallery and branches at 6 P.M. on July 28.[40]

As the last tamping material was patted into place, Pleasants and his men could look back at a remarkable achievement. No professional military engineers had been involved in their project. Burnside had touchy relations with the Army of the Potomac's commander, Maj. Gen. George G. Meade, and made it clear that he wanted no help from Meade's engineers. The famous Petersburg mine had been the creation of volunteer soldiers based on their experience as coal miners from Pennsylvania, although they obviously had learned something from the manuals about how to adapt their civilian knowledge to military purposes.[41]

But the mine had to be melded into an attack plan if it were to play a decisive role in breaking the stalemate at Petersburg, and that took a good deal of preparation. There was no sign that the Confederates were close to intercepting the gallery, so the Federals could afford to wait a couple of days before springing the charge. Grant planned and began executing an elaborate operation, first on the north side of the James River to draw enemy attention away from Burnside's sector, then authorizing the Ninth Corps to explode the mine and launch its assault early on the morning of July 30. Therein lay the trouble. Burnside had developed an attack plan that involved leading with his command's division of Black troops, which had not yet seen any combat. At the last minute Meade, with Grant's approval, rejected his proposal. He feared that if the Black division failed and was mauled, it would cause severe political repercussions in the North. Extremely frustrated and dejected, Burnside badly handled this change of plans, relying on drawing straws to select the lead division. Unfortunately, his worst division commander, Brig. Gen. James H. Ledlie, drew the assignment. Ledlie badly misunderstood his mission, believing his men were simply to hold the resulting crater and allow other troops to pass through to exploit the breach rather than taking the lead in doing this himself.[42]

Pleasants could not know anything of this as he arranged for the mine

Explosion of the Pleasants Mine, Petersburg, July 30, 1864. This is an accurate depiction of the dirt plume rising from the explosion. It appeared in northern publications soon after the Battle of the Crater. *Battles and Leaders of the Civil War*, 4:561.

to go up at 3:30 A.M. of July 30. A technical flaw delayed the explosion, and the colonel waited an hour before allowing Sergeant Reese and an officer to enter the gallery. They found that the fuze had stopped burning at a splice. After quickly fixing the problem, they exited before the charges exploded at 4:44 A.M. "A heavy jar, a dull thud, a big volcano-puff of smoke and dust," recalled Reese, "and up went the earth under and around that fort for a distance in the air of a hundred feet [30.48 meters] or more, carrying with it cannons, caissons, muskets and—men." Debris descended over a wide area, splattering the waiting Union infantrymen only 125 yards (114.30 meters) away. Many men described their impressions of the sight, noting that the plume resembled pictures of geysers they had seen or that it looked like a fountain in a city park. They noted red and yellow colors flashing at its heart from the powder explosion and huge blocks of hardened clay falling out and rolling along the ground.[43]

The massive explosion devastated Pegram's Salient, destroying most of the earthwork, taking out 150 yards (137.16 meters) of Confederate line, and killing 352 soldiers. The resulting crater measured 126 feet (38.40 meters) long at the surface and 69 feet (21.03 meters) long at the bottom. Its width measured 87 feet (26.51 meters) at the surface and 38 feet (11.58 meters) at the bottom. An estimated 100,000 cubic feet (2,831.68 cubic meters) of earth had been removed to make the crater 25 feet (7.62 meters) deep.[44]

The way was open for a campaign-winning attack that morning, but the battle of the Petersburg Crater that followed proved to be a classic example of an old problem throughout the history of military mining. Colonel Pleasants had engineered one of the most successful blow ups in history, and General Burnside followed up with one of the most mismanaged attacks. He deserved the lion's share of the blame for what happened. General Ledlie's men advanced but not with Ledlie; the division leader hid in a bombproof, where he prevailed on a surgeon to supply him with whiskey (to be fair, he suffered terribly from illness and relied on liquor as a sedative). His brigade leaders followed his instructions to simply hold the crater rather than advance. They crowded the constricted space of the breach and contributed to many problems other divisions encountered as they tried to get through the mess. In addition to the packed soldiers, a honeycomb of individual dugouts constructed by Confederates stationed in the salient presented a maze, and behind that maze loomed the secondary Confederate line, which remained intact and well manned.[45]

Federal commanders realized the problem and directed more troops to expand the breach right and left, achieving some success, but even those limited penetrations could not ensure a breakthrough. In three waves—the first right after the explosion, the next at 7 A.M., and the third (the Black division) at 9 A.M.—Burnside pushed 10,000 men into a breach of five hundred yards (457.20 meters), with more than 6,000 additional troops available to support the effort. Yet a combination of Union confusion and stiffening Confederate resistance prevented them from exploiting Pleasants's success. Beginning soon after 9 A.M., the Confederates began counterattacking and over time reduced Federal strength in the breach until driving all Union troops out of it by midafternoon. The Federals lost 3,798 out of 16,772 men engaged, while the Confederates suffered 1,140–1,612 men lost out of 9,400 engaged.[46]

The gaping crater was littered with dead and wounded men when the battle ended and the Confederates began the process of incorporating it into their defensive line. Details buried 177 dead in the floor of the crater, Union and Confederate alike. The number soared to over 200 if one counts the Confederates buried by the mine explosion and never found. A foot-deep (0.30 meters) layer of earth was all that separated this mass grave from the living soldiers who were forced to guard the crater for the next several months, emitting an awful odor that sickened the garrison. Confederate engineers tried to find the Pleasants gallery by digging two new shafts, although they had great difficulty dealing with the gases from the powder explosion and from the decaying bodies still trapped or buried in the earth. Not until late September did they find what was left of Pleasants's gallery and felt assured the Federals could no longer use it for further efforts. They were wise to worry because the Unionists did consider the possibility. Meade's engineers knew the gallery was intact up to the tamping, but a proposal to start a new gallery from the intact part was never approved. Meanwhile, by adding firing steps to the eastern rim and making other adjustments of its physical characteristics, the Confederates raised the floor of the crater and lessened the horrible odor until, by November, the hole was only ten feet (3.04 meters) deep.[47]

Although they had managed to save the line despite the spectacular explosion of Pleasants's mine, the Confederates were deeply affected by what happened on July 30. "I am becoming terribly demoralized at the thought of finding myself going up some fine morning or other," wrote John Walters of a Virginia artillery battery. His fear was shared by thousands of his comrades along the line, spurred more by the thought of being buried alive than by any other emotion.[48]

A frenzy of countermining resulted that carried through nearly to the end of the long confrontation at Petersburg. Confederate engineers accumulated a sizeable amount of material—rope, candles, carpenter tools, sandbags, lumber, and sheet iron. An Alabama private devised an auger that could be used to drill holes down through the earth in attempts to discover Union galleries. Oil-burning lamps were used by those men working at the head of countermine galleries, while candles illuminated the length of the excavation. Hand-cranked fans provided ventilation through wooden tubes. At the joints between sections of tubing, pitch-covered canvas provided a flexible union. Tools had to be adjusted to the nature of the soil, which at places was "very compact, pure clay of a grayish brown color," as Matthew Venable of the First Confederate Engineers recalled. This kind of soil was "so tenacious that the picks had to be made with short blades, with widened chisel-like edges and short handles, and we had to chip the material out by inches." Red clay was easier to dig, while sandy soil required a great deal of shoring.[49]

Mining details got to work at eight locations along the Confederate line. Generally, they dug shafts up to thirty feet (9.14 meters) deep in the bottom of the trenches, then started a gallery five feet (1.52 meters) above the bottom of the shaft so that water could fall into the bottom and not interfere with the gallery, bailing out the sump when necessary. Two or more galleries were extended forward at each site and connected by a transverse gallery. The Confederates usually drilled four-inch (10.16-centimeter) auger holes forward from the head of the gallery and prepared to release smoke into any Union tunnel they intercepted. Details nailed up shoring frames beforehand, using four-inch lumber, and prepared one-inch (2.54-centimeter) planks to connect each frame and to hold up the sides and ceilings of the gallery. Planks laid on timber runners provided a wooden track for wheelbarrows used to carry out spoil.[50]

The size and design of Confederate countermines varied from site to site. At Colquitt's Salient details dug eighty-three feet (25.29 meters) out and then constructed three branches. At Cooke's Salient they made three shafts twenty-six feet (7.92 meters) deep and extended galleries forty-three feet (13.10 meters) on the flanks and fifty feet (15.24 meters) in the center, with a connecting gallery at the end. The Confederates dug an inclined gallery rather than a shaft where the City Point Railroad crossed their line, beginning it 30 yards (27.43 meters) behind the entrenchment. Coarse sand caused the gallery to collapse, but they dug most of it out and applied heavy shoring. Confederate engineers dug their longest countermine gallery to reach a house owned by a Dr. Duval in no-man's-land. After digging reportedly 323 yards (295.35 meters) ten feet (3.04 meters) below the surface, they placed a powder charge under the house and planned to blow it only if the Federals occupied the structure.[51]

The largest Confederate countermine at Petersburg was located where the Jerusalem Plank Road crossed their line. Flat land and the close proximity of the Union position raised fears of offensive mining in this sector. The defenders started digging their main gallery on August 12 and extended it more than 146 feet (44.50 meters), with three branches and two connecting galleries, four listening chambers, and two chimneys for ventilation. The main gallery began horizontally in a ravine without a shaft. The complex totaled 1,088 feet (331.62 meters) of galleries and branches.[52]

At all but one of these eight countermine locations, the Confederates merely dug and listened, but at Gracie's Salient they exploded a charge. The Federals were advancing a sap toward this position, and Confederate engineers determined to stop them by blowing it up. After digging to a point he estimated was close enough, Douglas began constructing a chamber for 450 pounds (204.11 kilograms) of powder and placed the charge on the night of July 31, helped by two civilian miners. The lieutenant wanted to use lanyards to detonate the charge, but none of sufficient length were available, so he instead placed four strands of safety fuze, one to each of four barrels containing the powder. All four strands met at a powder train located near the mouth of the gallery. Tamping with sandbags and dirt completed the mine.[53]

When Douglas lit the fuzes at 11 A.M. on August 1, he waited for forty-five minutes as nothing happened. On investigating, a miner found that three pieces of the safety fuze had burned out while the fourth was still slowly

Arched Gallery of a Confederate Countermine, Petersburg, 1864. Part of a large countermine complex located at the Jerusalem Plank Road, it was discovered in the early 1920s and opened to tourists for a short while. Note the planked flooring and the electric lighting system, all applied by the twentieth-century owner of the countermine. Petersburg National Battlefield.

burning. Douglas extinguished that one and concluded that defective safety fuzes had been his problem. But he took this opportunity to improve the mine. His men removed the tamping and powder, pumped water out of the gallery, and dug two new branches, one of which the lieutenant believed was better aimed at the saphead. By August 4, he estimated that he had gone far enough and dug a chamber for eight powder barrels of 850 pounds (385.55 kilograms) total, placing four barrels in the chamber and four at the head of the branch. In a major error Douglas tamped the gallery only up to the top of the barrels rather than all the way to the ceiling. This time he procured a powder hose to be laid in a wooden trough at the mouth of the gallery, connecting it to the charges with more safety fuzes.[54]

When the countermine went off at 6:30 P.M. on August 5, it proved to be a severe disappointment for the Confederates. The charge happened to be forty yards (36.57 meters) from the Union sap head and did no damage to it. Even though the blast propelled earth one hundred feet (30.48 meters) into the air and scooped out a crater thirty feet (9.14 meters) in diameter, it had no effect on the Federals other than to draw their attention. The mine also collapsed the Confederate gallery all the way to the shaft. Moreover, according to at least one source, Douglas had started the gallery at too shallow a level and inclined it for drainage. By the time the tunnel reached its farthest extent, it was quite near the surface, and the lieutenant was forced to blow it prematurely or break through into the open. Finally, his failure to tamp properly led to much of the blast escaping along the gallery, which is why it collapsed all the way to the shaft. Douglas was blamed for the failure of the only Confederate charge to be touched off at Petersburg. Court-martial charges were filed against him, but he was allowed to resign his commission instead.[55]

If the Federals had intended to try another offensive mine at Petersburg, it is quite possible the Confederate listening galleries at eight prime locations would have detected and intercepted the effort. But Grant's men had no intention of doing so. Instead, Union miners dug countermines of their own at eight locations along the line just in case the enemy tried to approach any of them underground. They had already begun countermining at two sites a few days before the Battle of the Crater and extended protection to six more after the engagement. Four of those countermines were started in the first week of August, right after the dismal defeat at the Crater. In the ditch of Fort Stedman, miners began with a shaft that was 12 feet (3.65 meters) deep and 7 by 10 feet (2.13 by 3.04 meters) in dimension. By the end of the Petersburg Campaign, the gallery started from this shaft and its branches totaled 200 feet (60.96 meters). At Fort McGilvery nine companies of the Fiftieth New York Engineers extended a gallery 140 feet (42.67 meters) by November, using wheelbarrows to haul the spoil to a dump 400 feet (121.92 meters) from the entrance. So-called well holes dug into the glacis of Fort Burnham served as listening posts so as to avoid the labor of digging proper shafts and galleries.[56]

The Federals were more concerned about Fort Sedgwick than any other location. Positioned at the Jerusalem Plank Road and occupying ground that might tempt a Confederate underground offensive, they sank three shafts that were each twenty-two feet (6.70 meters) deep in the ditch. One method of listening consisted of shoving an iron ramrod into the earth and holding the outer end between the teeth, but that proved inconclusive. Someone had the idea to place an empty barrel at the bottom of the shaft and tamp the space around it with dirt so as to make a connection between the barrel and the surrounding earth. Then he tautly stretched a string across the open barrel top, believing that any digging within sixteen feet (4.87 meters) would cause the string to vibrate.[57]

In the end the confrontation at Petersburg was not settled by mining. Grant conducted a series of movements to stretch his position westward and flanked the Confederate line by early April 1865, forcing his enemy to abandon what was by then a continuous line of earthworks thirty-seven miles (59.54 kilometers) long. The relatively extensive underground digging at Petersburg was only slightly ahead of that at Vicksburg—the two campaigns combined represented the overwhelming majority of all mining and countermining ever conducted in the Western Hemisphere. At both places offensive mining certainly was applicable and justified, and it held a real potential for having a decisive influence on operations. But in both cases it failed to fulfill that potential not because of technical problems, but because of operational difficulties in coordinating infantry assaults with mine attacks.

At Vicksburg the Federals conducted classic siege approaches even though no one had any experience at this. They dug saps to cross no-man's-land and then started galleries from the head of those approaches when very close to target works, greatly reducing the length of their underground approaches. Except for Kossak's gallery at Stockade Redan, Union mines at Vicksburg were relatively short, proved quick and easy to dig, and needed little if any ventilation. Charges were relatively small compared to the Pleasants mine at Petersburg. Confederate countermining efforts at Vicksburg were largely ineffectual, mainly because the charges used were too small and ill placed to do much damage to Union saps and galleries.

At Petersburg the Pleasants mine has come to represent military mining in the Civil War for casual students. It was an interesting and impressive achievement, representing a technique to be widely adopted during World War I—that is, rather than sapping forward and mining at the head of the sap, Pleasants started from the main Union line and dug a tunnel completely under no-man's-land. Long-tunnel mining had occurred sporadically at other locations in previous wars, but it was not common and ran counter to Vaubanian doctrine. The Pleasants mine also has overshadowed the fact that a good deal of countermining on both sides took place at Petersburg after July 30. The only Confederate countermine explosion was ineptly planned and executed, but the listening galleries they dug at eight locations were capably designed.

Several other campaigns of the Civil War also represented static confrontations of two armies within striking distance of each other for extended periods. Often called "sieges" even though they did not have all the characteristics of a true siege, they nevertheless provided the opportunity for engineer officers to try their hand at mining. Yet at many of them, no one even attempted it. Even where siege approaches were dug, such as during the operations against Fort Wagner near Charleston, South Carolina, in the summer of 1863, mining was absent. But in the last of these types of campaigns, the Union offensive that captured Mobile, Alabama, in early April 1865, the Confederates began the last countermining effort in American history. Worried that the Federals might try to undermine the defenses of Spanish Fort, a heavily fortified position on the east side of Mobile Bay, Col. Samuel Lockett of Vicksburg fame ordered Confederate engineer troops to assemble mining material there. On April 8 they cut a passage under the parapet to gain access to the defensive ditch, where they intended to sink a shaft, but the order to evacuate Spanish Fort that night prevented them from starting it.[58]

Brig. Gen. Orlando B. Willcox, who commanded one of Burnside's divisions at the Crater, expressed it cogently when he said offensive mining was "generally considered as a series of partial, though desperate attempts in a siege, rather than counted on as finalities." He considered the aim of offensive mining to be the "gaining or destroying [of] some important part of the work" rather than opening up a hole for a successful assault.[59]

Willcox hit on an important point in the history of military mining—whether it should be used as a campaign-winning ploy or simply to gain small increments of advantage over the enemy. Many observers and commentators have debated that point over time. But Willcox was wrong to assume that his conclusion was generally held by those commentators—the debate would continue for the rest of mining history.

Emory Upton, a graduate of the U.S. Military Academy who served successively in the artillery, infantry, and cavalry arms of the Union army, illustrated two significant points when he commented on a visit he paid to the extensive coal mines near Pottsville, Pennsylvania, in December 1864. Upton was greatly impressed when entering the galleries and walking for one and three-quarter miles (2.81 kilometers) until exiting through an air hole. He thought that "all military men" should visit the diggings so "they may see what could be accomplished if necessary in the mining branch of their profession." Upton also "came to the conclusion that if General Grant can take Petersburg in no other way, he could easily undermine the enemy's entire works." His comment points out the vital link between military mining and civilian mining, how far advanced coal miners were compared to army engineers in that regard, and how much the latter could learn from their civilian counterparts. But his comment also indicates how little emphasis was placed on mining in the West Point curriculum. If an officer had to visit a coalmine to understand the possibilities of military mining rather than relying on his professional education, there is little wonder that the United States remained on the outer margin of the core area of siege mining.[60]

Nevertheless, the Civil War had made it clear that a shift was taking place in the application of mining in warfare. Along with what had already taken place at Sebastopol, observers noted that mining and countermining seemed just as applicable when dealing with temporary field fortifications as with permanent forts, which had been the primary targets of miners for centuries. This was a trend that would culminate along the Western Front of World War I barely sixty years after Sebastopol.[61]

But the Civil War failed to provide a test of electrical detonation for underground charges, the most important new technology available to military mining at the time. Vicksburg confirmed the continued viability of Vauban's siege principles and demonstrated that they could quickly be learned by soldiers otherwise untrained in such military endeavors. Although the Civil War was a showcase for the semisiege, a new kind of operation, it was in most ways an old-fashioned war as far as military mining was concerned.

9

Before the Great War

The American Civil War occurred at a time when improvements in the detonation of explosives were progressing in a marked way, but the military miners of that conflict failed to incorporate those new developments into their work. They also failed to use any of the new types of explosives that were being developed in the middle decades of the nineteenth century. But during the fifty years that separated the Civil War from World War I, both of these new methods of military mining became fully available to make of the Great War the greatest venue for the full exploitation of what was new and deadly in underground warfare.

Mining efforts that took place at the intense siege of Port Arthur during the Russo-Japanese War (1904–1905) finally demonstrated the usefulness of the new high explosives and electrical detonation systems for underground warfare. And yet the Japanese engineers who used that new war technology also were well aware of Vauban's principles. They combined the old principle of sapping toward the target before beginning to dig siege galleries and using the charges to blow in ditches to create ramps by which their infantrymen could attack Russian works. Port Arthur became a well-studied siege because of its lesson that mining was still relevant to modern warfare.

The application of electricity to the detonation of mines was linked to trends in civilian commercial mining. Old-fashioned powder trains, consisting of loose powder laid in wooden troughs, had given way to powder hoses, with the powder encased in linen tubes, long before the nineteenth century. Powder hoses, in turn, were being replaced by Englishman William Bickford's safety fuze of 1831. Bickford's device was surrounded by a covering of waterproofed jute and could burn at the rate of about a foot (0.30 meters) every thirty seconds. The key was the jute covering that held the gunpowder in place and protected it against the elements and any casual interference that could set it off accidentally. Bickford greatly increased the safety margin with his fuze, which was widely used in commercial mining throughout the mid-nineteenth century.[1]

Moving from the safety fuze to electricity represented a big step forward in terms of safety, but a source of electricity, insulated copper wires, and a device called a deflagrater to connect the wires with the powder charge were also necessary. Chemical and magnetic devices for creating electrical energy were in development for decades before the Civil War. Insulated wires were available commercially by 1848, and various types of deflagraters appeared by the 1860s.[2]

A handful of Americans participated in the work of developing electrical energy as a means of setting off underground charges, but they did not pioneer international efforts along that line. George W. Beardslee of Brooklyn followed the lead of others by developing an electromagnetic device with the coils remaining stationary and the magnets revolving by 1859. He increased the number of stationary coils to twelve and added a switchboard so the operator could "key in any number of coils to make up the generator's output." Beardslee's machine was not widely adopted for electroplating because firms involved in that business preferred chemical batteries. He then tried to interest telegraph companies in his work, filing patents in 1863 for devices useful in that industry as well as for exploding gunpowder. The Federal army tried his machine for powering field-telegraph lines, but it worked only for five miles (8.04 kilometers) instead of the promised twenty-five miles (40.23 kilometers). Beardslee then tried to interest the army in using his device to explode landmines and underwater mines, both of which were commonly called torpedoes at that time. He conducted a series of demonstrations at the U.S. Military Academy at West Point in January, February, and April 1865 that proved promising. The next step in Beardslee's tireless efforts to find an application for his device would have been to pitch it for military mining and countermining, but the end of the Civil War negated any opportunity for doing so.[3]

Beardslee had a close rival in Taliaferro P. Shaffner, a self-taught inventor and telegraph promoter who spent several years in Europe working on what he termed "the application of Torpedoes and Military Mines for Offensive and Defensive Measures in war." Most of his time had been spent in Russia, Denmark, and Sweden, and there is some indication that he was active for one side or the other in the Danish War of 1864. A fervent advocate of the Union cause despite his Virginia birth, Shaffner returned to the United States. "I abandoned my great enterprises in Europe and purchased about ten thousand dollars ($10,000) worth of materials suitable for Torpedo and Military Mining purposes," he wrote of his arrival early in 1865.[4]

Shaffner traveled to Petersburg to lobby General in Chief Ulysses S. Grant in late February 1865. Initially, the inventor pitched his work for the explosion of torpedoes on both land and water, guaranteeing that his deflagraters would ignite all the powder in each charge to produce the maximum effect. Grant appointed a board to examine Shaffner's system, and it filed a positive report by March 22. Shaffner then informed the general what he needed to equip the Federal army at Petersburg with detonating devices for exploding mines. He requested fifteen miles (24.14 kilometers) of copper wire insulated with india rubber, four electromagnetic generators, and one thousand deflagraters. Grant authorized the purchase of everything the same day, March 28. But a few days later the Petersburg Campaign came to an end. Grant's forces flanked the Confederate line and then broke through it with a massive frontal attack on April 2. A week later the surrender at Appomattox effectively brought the war to a close.[5]

Beardslee and Shaffner arrived too late on the scene to use their promising devices in the Civil War. Even when it came to antipersonnel landmines, all of those actually used during the war were self-activated devices triggered by contact with a person. Not a single torpedo was exploded on land by electricity. But electrical devices often were used during the war to explode underwater torpedoes. The Grove galvanic battery and the Wheatstone electromagnet machine were the commonly used devices for submarine mines during the conflict.

In the post–Civil War era, electrical detonation held great promise for both offensive and defensive siege mines. This was the area in which people like Beardslee and Shaffner could have made their inventions matter. Dispensing with the powder train, powder hose, and Bickford safety fuze alike, electricity might have made military mining as safe as it had made commercial mining.

The next significant war offered even fewer examples of mining and countermining than had occurred in the American Civil War. German doctrine prescribed rapid large-scale operations, as executed during the Franco-Prussian War (1870–1871), but their success in open-field movements and engagements overshadowed the fact that a number of sieges also took place. Prussian forces besieged several fortified cities with varying levels of success, digging siege approaches in the form of parallels and saps at some locations while relying on artillery bombardment or blockade at others. Several of the French garrisons held out until the end of the war, while others surrendered early in the siege. At Strasbourg (August 14 to September 28, 1870), the Prussians began to dig a mine gallery, but it did not extend very far before artillery opened an effective breach in the city defenses that led to the surrender of the garrison. This apparently was the only siege in which the aggressor contemplated mining, and there is no evidence of French countermining.[6]

In the United States a new generation of instructors at the U.S. Military Academy produced an updated manual of mining and countermining to supersede James Duane's book. Oswald H. Ernst, who graduated from the academy in June 1864 before joining General Sherman's engineer officers during the Atlanta Campaign, returned to teach at West Point in October of that year. From 1871 to 1878, he was primarily responsible for instruction in engineering and wrote the textbook used for those classes. The section on mining in his text clearly details all aspects of the process, including specially designed tools, the method of making frames for shoring galleries, ventilation, lighting, and tamping. His recipe for detecting enemy mining is decidedly old fashioned—placing a drum on the gallery floor with peas on the taut skin or using basins of water on the floor—but Ernst advocates a quick method of making a trench by digging a gallery at a shallow level, placing several small piles of powder thirteen yards (11.88 meters) apart along the floor, connecting them with a powder hose, and touching it off. This could instantly create a trench seven feet (2.13 meters) deep with at least the rudiments of a parapet and parados.[7]

Ernst is a bit dated in his recommendations for detonating underground charges. He tells cadets that gunpowder remained "the best explosive that can be used for mines." Without explaining his reasons, he believes that guncotton and dynamite were useful for "hasty demolitions, but they are not suitable for mines in earth." Ernst describes how to set off the charge with a linen powder hose laid in a wooden trough, which in turn is ignited by portfire stuck into the end.[8]

But Ernst was more progressive when it came to electrical detonation systems. He had been preceded in his teaching post by Dennis Hart Mahan, who had given limited credence to electrical detonation of mine charges. In 1867 Mahan had issued an updated edition of his *Elementary Course of Military Engineering*, in which he describes the European use of two methods of enhancing the surge of current through copper wires. One, the employment of a primary cell developed by the German scientist Robert Wilhelm Bunsen in 1841, was an improvement on the Grove battery. The other, an induction coil developed by another German scientist named Heinrich Daniel

Ruhmkorff in 1851, was an improvement on previous induction coils. But in describing those methods, Mahan does not fully endorse electricity as a means of firing mines.[9]

Ernst, therefore, became the first American writer to fully urge the use of electricity in mining and countermining. "The wires must be perfectly insulated, and should be thoroughly tested," he writes. They had to be laid in a wooden trough to protect them from the tamping. According to Ernst, Henry Julius Smith was developing the best detonating system for electrical explosions. Starting in 1868, Smith had crafted an electromagnetic device that, when perfected ten years later, was the prototype for all plunger detonators in succeeding decades. That improvement consisted of a device in which the electrical current was generated by pushing a plunger rather than turning a crank. But Ernst also liked Beardslee's crank-generated electromagnetic machine. "Each has exploders [deflagraters] particularly adapted to it," he observes. "Each has some defects." But Smith's system was better for exploding multiple charges.[10]

Two decades after the publication of Ernst's manual, James Mercur, professor of civil and military engineering at the academy, published a textbook for cadets on the subject in 1894. As all manual authors, Mercur covers all the bases of military mining. The term "globe of compression" had by now been replaced by the phrase "radius of rupture" to denote the extent of damage to the surrounding earth caused by an underground explosion. Mercur discusses in detail the construction of galleries and branches, urging engineers to make a map of everything that included the dimensions and directions of approach. He advises bringing deep galleries closer to the surface as they neared the target so as to reduce the line of least resistance, economize on the powder charge, and lessen the damage to the gallery and branches when the mine exploded. Like most commentators before him, Mercur warns that artificial ventilation remained necessary in galleries extending more than sixty feet (18.28 meters), but the only new twist he introduces is the use of masks or respirators by the miners.[11]

Mercur was the first manual writer in the United States to take seriously the new generation of explosives for underground warfare. Admitting that no good mining effort had yet taken place since the invention of dynamite in 1867, he was certain that this explosive could revolutionize military mining. Only half the amount of dynamite was as effective as the full amount of gunpowder. His text theorizes that poor tamping was less of a problem with dynamite because of its violent explosion, and one would need to dig only a small chamber for it. Mercur envisions a rapid series of digging, planting, and exploding episodes that could increase the pace of underground warfare with the use of this new explosive. Dynamite produced noxious gases, especially carbonic oxide, which were dangerous to miners, but then gun-powder explosions also produced harmful gases.[12]

Mercur's discussion of detonating underground charges is up to date as well. The older electromagnetic devices had given rise to a new genera-tion of blasting equipment with the development of the dynamo in several countries by 1866. It worked on some generally similar principles as the electromagnetic machine but was bigger, more complicated, and far more powerful. The dynamo produced very strong and, most importantly, very steady direct current, making it the first truly reliable electrical generator for widespread use in industry. It was not directly applicable to mine explosions, but the term "dynamo" became a synonym for any new device to produce effective electrical current. Mercur uses the term to refer to the new type of blasting machine that Smith had developed by 1878, which was small, portable, and worked by pushing down a plunger. The "electric-blasting ap-paratus is now in such common use," he observes, that its application to military mining was now taken for granted.[13]

But Mercur does not take into account that the new blasting system pro-duced some new problems as well. If a military engineer hooked up the new system to a charge for any length of time before exploding it, he had to test the system now and then to see if the connections remained unimpeded. This was done by sending a light charge through the wires, strong enough to register as working but not so strong as to set off the explosive. The older deflagraters could not easily be relied on to do this, so technicians developed new squibs that were less sensitive, calling them low-tension squibs in con-trast to the older, high-tension deflagraters.[14]

In addition to detailing the technical aspects, Mercur was the first Amer-ican mining author to discuss the tactics of underground warfare. Using explosions to create craters that could be converted into trenches or used to start another gallery for further mining was an important aspect of his chap-ter "Mine Tactics." Creating intermediate craters in no-man's-land could also be useful in countermining, as they easily served as the starting point of new galleries designed to block offensive mines. Permanent countermines were still highly relevant and had changed little if any in the past one hun-dred years.[15]

Mercur's manual represented a trend evident in all military manuals of the later nineteenth century that mixed technical information with tacti-cal advice. In short, they not only discussed how to dig galleries and plant charges but also delved a bit into the realm of doctrine. Moreover, Mer-cur's text addresses the human dimension of mine warfare, representing a real departure from past manuals. "Underground warfare is conducted in the dark," he writes, "in bad air, with constant danger of caving earth, suffocation by noxious gases, destruction of men and galleries by intentional explosions of hostile mines or accidental ones of our own, in addition to the usual dangers and difficulties of opening and supplying the mines under the close fire of the enemy."[16]

This brief glimpse into the world of the military miner, its dangers and discomforts, demonstrates the limits of mining as a tactical alternative in static military confrontations on the battlefield. Despite the increased advantages offered by the new technology of the late nineteenth century, many aspects of mine warfare had not changed, and there was only so much one could expect of miners. Perhaps this was one reason that no American forces undertook any mining projects after the Civil War.

There had always been an important link between civilian mining and military mining, with the former always leading the latter in terms of employing new developments in technique, equipment, and theory. That had been so during the long period of relatively little change before the nineteenth century and continued to be so during the short period of rapid new mining developments of the later nineteenth century. The application of electricity or the next generation of explosives and the use of new squibs had all been tried out in civilian mining and only slowly were applied to military mining. A big reason for this halting adaption is that, while civilian mining took place continuously all around the world, military mining only rarely occurred in a handful of conflicts. That is why Mercur had to theorize about the significance of new explosives rather than base his assessment on experiments, which he knew were impractical for budget-constrained armies, or on actual practice in the field.

In the civilian sector the greatest advances were seen in tunneling, the digging of permanent transit galleries, for railroads and subways. A huge increase in new tunnel construction took place from 1850 to 1880 in England, on the European continent, and in the United States. Those projects typically employed mechanical drills, advanced explosives, and iron sheathing. The galleries were so big that one could choose to dig them out by starting at the bottom and working up or starting at the top and working down. Shafts were typically dug upward to the surface at frequent intervals for ventilation, although some tunnels had systems of piped-in fresh air. While the purpose, size, and dimensions of those transit tunnels made them unfit as models for military mining, the new explosives and drills employed in their construction were potentially useful for army purposes. They would be tried out on the Great War's Western Front.[17]

Mining for coal and minerals had always tended to be more closely allied to military mining than transit tunneling. In comparison to the large and rather standard transit tunnels, the shafts and galleries of coal and ore mining varied a great deal in size, shape, and length, depending on many circumstances peculiar to the geology and the purpose of the diggings. In this, coal and ore miners shared many characteristics with military miners. Shafts could be "rectangular, circular, or elliptical," as noted in George L. Kerr's popular handbook of coal mining. Circular shafts better resisted "heavy pressure" and were more suitable for deep depths, while rectangular shafts were "more economical to sink, [and] easier lined and secured." Coal miners developed a technique called tubbing to negotiate water-laden strata while digging a shaft. This involved forcing cast-iron rings into the moist earth, digging out the dirt, and adding another ring to the top of the first one to continue pushing the shaft deeper. While this concept initially appeared in the late eighteenth century, it would come to play an important role in British military mining in Flanders during World War I.[18]

Kerr reports that gunpowder was still widely used in coal mining as of 1905 because it was "cheap, comparatively slow in action, and therefore suitable for coal and soft rocks, and less dangerous than some of the nitro-compounds." But he advocates blasting gelatin and dynamite in certain situations and mentions ammonal among other new explosives. Ammonal would play a huge role in British military mining on the Western Front. Safety fuzes were still used in many coal mining projects by 1905, although electricity was becoming more customary. Electrical devices tended to be the older crank-operated generator and dry batteries rather than the more modern plunger. "On the whole, firing by electricity is cheaper," Kerr advises his readers.[19]

Hand-held drills became more popular over the last four decades of the nineteenth century, especially in North America, where they dug out more than a quarter of all the coal extracted. But in Great Britain hand-held drilling machines were much less efficient than the American devices, and they accounted for less than 2 percent of the coal mined. Compressed air and electricity powered the hand-held drills. Larger machines designed to be stationed at the head of the gallery to dig away without manual labor were called heading machines. They were quite new in 1905, and only one successful device, called the Stanley Heading Machine, had been developed in Britain. Two revolving cutter bars entered the face of the gallery and dug out a core five feet (1.5 meters) in diameter and three feet (0.91 meters) deep. The machine needed to be anchored to the floor and sides "to maintain it in position, and to keep the cutters against the face." It had to be moved forward to continue the process after cutting each core into the face and clearing the debris. The Stanley machine ran on compressed air piped in from the surface, weighed three to four tons (3,000 to 4,000 kilograms), and worked at a rate of twelve to fifteen feet (3.65 to 4.57 meters) in eight hours.

It would be requisitioned by British engineers on the Western Front for use in military mines.[20]

To pump water out of coal-mine galleries, civil engineers used pipes made of wood, cast iron, wrought iron, and steel, although the last named was fast replacing all the others. Steel was cheaper and easier to handle, but Kerr has noted that cast-iron pipes could better deal with acidic or dirty water. The cause of bad air in coal galleries ranged from the exhalation of men and animals, to the effects of burning lamps and candles, to gases from the rock strata and those from exploding charges. Even the timber shoring decayed over time and gave off odors and effluvia, although taking off the bark and treating the wood before placing it in the shafts and galleries lessened this effect. There were no particularly new methods of ventilation better than pumping in fresh air and cutting ventilation shafts.[21]

Archaeological studies of a range of commercial mines in Great Britain reveal that the size, shape, depth, and length of shafts and galleries varied widely from site to site and from time period to time period. While candles and lamps were still placed in alcoves to provide light, electricity was becoming more common by the turn of the twentieth century. Power plants also were established deep underground, with whitewashing painted on the walls and ceilings of chambers to provide reflective lighting. Toilets for miners' use have been found in many commercial mines along with miners' graffiti and even the shoe and hoof prints of animals in some of the clay floors. But along with modernization came a noticeable increase in mining accidents. In Europe, Great Britain, and North America, the number of accidents that killed five or more miners steadily increased from 1875 to 1900, greatly accelerated more than 100 percent from 1901 to 1925, and then reduced to half of that peak by 1950. In 1907 alone at least 3,242 mining deaths occurred in those three areas of the world.[22]

Compared to major developments in tunneling and coal and ore mining during the later nineteenth century, military mining introduced new methods and technology in only a limited and halting way. Until military engineers were called upon to conduct extensive, intensive, and long-term siege mining, they were only loosely connected with developments in the commercial field. It would not be until 1914 that operational conditions on the battlefield called for a more intense copying of commercial mining technology in the military field. Meanwhile, some siege efforts around the turn of the twentieth century proceeded in an old-fashioned way, while others pushed the envelope of traditional military mining.

An example of a small, old-fashioned mining episode took place on the Indian subcontinent in 1895. When violence accompanied a succession crisis at Chitral in northern British India (modern north Pakistan), regional forces tried to overthrow the new ruler, who was supported by the British Raj. This led to the siege of a Sepoy garrison of 343 men of the Fourteenth Sikhs and some Kashmiri infantry in a fort at Chitral beginning on March 4. The besiegers made noise with drums and pipes at a small house near the fort, which raised suspicions among the garrison that these were intended to cover the sound of mining. The ploy did not work. One night a sentry heard the faint sounds of underground picking, and at 11 A.M. the next day, it was more distinctly heard, seeming to be only a few feet from the Gun Tower.[23]

The garrison mounted a large sortie of one hundred men that raided the house where the besiegers had been playing music at 4 P.M. They drove forty besiegers from the building and found the mouth of the mine inside. Thirty more Chitralis came out of the gallery and were bayonetted. The Sepoys then quickly tried to blow up the gallery, placing bags of gunpowder and tamping a bit. Before they finished, two Chitralis who had remained at the head of the gallery tried to escape, shots were fired, and the powder charge was ignited, killing the two Chitralis and injuring some members of the raiding party. Nearly a third of the sortie men were killed or injured, but they killed forty-five besiegers and destroyed the gallery. That evening the rest of the small garrison started digging two countermines under the Gun Tower in case another offensive tunnel appeared.[24]

The siege was lifted by a relieving column on April 19, allowing a closer examination of the offensive mine. It had been dug not only by Chitralis but also by Jandolis and extended for fifty yards (45.72 meters), with two turns in it that eventually directed the gallery straight toward the Gun Tower. "It was very shallow," wrote a British businessman–turned–newspaper correspondent who visited Chitral right after the siege. "Only eighteen inches [45.72 centimeters] below the ground, and the direction was most likely preserved by sticks driven through the surface at short intervals to serve as guides." The besiegers intended to use a gunpowder charge; they had already stockpiled it in the house and most likely intended to set if off with a powder train.[25]

In contrast to the low-tech methods at Chitral, the Japanese army employed dynamite and electrical detonation in its siege of Port Arthur during the Russo-Japanese War a decade later. The Japanese, however, used these new devices in a Vaubanian fashion, sapping forward to reach permanent and semipermanent fortifications and digging shafts from the head of those saps to blow in counterscarps to cross defensive ditches. While Chitral represented the full survival of old methods of military mining on the eve of the twentieth century, Port Arthur demonstrated a significant step toward using new technology during the first major war of that new century.

Formerly known as Lüshun, the port at the end of the Liaotung Peninsula in the southwestern corner of Manchuria had been captured by Japanese forces in the Sino-Japanese War of 1894 and later leased by the Russians, who changed its name to Port Arthur. By 1904, they had fortified the place with ten forts and twenty-five batteries a dozen miles (19.31 kilometers) from the port facilities. The forts were designed to be permanent, with moats thirty to forty feet (9.14 to 12.19 meters) wide, many of them dug into rock. The Russians constructed caponier galleries in the ditches, their roofs ten feet (3.04 meters) from the glacis and three feet (0.91 meters) thick as were their walls. Made of concrete, the caponier galleries had loopholes so those inside could fire along the length of the ditches in case attackers managed to occupy one. The Russians also applied a concrete glaze over the scarp (the ditch's inside wall nearest to the fort) to present "a smooth, perpendicular face" to anyone attempting to climb out of the ditch.[26]

The Japanese Third Army landed on the Liaotung Peninsula on May 31, 1904, and focused on capturing the outlying Russian defenses until the end of July. Its first attack on the line of permanent defenses in late August failed, so the Japanese began siege approaches on September 1. They started several saps to cross contested ground, aiming at three key positions—Erhlung Fort, Sungshu Fort, and North Fort. Those saps, designed on a zigzag pattern, were six feet (1.82 meters) deep and eight feet (2.43 meters) wide. The Japanese mostly worked on them at night, roofing sections that could be seen by the Russians. By October 23, they were 150 yards (137.16 meters) from Erhlung, 200 yards (182.88 meters) from Sungshu, and 47 yards (42.97 meters) from North Fort. For most of the distance thus far, sappers had encountered soft shale, which they easily dug, but then they found rock nearer the forts.[27]

At Erhlung, Japanese sappers reached the glacis on October 26 and began to sink three shafts near the counterscarp. "Every inch had to be chiseled out of hard rock," wrote English journalist David H. James, who accompanied the Third Army. With the high glacis, miners had to dig the shafts sixty feet (18.28 meters) down to reach the same level as the ditch bottom. One day James was allowed to inspect the work. "I saw the hair of the heads of some men sitting in a hole in the rock calmly chiseling away at the rock, while a party of infantrymen" tried to contend with hand grenades thrown over a sandbag parapet by the Russian garrison. After constant digging, the Japanese exploded dynamite with an electrical apparatus in all three shafts at 10 A.M., November 10, with mixed results. In two shafts the charges worked well, collapsing the counterscarp and filling the ditch up to six feet (1.82 meters) short of the top. In the third shaft the charge only partially collapsed the counterscarp and filled the ditch up to eighteen feet (5.48 meters) from

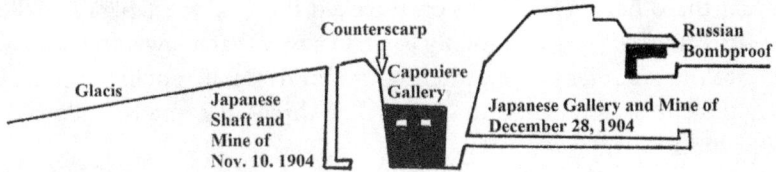

Japanese Attack, Siege of Port Arthur, 1904–1905. This schematic illustrates the use of mines to collapse a portion of the deep defensive ditch
of Erhlung Fort and to degrade part of its defenses on November 10 and
December 28, 1904. The Japanese employed Vaubanian principles (mining at the end of a sap and using explosions to prepare entry for infantry
assault) as well as modern technology (dynamite) while digging through
solid rock. Based on diagram in James, *Siege of Port Arthur*, 62–63.

the top. None of the three mines had created the anticipated ramp that assaulting infantry could use.[28]

When other Japanese soldiers reached the glacis of Sungshu Fort, they
dug seven shafts. Four of them hit the concrete roof of a caponier gallery,
and the miners had to diverge outward to bypass it. Planned as a group, the
four had been started "equidistant along the glacis." The other three shafts
also were grouped so that the two collections would have maximum effect
in collapsing the counterscarp. The besiegers also dug a transverse gallery
when the shafts neared the base level of the ditch.[29]

The Japanese planted dynamite and rigged an electrical detonation system, setting off the charges in all seven shafts at Sungshu Fort at 3 P.M. on
November 17. James witnessed "a great upheaval of rock and concrete,
mingling in a dense column of whitish smoke." When it cleared off, the
Japanese found that the blasts had collapsed the counterscarp and mostly
filled in the ditch while also destroying half of the caponier gallery. For the
next several days, they constructed a covered passage across the ditch and
replaced the saps that had been damaged by the multiple blasts, all to open
an avenue of infantry attack.[30]

The level of intense underground fighting rose enormously when the
Japanese approached North Fort. Their engineers thought it expedient to
begin a gallery forty-seven yards (42.97 meters) from the work and made
good progress until miners began to smell coal tar, which they took to mean
the near presence of a Russian countermine. Upon investigation they discovered a small hole in the wall of their gallery thirty-four yards (31.08 meters) from the shaft. Six volunteers tried to enlarge the hole and dismantle
the Russian charge, but a half hour later it went off, killing three and badly

injuring the others. The explosion wrecked most of the Japanese gallery, yet it also cracked the surrounding earth in a way that allowed them to gain access to the caponier gallery, where they found a small rent in the concrete. Using guncotton to enlarge this hole, they threw in grenades followed by infantrymen.[31]

This started a bitter fight for the caponier gallery at North Fort on October 30, a bizarre section-by-section struggle that resembled room-to-room fighting in a house during urban combat. As this combat lengthened into days, the Japanese also began to mine the right flank of the ditch near another caponier gallery. A Russian countermine charge exploded on November 19, which did little harm to the Japanese gallery but damaged part of their own caponier. The enterprising besiegers again used guncotton to widen the damage and captured three-fourths of this caponier gallery by November 23.[32]

With enough of this second caponier gallery in their control, Japanese miners began to dig two galleries directly into the scarp of North Fort from the ditch, forty feet (12.19 meters) below the surface. They dug branches to the right and the left from both galleries and planted 4,409 pounds (1,999.88 kilograms) of dynamite in seven charges. They laid a wire system attached to a switchboard to set off all seven simultaneously after tamping the galleries. Those charges exploded on December 18, and the follow-up infantry assault captured North Fort, the first of the permanent forts to fall into Japanese hands.[33]

This success occurred two weeks after other Japanese forces had captured 203 Meter Hill, a key Russian position that overlooked the harbor where a fleet of warships was bottled up by the Japanese navy. The fall of 203 Meter Hill, which was accomplished by heavy artillery bombardment and fierce infantry fighting, was the beginning of the end for the Russian garrison. After North Fort also fell, mainly through sapping, mining, and infantry combat, Erhlung Fort was next. The Japanese set off another group of mines on December 28 that buried fifty Russians. The other three hundred left in the fort resisted the follow-up attack for several hours before evacuating the work. At Sungshu Fort two Japanese mines went up on December 31 and the following infantry assault cleared that work in a few hours. As a result, the Russian garrison commander entered into negotiations for surrender, finalizing the terms on January 2, 1905, to bring the siege of Port Arthur to a close.[34]

Mining had played an important role during this bloody encounter in Manchuria, and it represented a mixture of old concepts and new technology. The foundation of mine warfare at Port Arthur was traditional and would have been quite familiar to Vauban. The Japanese sapped their way toward selected works in the long Russian defense system, digging more than twenty miles (32.18 kilometers) of saps in about six months and using 1.2 million sandbags plus "thousands of rice-bags . . . to bring the infantry in touch with the forts." The idea of then mining at the head of those saps in order to collapse the counterscarp and prepare a way for the infantry to cross the ditch also was Vaubanian.[35]

But the Japanese were the first to use any of the new high explosives in actual mining operations. They were not the first to use electrical detonation—the Russians had preceded them with an earlier electrical system at Sebastopol—but they used the more advanced equipment available by the turn of the twentieth century.[36] The intensity of their mining effort was reminiscent of that found at Schweidnitz and Sebastopol, although the number of charges exploded at Port Arthur did not equal the number exploded by either side at Sebastopol.[37] But the key role played by mining at Erhlung Fort, Sungshu Fort, and North Fort was reminiscent of many previous sieges. As a test case in mixing new technology with previous doctrinal and tactical concepts, the Japanese siege of Port Arthur ushered in the context for what was to happen on the Western Front only a decade later.

The main difference, however, between the mining at Port Arthur and along the Western Front was that the former took place against new permanent forts made of concrete, while the latter was conducted against temporary or semipermanent earthworks. But the Russo-Japanese War also witnessed some examples of mining and countermining in field operations that involved temporary earthworks. While some Japanese forces were industriously sapping and mining at Port Arthur, others were holding a bridgehead protecting a railroad crossing of the Hun River near Mukden that was defended by earthworks from October 1904 to February 1905. The Russians mined toward those defenses until confronted by two Japanese countermines, which stymied their efforts. Another example took place along static lines near the town of Linschinpu, where in mid-January 1905 a Russian offensive mine extended 164 yards (149.96 meters) before meeting a Japanese countermine. The Russians hurriedly set off a charge of 1,984 pounds (899.92 kilograms) of gunpowder on February 28, but the subsequent infantry attack failed. Later, when the Russians evacuated their position, the Japanese found several mine shafts at various stages of construction.[38]

Whether at Port Arthur or in the field, operations during the Russo-Japanese War provided fit venues for the employment of siege approaches and offensive mining of a type reminiscent of past doctrine yet employing new technology. The war provided an important lesson that was taken seriously, especially by the Royal Engineers in England. They mounted a massive practice in military mining at Chatham in 1907 that was heavily based on the Japanese experience at Port Arthur. This was by no means the first English mining exercise. The Royal Engineers had conducted much smaller exercises in 1844, 1848, and 1868, which involved the explosion of nine charges; 1877, which involved the setting off of 11,000 pounds (4,989.51 kilograms) of powder; and 1878.[39]

But the exercise of 1907 was the biggest in the country's history. It involved a total of twenty-six charges exploded from July 1 to August 20. The primary targets were two masonry fortifications that had been constructed after 1860. A total of 3,575 men were involved in digging the offensive tunnels, which began relatively close to Fort Bridgewoods and Fort Luton because the government did not own enough land to allow the miners to start at a more realistic distance. Some engineers were assigned to dig countermines starting in the ditch and entering the counterscarp. Both sides had relatively easy digging thanks to the mostly chalk nature of the ground. Their shafts and galleries needed little shoring, and they could hear their opponents working up to 150 feet (45.72 meters) away, as the sound of picks and spades resonated widely in the chalk. The miners mostly used candles but experimented with electric torches, which did not work very well due to defects in their design. They also tested an artificial respirator that, if it had worked well, could have reduced the need for a formal ventilation system. The engineers used a variety of the new high explosives then available as well as different detonation systems.[40]

The 1907 experience at Chatham was the most extensive practice mining exercise in any nation before the onset of World War I. The techniques used there proved highly successful in opening breaches to attack forts protected by concrete ramparts. The engineers employed the same method as the Japanese—sapping forward until reaching the glacis, then digging shafts and exploding charges to collapse the counterscarp, which filled in the ditch and created breaches in the defensive work. No one could have anticipated that the next major conflict would hardly involve offensive mining against permanent fortifications. But the principles of attacking temporary earthen field fortifications were essentially the same as those for attacking permanent concrete defenses.[41]

By the time of the Great War, a decades-long series of developments had given military miners a new generation of explosives far more powerful and easier to use than gunpowder. Along with new ways of setting off charges, military mining was poised to enter a new phase in its long history. Japanese mining at Port Arthur proved to be a significant pivot in the history of underground siege approaches. It pointed out the utility of combining older doctrine with new technology, indicated that mining could play a decisive role in determining the outcome of sieges, and prompted the huge mining exercise at Chatham, the experience of which would play a role in British mining efforts in World War I. The Western Front provided an apt setting for the full employment of those improved tools as well as for expanding the scope and doctrine of underground warfare.

10

Underground Warfare on the Western Front

If anyone had told the men and officers who were mobilizing for war in the summer of 1914 that they would soon be engaged in the greatest military mining era in world history, they would not have believed it. While every major military force in Europe retained the knowledge, doctrine, and technical expertise to construct shafts, galleries, and explosive chambers, all of them did so with limited expectation of having to employ this method of siege warfare. A continuous line of field fortifications demarking an international border was never anticipated, nor a static confrontation between armies heavily entrenched opposite each other that would last for years. There were many unexpected developments to come as the armies marched confidently off to war late that summer, and military mining conducted on a scale that would dwarf anything seen in history was just one of them.

In the British Army the Royal Engineers had conducted several mine exercises, the biggest and most impressive seven years before the creation of the Western Front. Yet they had practiced only the basics of military mining, giving engineers elementary instruction in the process and relying on old rotary blowers for ventilation, and had no listening instruments or surveying instruments adapted to their needs. The German army, for its part, had essentially given up on mining as unlikely to occur in a future war of maneuver. The expectation was that modern infantry weapons and artillery would decide campaigns. Fortress warfare, which was still heavily associated with military mining, had long ceased to be the cornerstone of national strategy.[1]

But as soon as the German effort to knock France out of the war failed, the military situation provided a perfect venue for the growth of military mining. During October 1914, the French, British, and Germans knitted together a continuous front as they sought to flank opposing forces by extending their lines across northeastern France and western Belgium, soon constructing a continuous line of field fortifications known as the Western Front. Within a few weeks, the opposing armies were shielded by trench systems that stretched from the Swiss border to the North Sea coast across varied terrain. Millions of soldiers began to learn how to live, fight, and survive in a bizarre world of ditches and underground shelters, pummeled by artillery fire and gazing at exposed ground separating them from the enemy, which appropriately came to be called no-man's-land.

The French started mining almost immediately. On October 15 in the Argonne sector, they began to dig shallow underground shelters and on October 25 started constructing countermine galleries to protect selected points. South of the Somme River, the Twenty-Eighth Division began digging an offensive gallery that same month. The miners began it 328 yards (217.62 meters) from the target, and the Germans began countermining after learning of it from prisoners of war. Offensive and defensive galleries began to appear all along the front during the next several months, with the Germans springing the first siege mine on November 13 in the Argonne sector.[2]

From the beginning of underground operations in World War I, the pace and scale of these efforts exploded to levels never seen before. With large numbers of men packed in static positions apparently unable to break through opposing field works, military mining was set to expand into an operational vacuum. Instead of one gallery, both the French and the Germans dug multiple galleries in one restricted area and touched off several charges simultaneously. In the Argonne sector, for example, the French fired two mines on December 8. The Germans set off nine mines on December 12 and six on the seventeenth. This pace increased into the early months of 1915, as the number of countermine explosions picked up and both offensive and defensive galleries were dug at many new locations. In some cases those mines and countermines were associated with the digging of saps to carry troops at least part way across no-man's-land, as had been done at Schweidnitz, Sebastopol, Vicksburg, and Port Arthur. But in the majority of cases, galleries were begun behind the front line of trenches as had been done at Petersburg.[3]

The British were targeted for the first time by a mine attack on December 20, 1914, at Festubert. The Germans exploded ten charges along a sector of 1,000 yards (914.40 meters) and took possession of that area but could go no farther. In some cases along the front, French, British, and German follow-up attacks managed to capture part of an enemy trench but never achieved a breakthrough. The basic trench system in use by all belligerents consisted of multiple lines of trenches, usually three, each one two or three hundred yards (182.88 or 274.32 meters) from the other, making it difficult to break through all of them no matter what method of attack was employed.[4]

While the French and Germans relied mostly on military engineers to conduct their mining and countermining operations, the British experimented with an old idea—recruiting civilian miners. This was initiated by John Norton-Griffiths, a civil engineer with military experience. He had proposed recruiting men from the business he operated to dig sewers under London. Believing that the clay to be found under the city was similar to that to be found in western Belgium, Norton-Griffiths thought his "moles," as he called them, knew the right techniques to rapidly dig tunnels in this medium. His proposal was submitted in November 1914, but nothing came of it until the pace of military mining increased to an alarming degree in February 1915. Called to discuss his proposal at the War Office, Norton-Griffiths explained to Lord Kitchener, the British secretary of state for war, how his miners had developed an effective digging method called clay-kicking. They constructed a wooden frame that allowed one man to sit and lean back, using his feet and legs to manipulate a spade for breaking off large sections of clay from the face of a mine gallery while others carried away the spoil. Norton-Griffiths actually demonstrated the technique by lying on the floor with a shovel from the fireplace. This method could give the British a decisive edge in digging offensive mines, he argued. Kitchener was so impressed he immediately adopted the plan.[5]

Norton-Griffiths went to France to inspect the ground and explain his plan to high-level commanders, also setting into motion the recruitment of tunnelers from London. He told Robert Napier Harvey, who soon became his right-hand man, that "the nature of the clay made his mouth water" for the work to come. On February 20, 1915, only five days after demonstrating the clay-kicking technique to Kitchener, the first batch of tunnelers arrived on the Western Front. They had no military training but formed the nucleus of two new companies of Royal Engineers. As these and other new tunneling companies were organized, 80 percent of their officers were civilian mining engineers recruited from Britain's global empire. Such men "were inured to danger, had any amount of pluck and resource, and had been used to thinking and acting for themselves and to handling men." They made excellent military miners.[6]

CLAY-KICKER, WITH GRAFTING TOOL (LEFT)
AND "CROSS" (RIGHT)

Clay-Kicking. John Norton-Griffiths introduced this method for faster digging through clay from his work as a civil engineer creating tunnels under
London. This gave British tunnelers, famously referred to as "moles"
by Norton-Griffiths, a decisive edge over German miners in the pace of
gallery construction. It also kept the dimensions of the gallery as small as
possible while shoring it as little as possible. From Grieve and Newman,
Tunnellers, 34.

The rank and file of what would be designated as tunneling companies in
the Royal Engineers consisted of many civilian miners. In the early history
of these units, their men were fresh from the sewers and of all ages. The
185th Tunnelling Company had a fifty-year-old member and another who
was over sixty and had five sons already in the war effort. But as the conflict
continued, soldiers in other branches of the service transferred to the tunneling companies and men were recruited from nonmining segments of civilian society, diluting the original component. In addition, officers regularly
detailed infantrymen to work with miners for short periods of time to haul
away spoil, prepare sandbags, and do the other drudge labor that the miners
preferred to avoid. Later in the war entire infantry companies were assigned
to such work for a month or more. This allowed infantry officers to control
their own men and offered the infantrymen better food, accommodations,
and regular hours with breaks. Still, many detailed infantrymen did not like
to venture underground. Bernard Newman, an infantry officer, recalled that
he went below only once "and was mighty thankful" he did not need to go
again. "The low galleries alone were enough to terrify an ordinary man—
mere burrows, apparently about to collapse at any moment. I came to the
conclusion that tunneling demanded a different brand of courage to that
which I possessed."[7]

With a worldwide empire to draw from and inspired by Norton-Griffiths's
driving energy, the Royal Engineers organized many tunneling companies

in 1915 and 1916. From the empire they tended to recruit mineral miners, while from Britain they drew on coal miners and sewer diggers. Australia organized three companies that reached the Western Front in May 1916. The Australian units operated independently at various parts of the front. Originally recruited with 360 men in each company, that troop strength was later raised to 750 men of all ranks. Sections consisted of 170 men each. Later, 30 men were taken from each of the original units to form a fourth one called the Australian Electrical and Mechanical Mining and Boring Company.[8]

Canada contributed two tunneling companies, one recruited from among miners in the eastern part of the country and another from British Columbia and Alberta. Both began organizing in the fall of 1915 and appeared on the Western Front early the next year. A third company formed later from men already serving in the first two. New Zealand also began organizing a tunneling company in the fall of 1915 that reached the theater of war in March the next year.[9]

Within a relatively short time, Norton-Griffiths's idea had sparked the creation of a large number of special units designed to apply civilian methods of mining to military use. Seven tunneling companies represented the British Commonwealth nations (three Australian, three Canadian, and one New Zealand), while twenty six were organized directly by the Royal Engineers and referred to as imperial companies because they consisted of men not only from Great Britain but also from colonial holdings around the world. The thirty-three tunneling companies held about 25,000 men at any one time.[10]

Engaged in heavy mining and countermining, the loss rate of the companies was substantial, although only about half of that suffered by infantry units on the Western Front. The First Australian Tunnelling Company, for example, lost 67 killed, 166 wounded, and 37 invalided out of the 1,220 men who served during the course of the war, a loss rate of 22.13 percent. The Third Australian Tunnelling Company's loss rate was double this, but in comparison the Australian Imperial Force overall endured 66 percent casualties during the war.[11]

While detailed information on mining units in other armies on the Western Front is far less available than for the British forces, it is clear that no other belligerent moved forward with enlisting civilian miners on the scale to be seen in Norton-Griffiths's effort. The French and Germans to a limited degree recruited civilian miners but never relied on them as a main resource and did not organize them into special units. Mining for them remained the preserve of professional military engineers, who had a smattering of civilian miners in their commands. In fact, the German army was forced to release most of the civilian miners in its ranks partway through the war so they could return to commercial mines and keep up the supply of coal to the home front. There is no doubt that the specialized tunneling companies gave the British a decisive advantage in the developing underground warfare on the Western Front.[12]

When the United States entered the war, the staff of Gen. John J. Pershing was impressed by the extensive mine warfare and pushed through efforts to create an American engineer regiment dedicated to the work. As a result, the Twenty-Seventh Engineer Regiment was recruited from civilian miners and commanded by an accomplished civilian mining engineer. By the time the regiment reached France in March 1918, mine warfare had waned so much that there was no need for its specialized services. Pershing's chief engineer appointed Lt. Col. Alfred H. Brooks as chief geologist for the American Expeditionary Forces in the late summer of 1917. Brooks intensively studied mining operations, but no Americans engaged in mine work. Other than Brooks, few American engineer officers gained substantial knowledge of the mining operations conducted by the French or the British.[13]

The American lack of experience stands in stark contrast to that of the British. Not only did the Royal Engineers engage in aggressive and extensive mining and countermining but also their work has been thoroughly documented in English-language publications. While in most areas of detail British mining resembled closely that of the other two national armies, in some aspects it was unique.

The British, for example, pushed forward with a more sophisticated administrative structure to monitor, control, and document mining projects along their sector of the Western Front than did the French or the Germans. The energy with which tunneling companies worked while reporting to brigade, division, or corps commanders was astounding, but gradually it became apparent that such a large effort needed higher direction. After Sir Douglas Haig replaced Sir John French as commander of the British Expeditionary Force, he created a higher-level administrative structure to meet this need. Beginning January 1, 1916, a controller of mines at each field army headquarters assumed responsibility for documenting and reporting on all mining activity. They reported to Gen. Robert Napier Harvey, who was appointed inspector of mines for the British Expeditionary Force, a post created on January 4. Harvey collected all mining reports, summarized the important information to be passed on to general headquarters, and issued advice to tunneling companies through the publication of "Mining Notes" and other kinds of documents. Harvey had a staff consisting of two officers, a geologist, a medical officer, and a mechanical engineer along with a number of clerks and draftsmen. He was primarily responsible for evaluating suggestions for mining projects coming from the commanders of tunneling companies and line officers, for recommending what projects should be pursued and which dropped, and for shifting around the tunneling companies to places where they were most needed. He reported directly to the commander of the British Expeditionary Force.[14]

This administrative structure was unique in the history of military mining. It was necessary only because of the gigantic scale of mine warfare that was fast developing in 1915 and 1916. The administrative structure also spurred an increase in the accumulation of information about the context and details of mining operations. Officers collected information from a variety of sources, including reconnaissance on the surface, mapping German trench systems, and aerial photography. They filed detailed reports on mining projects on a regular basis. Harvey had to sift through all this information as he made tough decisions about which projects seemed to hold more promise than others and, above all, how each could be incorporated into larger operational plans for breaking the trench stalemate or achieving lesser but important operational goals.[15]

About the same time that they created an inspector of mines, the British began developing a system of in-theater schools to train personnel near the scene of their operations based on the latest information about conditions in the field. It started in the summer of 1915 with the Army Mine Rescue Schools, organized to train men in extracting miners from galleries affected by cave-ins and gas poisoning. From this start, the engineers developed Mine Listening Schools early in 1916. This was followed by an overarching concept, the First Army Mine School, which opened on July 1 and embraced lessons in all aspects of mining, from rescue to listening and other activities. The army-level schools trained both officers and enlisted men. From this the British developed mining war games, in which two teams of miners received instructions and competed with each other. Students also blew practice mines nearby. This level of training processed quite a few personnel. The First Army Mine School alone graduated more than 500 officers and 5,500 other ranks in two and a half years of operation.[16]

Despite the elevated management scheme and the many-layered school system, how to make mining work within the general operational plan proved to be the toughest part. None of the many underground projects had yet produced a real hope for a breakthrough in the static nature of the Western Front. Many British explosions had been followed by infantry attacks that failed to achieve much in the way of results.

At Hill 60 south of Ypres, Belgium, the British blew a mine on February 17, 1915, which led to retaliatory German charges. In the give and take, both sides lost men until the British set off six mines with about 1,000 pounds (453.59 kilograms) of gunpowder in each and attacked with infantry on

April 17, capturing the small terrain feature. But the Germans used chlorine gas to recapture Hill 60 only nineteen days later and held it for the next two years. Farther south a similar tactical scenario developed at the village of St. Eloi when a German mine-infantry attack captured the place and a British counterattack reclaimed it. Both sides mined each other's position to produce casualties but with no larger aims accomplished. Intensive mining and countermining took place around Ypres. A British shaft descended 35 feet (10.66 meters) through sand before hitting clay, and the subsequent gallery extended 190 feet (57.91 meters) out. Miners placed a charge of 4,900 pounds (2,222.60 kilograms), mostly consisting of ammonal, which created a large crater. While British troops took possession of the crater, German infantry counterattacked with flamethrowers to limit their gains. At the Brickstacks, two British offensive galleries blew up a section of the German first-line trench on April 2, 1915, but the troops could not exploit the advantage. At Loos miners exploded sixteen charges on September 25 and captured the German first line. But the Germans had begun evacuating that line just before the assault, and continued to hold on one hundred yards (91.44 meters) behind it.[17]

Much the same could be said of French mining efforts farther south of the British sector. At Carency, a German-held town at the north end of Vimy Ridge, French troops advanced five saps that aimed at three German salients beginning January 1, 1915. They connected them to form a parallel close to the target in classic Vaubanian fashion and then began a tunnel toward one salient on January 6. On hearing the sound of German countermining, the French blew a charge consisting of twenty-two pounds (9.97 kilograms) of cheddite, a commercial explosive introduced into the army the previous November, and stalled the German defensive effort. But further offensive work by the French was in turn stalled by more enemy countermines.[18]

Like Totleben at Sebastopol, the French at Carency now tried to dig deeper in order to blow out the countermines. They began new galleries from behind their front line, employing inclined entrances rather than shafts so as to remove spoil faster. This underground confrontation heated up in March 1915 as both sides exploded camouflets to collapse opposing galleries, with losses of miners trapped or killed in demolished tunnels. The French began nine new galleries at a third, deeper level and connected them with transverse galleries for good ventilation. While offensive mines now and then met defensive mines at the two upper levels, the French managed to bypass all that in their third level and planted seventeen charges with a total of nearly twelve tons (10,886.20 kilograms) of explosives. They blew them as part of the Second Battle of Artois on May 9, 1915, exploding them in batches over a period of twenty minutes. The blasts played a role in the

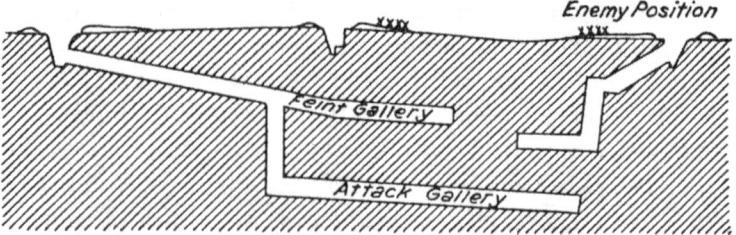

Fig 937 A. Feint and Attack Galleries

Attack Plan of Offensive Galleries, Western Front. This two-pronged approach, with a feint gallery dug shallow to distract enemy attention and the true attack tunnel much deeper to undercut the defender, represented a typical strategy in the underground warfare of the Western Front. Mitchell, *Army Engineering*, 230.

infantry assault that captured the first layer of German fortifications, and Carency fell three days later. Still, there was no breakthrough or larger strategic result.[19]

The interplay of offensive and defensive mining efforts at Carency during the first six months of 1915 was on a level with that at Sebastopol sixty years earlier. It represented only one of hundreds of similar examples along the Western Front. In terms of intensity of effort, the push to beat the opponent, outsmart him, and blast his efforts to attack or impede offensives, miners on both sides of the long front were sucked into a natural process of escalation that involved tens of thousands of men and mountains of supplies and equipment.

Offensive mining came first, soon to be followed by defensive efforts, or what some observers called protective mining. Many of the British tunneling companies were forced into defensive efforts when they initially deployed along the front in 1915. The First Canadian Tunnelling Company, for example, moved into a salient at Le Buzet and Le Touquet near Houplines in the Lys valley. It was "very quiet" on the surface, but there were already four shafts eight feet (2.43 meters) deep with five hundred feet (152.40 meters) of defensive galleries. The Canadians expanded the system for the next three months. The miners of the Second Canadian Tunnelling Company, deployed between Hooge and Armentières, dug shafts at one-hundred-foot (30.48-meter) intervals along the first line and connected them with transverse galleries. Then the men ran listening posts forward thirty feet (9.14 meters) from each other. At thirty feet out from the transverse, they dug a right angle of ten feet (3.04 meters) in order to break the force of any German camouflet and planted a listening post. Boring holes twenty feet

(6.09 meters) forward from the listening post, they were able to quickly plant camouflets of their own if needed.[20]

Farther south, around Fricourt in the Somme sector, British troops replaced French forces in the surface earthworks, accompanied by their miners. Blow and counterblow were traded between Norton-Griffiths's men and the German defenders as a vigorous underground war developed late in 1915. On December 21 the Germans exploded a countermine that killed nineteen and injured twenty-two miners, the largest loss of tunnelers in a single incident on the British side of the war. The German defensive system was twenty to thirty feet (6.09 to 9.14 meters) deep and one hundred feet (30.48 meters) forward with a transverse gallery; it also was equipped with electrical microphones and a central listening station. The British decided to sink a new offensive mine system one hundred feet deep to undercut their opponent but encountered water, which greatly impeded progress.[21]

The growing sophistication of offensive mining techniques and designs sparked a corresponding increase in the methods of preventive mining. Alfred H. Brooks, the American chief geologist, described countermine systems he examined on the Western Front late in the war. At one location he found a lateral gallery one hundred feet in front of the first trench line with listening galleries extending toward the Germans at intervals of sixty to one hundred feet(18.28 to 30.47 meters). Listening posts had electrical wires that connected the head of the gallery with a central station manned twenty-four hours a day. There, the operator could tell from which gallery suspicious noises originated. First-aid stations had been set up at frequent intervals, and those galleries that lay above the water level were provided with electric lights. Gas engines powered a dynamo at each flank of the system to provide power for the lights and for electric pumps to dispose of seepage, while a well provided water for human consumption.[22]

What Brooks saw was something far advanced compared to anything that had been constructed previously in the history of military mining. It was a complete system of semipermanent countermines for semipermanent earthworks, a Western-Front counterpart to the permanent countermine systems that had developed from the sixteenth through the eighteenth centuries at masonry fortifications, but with many new technologies incorporated into its design and daily use.

Brooks discussed two basic types of mine designs that were used not only for semipermanent countermine systems but for offensive mines as well. One was the herringbone (or as he called it the fishbone) type, and the other came to be known as the lateral, or transverse, type. The French preferred the fishbone system and the Germans usually adopted it as well, according to Brooks. This system, with independent galleries that had

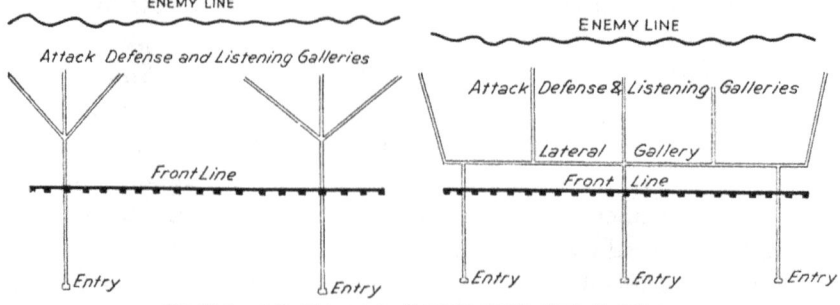

Figs 936 A and B. Listeners for Detecting Hostile Flank Operations

Two Gallery Designs, Western Front. On the left the herringbone, or fishbone, design competed with the lateral, or transverse, design on the right in the construction of offensive, countermine, and listening galleries by the belligerents. Mitchell, *Army Engineering*, 228.

branches shooting off from it for attack, flank protection, or listening, was more widespread than the lateral system. The transverse system began with a lateral gallery parallel to the friendly trench line from which the galleries began toward the enemy. Farther out from the starting point, miners dug other lateral (transverse) galleries to create an interconnected system that promoted natural air flow.[23]

The French criticized the lateral system for several reasons. They pointed out that the interconnected nature of the galleries meant that if gas was introduced into one, it would communicate to the others because of the natural flow of air through the system. Brooks, after thorough study of the designs, argued that this could be dealt with by placing gas-proof doors at key junctions, although this would impede foot traffic through those junctions as well. The French also argued that disposing of spoil would be slowed and complicated because of the need to haul it through a system with several sharp corners to turn. But Brooks countered that the transverse system had more entrances so that many galleries were connected to the surface, allowing details to have more straight paths to remove the dirt. If a counterattack allowed enemy troops to enter one gallery, said the French, the entire system could be compromised. Brooks argued that the starting parallel could be located at least one hundred feet behind the forward line. If enemy troops penetrated that far, they would be exposed to detection and counterattack before they could consolidate their control of the system.[24]

Brooks sided with the British, favoring the lateral (transverse) system of mine design. He found a connected system better than a series of independent galleries but recognized that both types to a greater or lesser degree served the purpose of offensive mining. That is why he admitted, after his extensive survey of French and British mining efforts in the latter part of the war, "that in any extensive mining system, both types will be more or less used."[25]

With so many miners working along the line, infantrymen on the surface could not help but be aware that underground work was at a fever pitch. The reverberation from a big blow could be felt on the surface up to a mile (1.60 kilometers) away, and closer to it "dugouts would shake badly," wrote Lt. Harry Davis Trounce of the 181st Tunnelling Company. "Timbers would be loosened and many men buried," leading to hasty rescue efforts.[26]

Most infantrymen were concerned about the dangers of underground warfare, and often there was a touchy relationship between them and the miners. Capt. Herbert W. Graham of the 185th Tunnelling Company recalled that when his men replaced the New Zealand company at Vimy Ridge on April 1, 1916, they found a dangerous situation. Judging from listening reports, German offensive mines were very close, so the tunnelers worked on building an effective system of countermines for some time. "Many can still recall that unpleasant feeling," Graham mused in 1927, "one of the most uncanny I know of, and one that sticks persistently behind one's mind, of the possibility of being blown to eternity at any moment." That feeling was especially sharp among the infantrymen. Once when an infantry officer reported to Graham that his men heard sounds of pick work near their position, the mining officer lied and said it came from a British gallery when he knew that it was a German mine. Graham felt sure that his company's countermining efforts would protect the position and so did not want to increase the worry level of the infantryman.[27]

Nevertheless, the Germans managed to set off a blast now and then, leading to losses, for which the infantrymen tended to blame the 185th Tunnelling Company. In Graham's view the enemy "never seriously endangered our line," and the loss of a few men now and then was a natural price to pay for being in a war zone. With a few weeks of hard work, the 185th managed to bring the underground confrontation to a stalemate.[28]

Not surprisingly, it became obvious that defending against enemy offensive mining was not enough. "Not long after mining was commenced," recalled H. Standish Ball, the assistant inspector of mines, "it was recognized by both sides that the 'weakest form of mining was a pure defence,' and that to be successful one's policy had to be as offensive as possible in character." Harvey saw most of the mining efforts of 1915 as "ornamental destruction" because they were focused on very limited tactical goals. "This sort of mining is perfectly useless for ending a war." To make offensive mining felt on the operational level, it needed to be tied together by a coordinated plan, and that was the primary purpose of creating the administrative structure he now led. Harvey worked to tie together all offensive mining efforts to the needs of infantry operations. He also mandated that even defensive mining schemes had to be considered with a view toward promoting offensive mining as well.[29]

Those ideas were, of course, not new in the global history of military mining. Running galleries forward to undermine a defensive work had always been at the heart of this military craft, and coordinating it with surface attacks had always been difficult. The difference between World War I and all mining efforts of the previous two thousand years was a matter of scale. Mining had expanded so far and so rapidly that literally hundreds of galleries were in operation at any given time rather than just a handful, and all three belligerents on the Western Front struggled with the effort needed to manage them. As part of that struggle, Harvey had to impart an overall sense of purpose and direction. He asserted what Totleben had so well discussed about the lessons to be derived from mining at Sebastopol—an offensive spirit was essential to success, whether one dug an attack tunnel or a preventive gallery.

We know more about the British effort to manage mining along their sector than we do about the French or Germans simply because there are many more English-language publications on the subject. But the French experienced the stress of managing large-scale mining on their sector. A pamphlet issued by the headquarters of Joseph Joffre, commander in chief of French forces, in March 1915 reflected prewar mining doctrine. It envisioned the digging of two offensive galleries in tandem, the use of inclines, and the aim of reducing strongpoints in the enemy position. Nothing in this deviated from the heritage of mining doctrine. But by December 1915, about the time the British were constructing their higher-level mine management, Joffre circulated a dispatch to commanders of army groups urging better coordination of mining efforts. Thus far, local commanders controlled them, and the tactical results were uneven at best. Joffre required more thorough reports on what was being done and asked for information on any special equipment needed. Gen. Charles Ebener of the Thirty-Fifth Corps and Gen. Robert Nivelle of the Sixty-First Division responded to this circular by arguing that tight controls had to be exerted on mining efforts. Higher levels of authority had to select key points and use mining to support surface attacks. As a result, French mining efforts were streamlined in 1916 and continued only in sectors where the Germans refused to ratchet down their own mining. "It is necessary to weigh the effect and results" when planning underground efforts, wrote a French engineer officer in a widely circulated pamphlet that was translated into English.[30]

The spirit of offense seemed to be more intense among British miners

In Fig 931 A, a cross-section of a common crater:
OE = line of least resistance = LLR.
AE = crater radius = OE = LLR.
AO = radius of explosion.
OF = horizontal radius of rupture = 1.7 LLR.
OH = vertical radius of rupture = 1.1 LLR.
AOB = real crater.
ACB = apparent crater.
EC = apparent depth = 0.33 LLR.

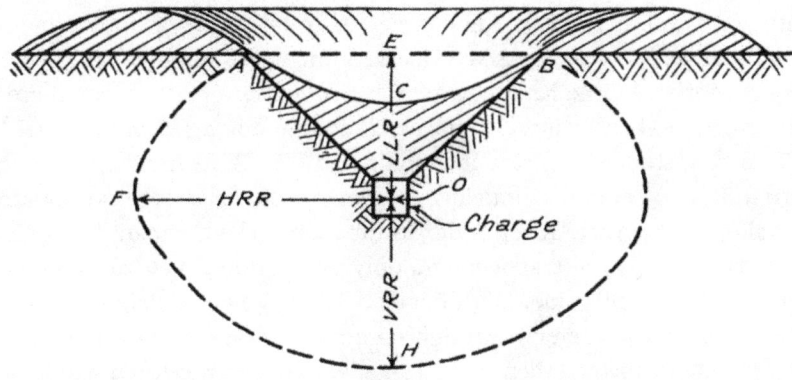

Fig 931 A. Mine Crater

Crater on the Western Front. This diagram clearly shows the dimensions and the technical terms of a typical crater while delineating the effect of the blast both on the surface and underground. Mitchell, *Army Engineering*, 224.

than among their French or German counterparts, and it is mostly due to the influence of Norton-Griffiths and Harvey. The former held a sort of roving commission to examine mining efforts and make recommendations, continuing to exert an important influence on its shape and spirit for most of the war. The latter was of a like mind with Norton-Griffiths. The Royal Engineers establishment easily agreed with the offensive doctrine of mine warfare, reproducing almost verbatim Harvey's postwar words about it in their official history of engineering work in the Great War.[31]

That official history also summarized the offensive doctrine of mine warfare better than Harvey did in his postwar writings. "The best form of offense is attack," it begins. "Surface and underground fighting must go together." The official history also mandated thorough study of the sur-face effects of underground explosions so those effects could be accurately predicted "and used to the best advantage" by troops after the blast. Every advantage in the underground war was on the side that could dig deeper than its opponent because it was easier to damage an enemy gallery with explosives from below than from above. No two offensive galleries should be so close to each other that one defensive charge could disrupt both of them. "Speed and silence" were key ingredients of success in offensive mining. In contrast, countermine galleries had to be close together so that an offensive gallery could not sneak in between them without detection. As a brief statement of offensive mining doctrine, the Royal Engineers official history could not be surpassed.[32]

Miners on all levels of the chain of command supported an aggressive attitude toward their work. They preferred to take the war to the enemy rather than act on the defensive. That sentiment was bluntly expressed by George Morley, a Canadian tunneler, right after the war ended, and there is every reason to believe it was a universal thought among British miners.[33]

Two case studies illustrate this point. Capt. L. B. Reynolds related the tactical situation after his tunneling company arrived at a ridge overlooking the valley of the Ancre River in the Somme area. The Germans had galleries thirty feet (9.14 meters) deep and only one hundred feet (30.48 meters) from the British first line. With detailed infantrymen to help carry spoil 1,000 yards (914.40 meters) through communication trenches, Reynolds's men dug an incline and then started a gallery, stopping to listen when they estimated the head of their tunnel was thirty feet from the German gallery. The sound of continued digging was reassuring, but the sound of "walking, shuffling, [and] thumping" meant the opponent was charging a mine. The British set a camouflet and sprung the charge. "Air pumps are rushed in" immediately after the blow, Reynolds dramatically writes, "and, as soon as the air is free from gas, men begin work clearing the wrecked gallery to the point where it is completely smashed up. A new gallery is then started off abruptly, and this is pushed with might and main. The men strip to the waist and dig for all they are worth, for now it is a race." Because the Germans heavily shored up their galleries and dug by the book, British moles usually beat them in such a race. "We could gain 20 to 50 feet [6.09 to 15.24 meters] on him, in the rush back to the old conditions. So gradually we drove him back, step by step, never blowing unless he refused to stop, but endeavoring always to force him to blow first."[34]

In another example, George H. Morley related action near Ypres in which his company contended with two German offensive galleries. They sank a shaft behind the first line to a depth of ninety feet (27.43 meters) and extended a gallery until estimating that they had reached the deeper German gallery. Then the Canadians dug branches to right and left to form a T and set four charges fifty feet (15.24 meters) apart, with six hundred to eight hundred pounds (272.15 to 362.87 kilograms) of ammonal in each. The resulting blast caved in the deep German gallery. While all this was going

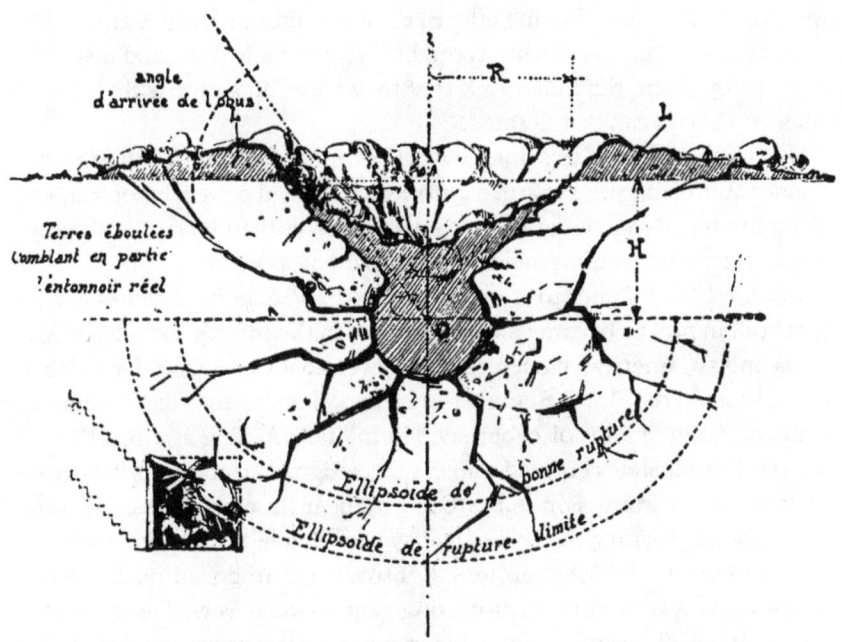

Zone of Disruption of a Mine Explosion, Western Front. A French illustration of the effect created by a detonated mine charge underground and on the surface. Called the globe of compression in the seventeenth century, it was by World War I more commonly called the zone of disruption. This blast effect created not only a crater on the surface but also fissures through the underground strata, crushing any gallery within that zone. "German Shelters on the Somme," 403.

on, other miners were listening on the surface in no-man's-land during the night and discovered a shallow German gallery. The Canadians exploded twenty-five pounds (11.33 kilograms) of ammonal on the surface to create a small crater into which they then put one hundred pounds (45.35 kilograms) of ammonal, exploded it at night, and crushed the shallow gallery below. This shallow gallery had made it to a point very near the British first line, but the deep gallery had been destroyed quite near the German line.[35]

The tone of underground combat on the Western Front impelled British miners to extreme exertions, perhaps more so than among the French or Germans. Perhaps this was because of the army's heavy investment in civilian miners, who brought to the war a heightened spirit of competitiveness. But British efforts are attributable not solely to the influence of Norton-Griffiths, but also to a tendency of the British Expeditionary Force to focus on specialty arms, which instilled a strong sense of purpose among the men. A New Zealand miner described the charging of a mine, the capstone of all gallery-digging efforts, as "a continuous feverish rush." Propelled by a fear that the Germans might discover their work and take measures to stop them, the miners felt that they were "in a race with death" in which "no delays can be tolerated."[36]

The development of this unique world of military mining, full of technical innovations and supercharged emotions, received quite a bit of publicity during the war. No one did more to popularize it than Harry D. Trounce. Born in England, Trounce had moved to Canada as a young man and soon after to the United States to study at the Colorado School of Mines. He left the school in 1910 to become a civil engineer in the mining industry of California and an American citizen. Trounce went back to England in October 1915 and joined the Royal Engineers as a lieutenant in the 181st Tunnelling Company, gaining a lot of experience in military mining as a result. Soon after the United States entered the war, he resigned from the British Army and received a commission in the U.S. Engineer Reserve Corps. He wrote his first book, *Fighting the Boche Underground*, while waiting for orders to return to France. Published in 1918, it introduced American readers to the underground war in Europe, discussing general ideas as well as illustrating them with specific examples from his own personal experience. He followed that up with *Notes on Military Mining*, published by the U.S. Engineer School in 1918. "The development of military mining and countermining is very marked in the trench warfare operations of this war," Trounce declares. But he felt that its importance was "not generally fully understood, and a study of the subject is essential." The eighty-eight page *Notes* was mostly directed to military personnel rather than the general reader, who had been the main audience for *Fighting the Boche Underground*. For the first time in history, the Great War witnessed efforts to popularize mine warfare outside the military community.[37]

During 1914 and 1915, the foundations of mine warfare were laid for all three armies along the Western Front. In many ways the character of that warfare emerged as quite different from its historic precedents. Scale was among the more important characteristics. Covering a front of semipermanent earthworks that stretched for 475 miles (764.43 kilometers) from the Swiss border to the North Sea, the Western Front greatly overshadowed any siege or semisiege lines that had previously involved military mining. The continuous lines at Sebastopol—35 miles (56.32 kilometers) long—and at Petersburg-Richmond—37 miles (59.54 kilometers) long—paled in comparison. This difference in length allowed for a hugely expanded mining effort against many different points at the same time, using methods of approach that differed markedly. Vauban had little show along the Western Front because above-ground sapping was too vulnerable to heavy artillery fire to be effective. Miners overwhelmingly chose to start behind friendly lines, leading in many cases to unusually long galleries that required sophisticated artificial ventilation and innovative ways to extract spoil.

Mining operations along the Western Front fully employed a wide range of new high explosives to replace gunpowder and made full use of electrical detonation systems to replace powder hoses and safety fuzes. Those innovative technologies had been developing for decades before 1914 but had never been fully tested in the field.

The high level of mining management instituted by the British was an innovation as well. It produced an enormous amount of detailed information about individual mining projects that was sifted by army administrators to determine how best to conduct underground operations. Despite the new approaches to mine warfare, all belligerents to a greater or lesser degree practiced the age-old policy of enlisting civilian miners for military purposes. No one did this more enthusiastically than the British. Their famous tunneling companies mixed civilian with military personnel to a greater degree than did the French and Germans, and it paid dividends in an unusually aggressive, competitive atmosphere among British miners.

Military mining in the Great War had already come of age by the end of 1915 and was poised to reach a historic level in the period 1916–1917. But first, one must consider the many new technologies and methods of digging shafts and galleries, blowing up charges, and dealing with surface operations used by the military miners of the Western Front.

11

The Materials and Methods of Mining in World War I

During the Great War of 1914–1918, literally hundreds of offensive and defensive galleries were dug along the Western Front, representing the largest accumulation of military mining projects in global history. For the most part, the activity was thoroughly documented so that we have access to the largest amount of information concerning the techniques and methods of military mining. Thus, the full use of the most advanced material, ranging from high explosives to rescue apparatus, can be appreciated, linking military mining more firmly with commercial mining and reinforcing the modern nature of military operations generally in the Great War.

The methods employed in siege mining along the Western Front also characterized the Great War as unique in world history. Traditional Vaubanian principles were largely abandoned as miners surged forward to develop their own way of digging and using underground approaches. They mostly gave up hope of sapping forward and started shafts behind friendly lines, learning how to deal with the unusually long galleries that resulted. Western-front miners selected traditional targets, such as enemy strongpoints and salients, that often were located on hills and ridges. But they much more often employed attack galleries to achieve short-term, incremental goals in no-man's-land than had been common before 1914. They did not shy away from attempting huge mining projects designed to shatter opposing trench lines, however, mounting by far the largest such efforts in mining history.

Great War military miners more heavily borrowed ideas, methods, and especially technology from the civilian mining industry than ever before. They used tubbing, a method of sinking shafts through sand layers; experimented with large coal-mining machinery in an attempt to dig galleries faster; made a careful study of geological layers because they dug much deeper than previous military miners; borrowed listening technology from the civilian sector in an effort to better detect enemy mining activity; and learned from civilians how to use canaries and mice as air sentries to alert them to the danger of gas poisoning for the first time in military mining. They employed the latest civilian oxygen devices to rescue military miners caught by gas in the galleries and recruited medical personnel experienced in treating gas-poisoned coal miners into the army. Charging and exploding their mines, Great War personnel used a variety of new high explosives that had been developed for civilian mining, touching it off with sophisticated electrical systems taken from commercial industries.

First among these many topics is how far Great War miners dug their galleries. Throughout history, gallery length had been governed by the relative position of the opposing forces. On the Western Front the width of no-man's-land was the decisive element. It varied from a few yards to hundreds of yards, with varied types of soil and differences in topography from one sector to another. The preferences and judgments of military engineers also came into play in determining exactly where to start an offensive mine now that beginning at the end of a sap near the target was largely a thing of the past. Approach trenches dug on the surface were simply too vulnerable to the heavy artillery utilized by both sides along the Western Front, compelling miners to dig under the ground between the lines rather than sap across it.

Alfred Brooks reported that the French established a general rule that a gallery should not be started if the opposing lines were more than five hundred feet (152.40 meters) apart, while the Germans reportedly also used that estimate. The British, on the other hand, had no difficulty with starting operations from points much farther from the target. Given the relative role played by the three belligerents, Brooks concluded that, on average, attack galleries started from less than five hundred feet from the objective, and he probably was right. In comparison, the Pleasants mine at Petersburg was a little more than five hundred feet long.[1]

After determining where to start the gallery, how to dig the shaft in order to conceal its location from enemy reconnaissance was the next

Digging at the Face of a Gallery, Western Front. This photograph was taken by Lt. Col. Robert D. Percival-Maxwell of the Royal Irish Rifles. It is perhaps the only authentic photograph of a sapper working at the face of a gallery on the Western Front, in this case the Hawthorn Ridge Redoubt mine, as it is the only photo I know of anywhere. Note the cramped dimensions, the rough-edged face, and the fatigued and dirty appearance of the sapper working hard with a small pick. Imperial War Museum, HU087951.

consideration. Miners needed to have easy access to the entrance, and materials had to be funneled through it and spoil taken out of it, compounding the problem of concealment. Support trenches behind the first line proved to be good locations for shafts, as they were relatively shielded from enemy observation. This fact played a large role in lengthening Western-Front galleries because supporting trenches could be placed at any distance behind the first line.[2]

Military miners had two choices in the type of entrances to galleries, and both had deep historical roots. Shafts had the advantages of being relatively easy to make; they gave quick access to deep levels, and there were ways to dig then through wet soil. Disadvantages were that they were vulnerable to well-aimed artillery fire; it was difficult for men to get in and out of them; they needed special framing; and they provided less effective natural ventilation. In contrast, an inclined entrance provided much better ventilation and easy removal of spoil but took some time to reach the desired depth. Ironically, this type was even more vulnerable to artillery fire because the tunnel was at a shallow depth for some distance until gradually descending to a point where shelling proved less destructive. The ideal situation was to start a gallery in the face of an escarpment in order to begin at a deep level and avoid utilizing either the shaft or the incline, but such opportunities were rare. A conveniently located quarry, a commodious mine crater, or as in the Saint-Mihiel sector, a partly demolished church could be used, but such sites were not readily available. Thus, the majority of all galleries began with either a shaft or an inclined entrance.[3]

Protecting the entrance became a priority. The Germans, renowned for thoroughness in construction of their mines, dispensed with wooden sheathing in at least one of their shafts and lined it with concrete instead so that artillery rounds would not collapse it. They could not afford to do this with every shaft, but this one descended ninety-three feet (28.34 meters) and was eight feet (2.43 meters) in diameter. It became common to employ camouflage to conceal the location of shafts from aerial photography, and setting up a dummy shaft at another location was a good ploy.[4]

Geology heavily influenced mining technique and method. In Belgium and French Flanders, layers of sand alternated with layers of clay in many areas. The British found it extremely difficult to sink shafts through the sand layers by the traditional method of digging a couple of feet before erecting timber framing. In May 1915 they borrowed the concept of tubbing from the civilian mining industry. This involved pushing a ring of steel 2.5 feet (0.76 meters) wide and 5–6.5 feet (1.52–1.98 meters) in diameter, open at both ends, into the ground. Engineers erected a heavy roof over the shaft and attached screw jacks to it, which they used to push the tube into the sand as far as possible, then dug out the sand from inside the circular perimeter. Adding more rings enabled them to push the excavation to a layer of clay beneath the sand where they could dispense with the steel tubes and revert to timber framing. Tubbing was a perfect example of how civilian mining influenced military mining to deal with geological challenges to shaft construction.[5]

Geology also played a significant role in the digging of galleries because shoring was necessary in clay but hardly necessary in chalk. In turn, the need to shore up galleries influenced their size. Miners required enough room to work and allow traffic through the excavation (spoil to the top, materials and men to the head of the gallery). But the larger the tunnel, the more shoring lumber it required. The British settled on a standard-size gallery of four feet (1.21 meters) by two feet, three inches (0.70 meters) but allowed officers to go larger if needed. They fashioned the shoring in sets above ground for easier installation. The frames were held together by slats spaced eighteen inches (45.72 centimeters) apart in dry ground but only one-half to

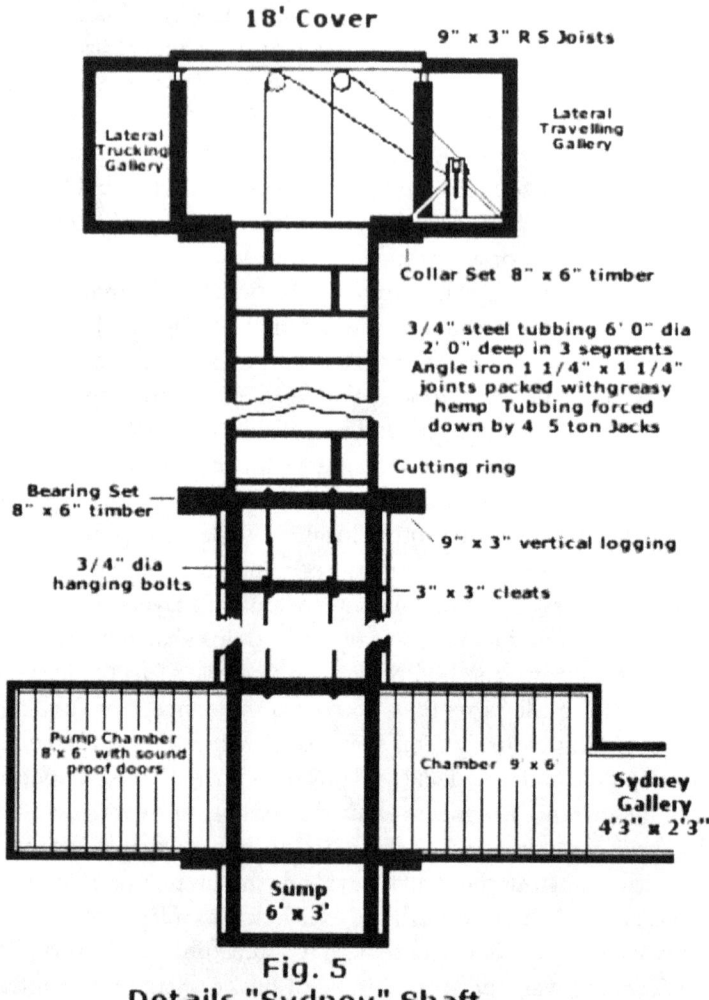

18' Cover

9" x 3" R S Joists

Lateral
Trucking
Gallery

Lateral
Travelling
Gallery

Collar Set 8" x 6" timber

3/4" steel tubbing 6' 0" dia
2' 0" deep in 3 segments
Angle iron 1 1/4" x 1 1/4"
joints packed with greasy
hemp Tubbing forced
down by 4 5 ton Jacks

Cutting ring

Bearing Set
8" x 6" timber

9" x 3" vertical logging

3/4" dia
hanging bolts

3" x 3" cleats

Pump Chamber
8'x 6' with sound
proof doors

Chamber 9 x 6

Sydney
Gallery
4'3" x 2'3"

Sump
6' x 3'

Fig. 5
Details "Sydney" Shaft
Scale - 1/8 in = 1 ft.

Sydney Shaft, Western Front. This diagram shows in detail the design and construction of a typical shaft British miners used in World War I. Note the casemate covering over the top, with eighteen feet (5.48 meters) of space for two winches used in lowering and raising personnel and spoil. The shaft was constructed with much tubbing through sand layers, and a spacious sump was dug at its bottom for the collection of water. Woodward, "Notes on the Work of an Australian Tunnelling Company," 9.

one inch (1.27 to 2.54 centimeters) in wet places. They tried always to have at least some space between slats even in the wettest ground to allow for the clay to swell without breaking the shoring.[6]

When capturing German galleries, the British were surprised to discover that their opponents tended to be lavish with shoring. "Even in hard chalk all galleries were close timbered." A German mining regulation captured at Fricourt in July 1916 indicated that closely shored galleries muffled the sound of digging better than those loosely framed. While this probably made it more difficult for the British to hear their mining, it also lessened the Germans' ability to hear British mining and slowed the pace of their own digging.[7]

With the ability to dig almost as deep as they desired, Western-Front miners utilized all levels within their reach. At times they dug deep in order to avoid bad geologic conditions, such as underground water, but at other times they deliberately dug galleries at several different levels as part of a tactical plan. Upper levels often were designed to draw enemy attention away from deeper work, giving advantage to the side that could deceive the other. The massive number of men assigned to mining work by the three belligerents also made this multiple-layer tactic possible. At combined peak strength in the three armies, roughly 75,000 men worked on mines and countermines in France and Belgium. Far more than in any previous war, mining systems became three dimensional on the Western Front, expanding down as well as sideways under the surface of the battlefield.[8]

Brooks, following his extensive study of mining in Belgium and France, concluded that galleries needed to be at least 20–30 feet (6.09–9.14 meters) below the surface to avoid all damage from shelling. If they were shallower, heavier shoring was required. It was best to dig above the water level, but if that was not possible, constant pumping was needed. Brooks also noted that the British tended to go deeper, between 20 and 100 feet (6.09 and 30.48 meters) below the surface and as much as 200 feet (60.96 meters) in some cases. The famous galleries that produced the biggest mine attack in history at Messines in June 1917 were on average 100 feet deep. French captain Cussenot, author of a pamphlet on mining that was translated into English and widely distributed by the Americans in 1917, noted that there was no real rule about depth but believed that galleries were normally from 80 to 160 feet (24.38 to 48.76 meters) below the surface.[9]

All tactical considerations aside, there was an emotional element that influenced the depth of galleries. R. W. Coulthard, a member of the Second Canadian Tunnelling Company, ably expressed the underground warrior's sense of purpose and professional pride when he argued that the true miner wanted to dig as deep as possible. He writes of "the deep mine—the real mine" set at 130 feet (39.62 meters) below the surface, where "30 to 40 tons [27.21 to 36.28 metric tons] of canned hell lie sleeping" like "some huge monster." This sort of mining operation was, in his view, "the *raison d'etre*, of the tunneller."[10]

Coulthard expressed the offensive nature of mine warfare, which Canadian engineer Capt. L. B. Reynolds also described when writing about the underground contest. He noted that the Germans tended to run two galleries in tandem at an upper and a lower level, pushing the upper one farther than the lower to draw British attention. They dug as far as possible and blew a charge. The lower gallery, deep enough to avoid the area of rupture of that blow, would then be pushed rapidly forward while British efforts were focused on dealing with the work at the shallow level. The British learned to anticipate such tactics and to be prepared for them. In the process, tunneling dove deeper and deeper into the earth as both sides tried to descend below the other.[11]

Running galleries out from the shafts involved creative solutions to problems. According to the official history of the Royal Engineers, galleries were started three to four feet (0.91 to 1.21 meters) up from the bottom of the shaft to allow water seepage to collect below the entrance and lessen the chance of its running into the tunnel. Then there was the age-old problem of navigating underground. "Straight driving is essential for quick work and good ventilation underground," notes the official history. The easiest way to achieve this dated back centuries. It was to hang weighted lines from the roof of the gallery, placing them every ten to fifteen feet (3.04 to 4.57 meters) apart "on the bearing required." Then, holding a light on the center of the face of the gallery and seeing if it lined up precisely with the weighted lines gave one an accurate picture of whether the tunnel was running straight.[12]

This method worked very well underground, but newer technology was needed for above-ground surveys, all of which was borrowed from civilian mining. The theodolite was perfect, but for military purposes, shorter legs on the tripod were necessary to avoid exposure to enemy fire. The miner's dial was very useful for taking bearings underground but was a bit large to be used in smaller galleries. In those cases, the prismatic compass sufficed, although the presence of iron in the soil affected its performance. To measure inclines, the Dumpy Level was effective in large galleries, while the Abney Level was used in smaller ones.[13]

As had countless miners before them, World War I engineers realized that some means of ventilation was necessary in galleries that extended more than sixty feet (18.28 meters). It was possible to create a natural flow of air with a system of galleries connected by a transverse. If there were two shafts, one could become the upcast to vent bad air while the other became the downcast that brought fresh air into the system. One could manipulate the flow of air by placing doors at strategic points to close off part of the system and direct air to other areas. Gas curtains made of "blankets soaked in anti-gas mixture" also could be placed at strategic points to prevent enemy gas from entering.[14]

Given the many large systems of mines and numerous independent galleries that appeared along the Western Front, artificial means of ventilation became common. The longer the gallery, the less a natural flow of air would suffice. Miners experimented with a variety of devices to pump fresh air into their works. Ordinary blacksmith bellows attached to tin pipes six to nine inches (15.24 to 22.86 centimeters) in diameter with canvas joints to link one section with another worked well enough, but they were cumbersome to set up. India-rubber pipes could substitute for tin. Bellows were useful to ventilate up to 1,200 feet (365.76 meters) of gallery, although some tunnelers recalled hand-powered air pumps with disdain. If the pumps were small, even hard pumping could barely provide enough fresh oxygen to light a candle at a distance. Compressors powered by electricity or gasoline were needed to ventilate long galleries. Holman Brothers, a mining-equipment manufacturer in Camborne, Cornwall, England, had been making a compressor since 1894. Attached to a three-inch (7.62 centimeters) hose, it was a great improvement over hand-operated bellows but, along with other air compressors, created a good deal of noise and often broke down when most needed. A small air compressor powered by a gasoline engine was placed "in a fairly deep dugout that muffled the sound" at one location along the front.[15]

Ventilation could also be used as a weapon if conditions were right. In the area of Vimy Ridge, the Germans dug into one of the coal-mining systems that happened to be near no-man's-land and flooded part of it with water. British engineers came up with a scheme to retaliate by flooding the system with poison gas, assuming that it would enter the German gallery. The plan was dropped because the French worried that a large civilian population might be located just behind the enemy line and no one assumed the Germans would protect them if the gas escaped to the surface.[16]

British miners commonly used candles to illuminate their galleries, although some electric light systems were installed. The Germans employed gas-powered electricity more often than their opponents. Both electricity and gasoline lights, borrowed from the civilian mining industry, were used for the first time in military mining during World War I.[17]

When driving a gallery forward, the work regimen was demanding. In the Canadian companies the tunnelers were organized into four shifts of six hours each day. Individual miners worked up to two shifts, or twelve hours, a day and slept nine hours, which allowed no time for anything else. The man in charge at the face held, as George Morley called it, "the post of honor." He typically wore only a pair of boots and a pair of pants, leaving his upper torso free. The face man broke clay, chalk, and earth from the front of the gallery, which other miners placed in sandbags. Then they either dragged the bags or moved them on carts with rubber-tired wheels. Part way back at intermediate stations, they often transferred the bags to larger trucks that ran along wooden or steel tracks. If the entrance was inclined, the track could go all the way to the surface and to the spoil dump.[18]

Getting spoil out of the mine was much more difficult through a shaft. The only way to do so was by a windlass, operated either by hand or by a small engine, which lifted a platform containing a few sandbags at a time. The Royal Engineer official history reported on work at a shaft eighty to one hundred feet (24.38 to 30.48 meters) deep, using a windlass drum of twelve inches (30.48 centimeters) in diameter with fourteen-inch (35.56-centimeter) handles. Two men could bring up 2 sandbags per haul in one minute, making for a steady rate of 1,200 sandbags per day. This rate of spoil removal supported an estimated driving rate of thirty feet (9.14 meters) per day in earth suited for gallery digging.[19]

What to do with the spoil once it reached the surface was a puzzle. Secrecy was the objective, so a number of expedients were tried. Erecting camouflage netting behind the trench line and piling up spoil underneath it could fool aerial observers. A more common trick was to place the spoil bags in the lines of earthworks along with sandbags filled with surface earth. Other tricks included dumping spoil in shell craters and unused trenches. Given the proximity of the opposing lines, the static nature of operations, and especially the use of airplanes for reconnaissance, disposing of spoil had become "one of the most difficult problems in military mining," as Brooks put it.[20]

The rate of digging varied widely from unit to unit, sector to sector. The most important factors were the type of soil, the water conditions underground, and how much shoring was needed. But the officers in charge also had an influence; much depended on how hard they pushed their men. The French accepted as a rule of thumb a digging rate of ten feet (3.04 meters) per day, while the British generally agreed they often exceeded that rate. Brooks reported progress of fifty-seven feet (17.37 meters) in one day under unusual conditions that included digging through solid chalk, which needed very little shoring, and dumping the spoil near the entrance. Not even the most aggressive of John Norton-Griffiths's moles could expect to dig that much for more than a day or two at a time. Brooks also noted that French miners managed thirty-feet (9.44 meters) in one day in similar conditions but with a pneumatic drill to cut away at the face of the gallery (another borrowing from civilian industry). As far as vertical digging was concerned, Canadian tunnelers once dug a shaft one hundred feet (30.48 meters) down in fifteen days, achieving a rate of more than six feet (1.82 meters) per day.[21]

Digging galleries at a daily rate of ten feet (3.04 meters) or more involved rigorous application of manpower for lengthy periods of time. The history of the New Zealand company laid out the weary routine that tunnelers accepted for months at a time. "Eight full hours hard slogging in the solid chalk and flints, till relieved at three, eleven or seven o'clock, then a good hour's plod back to billets along the trench, a rum ration, a hot meal and to sleep till time for the next shift; each day of the week or month exactly alike, the only variations being in whether it rained or snowed, whether the mud was liquid or merely sticky or in 'Jerry's' supply of ammunition for 'hate' purposes."[22]

To a limited extent, all three armies on the Western Front used mechanical devices to help them dig tunnels. Such machines, taken from the civilian mining industry, failed to prove effective enough to play a significant role in military digging. Everything had to be taken through a maze of communication trenches to reach the work site, and thus it was preferable that the machines be relatively small, not over eighty pounds (36.28 kilograms) in weight, and reliable under harsh conditions. They were powered either by electricity or air compression and made a lot of noise, a factor that could play havoc with secrecy. While the British and Germans were not enthusiastic about the use of mechanical diggers, the French pushed forward more widely with their use. Brooks thought this was because the French had more difficulty incorporating civilian miners into their military force due to the need to keep the supply of coal to the national economy moving forward. They also dismissed the noise issue by noting that digging by hand also was noisy, especially at close range, and using machines gave them more speed to compensate for that disadvantage. It was possible to place the necessary electrical generator or air compressor at great distances from the worksite. Miners had at their disposal, for example, air compressors that could push air for ventilation purposes through more than 3,000 feet (914.40 meters) of tubing. But it also was smart to place the generating unit as close to the worksite as possible to lessen the danger of having the tubing or electrical line cut by enemy artillery fire.[23]

Small diggers, meant to bore narrow holes forward from the gallery face or straight up to the surface through the roof of the tunnel, were more successfully used than larger pieces of equipment. The British Wombat was capable of drilling a hole six and a half inches (16.51 centimeters) in diameter up to two hundred feet (60.96 meters) long at the rate of three to four feet (0.91 to 1.21 meters) per hour in chalk. Charges consisting of ten pounds (4.53 kilograms) of high explosives in a tin tube could be inserted into such bore holes to blast out surrounding earth or blow out crude trenches on the surface. While the Wombat was the most commonly used small borer, the British also employed the Burnside borer to a lesser extent. One day a miner hit something solid when boring a hole with this machine that turned out to be the timber casing of a German gallery. The British pushed forward and got into the enemy tunnel to do mischief.[24]

The British experimented with large machines designed for use in coal mining and subway construction, but they proved inadequate on the Western Front. The Stanley Heading Machine operated on a rail track and ran on compressed air. It rotated two blades to cut a circular tunnel and weighed up to four tons (3.62 metric tons). While it took a massive logistical effort to get such a machine into a gallery, the Stanley worked fine at first. But gradually the surrounding clay swelled to such an extent that the unit became trapped and was left in place 118 feet (35.96 meters) below the surface, where its remnants still remain. The British War Office commissioned another header from Douglas Whitaker, mostly constructed in Glasgow, Scotland, but it was still undergoing trials in Great Britain when the war came to an end.[25]

Geology played nearly as large a role in negating the value of large digging machines in military mines as did the depth, narrow confines, and difficult access of galleries. In every way, as miners worked forward underground, they came to experience the importance of dealing with the character of the earth they invaded. It was a very personal experience because those conditions literally could spell life or death for the men. Water conditions often dominated that experience. In places the galleries "were generally damp and smelly," according to Canadian George H. Morley, and the need to find out where the water level lay, how to avoid it, and how to keep the galleries reasonably dry assumed large proportions. By April 1915, the British found a trained geologist in the infantry arm, Lt. William B. R. King, and assigned him to work full time as the army's chief geologist. King possessed a college degree and had worked for the Geological Survey of Great Britain before the war. The Australians relied on Maj. Tannatt W. Edgeworth David, a Welsh-born professor at the University of Sydney, as their chief geologist by May 1916. King mostly worked on borings to chart water levels along the Western Front, while David concentrated on mining and deep shelters, although both men spread their talents across many other functions, too. Systematic boring began in May 1916 among the British forces and reached a crescendo the next year, with as many as 100 bores drilled per week. Some estimate that more than 1,000 bore samples were extracted during the war by British geologists. The Germans began consulting geologists early in 1915 and employed twenty of them along the Western Front by February 1916, with one hundred at work by the end of the war. Even though the French did not "realize fully the application of geology to military mining," according to Brooks, they had many officers who had earned academic degrees in geology before the war. Brooks admitted that the French pulled off "some of the most brilliant mine attacks of the war" and that their "methods and equipment were used by all the belligerents." The Americans deployed eighteen geologists during their short time on the front, but none of them worked on mining projects. Of course, even among the British, French, and Germans, geologists were as busy if not more so in dealing with nonmining issues, but the Great War witnessed the first use of professionally trained geologists in mine warfare.[26]

Miners throughout military history had always known that their underground work was intimately tied to geology even if the word was unfamiliar to them. Burrowing through the upper layers of the earth, they entered a three-dimensional world of darkness, varied layers of soil, water, and gases. But while the shallow mining of the past was relatively safe to do, the deep mining of World War I called for the assistance of professional geologists. Yet one must also consider that the Great War was the first major conflict to occur after the creation of geology as a scientific discipline in Western academic circles. Both King and David held degrees in the field. "The first requisite for success in military mining is to secure the services of experienced geologists," concluded Brooks after studying British and French operations. "Mine warfare calls for the most detailed geologic information."[27]

Water posed the most serious problem that a geologist could address for the military miner. As Brooks has noted, it affected operations in three ways. First, surface water falling as rain penetrates the upper layers to a greater or lesser degree to affect shallow mining. Second, the water level settles at varied depths underground according to the geological and topographical circumstances of any given site; the geologist has to study each area to understand where miners could expect to hit trouble. Third, the nature of each underground strata is either resistant to holding water or holds it in great quantities. Some layers had to be avoided by tunnelers, while others were conducive to effective mining.[28]

The chief method of finding out about such factors lay in drilling boreholes from the surface down as far as mining operations were intended to go. Examining the samples gave a trained geologist a great deal of information on how to advise tunnelers about planning their systems. Existing civilian wells also could be examined, but that involved lowering the geologist down the hole. As Brooks has noted, surface water rarely caused much of a problem for most mining systems because of their depth. Heavy sand layers near the surface, however, as often existed in Flanders, posed a serious problem for sinking shafts; tubbing was the best way to address this issue.[29]

The British sector of the Western Front, which stretched from the North Sea across western Belgium and into northeastern France, displayed a variety of geologic zones to the military miner. A dune belt along the coast protected the inland polder country from flooding. Three-quarters to two miles (1.20 to 3.21 kilometers) wide, its sand dunes averaged forty feet (12.19 meters) tall above the high-water mark; one dune rose to one hundred feet (30.48 meters). This belt was unfriendly to mining, and if one wanted to dig a shaft, it required tubbing all the way. It was possible to begin a gallery by digging horizontally into a large sand dune, but that required instant shoring. The "perfectly dry silver sand" ran quickly "like water," recalled one miner. He reported an incident in which a small hole no larger than a fist developed in the wall of a gallery, and in less than a minute it had enlarged and disgorged so much sand that a man "was knocked down and buried over his head," requiring rescue. These problems allowed for very little military mining in the sand-dune belt.[30]

Moving south along the front, the entire western portion of Belgium is undergirded by large belts of clay, the same clay layer that Norton-Griffiths's moles dug subways through under London. It holds rainwater close to the flat surface of this region, while underneath it is a layer of fine sand laden with water. At times heavy artillery shells penetrated down to this sand layer and released the water to the surface. The clay layer is more substantial in the east, as much as 130 meters (426.50 feet) thick, but much thinner in the west, around 50 meters (164.04 feet). It proved an almost perfect medium for military mining, lending itself to easy and quiet digging. One of the few problems was that the moisture-laden clay tended to expand and burst the shoring of timbers, but one could compensate for this by creating small gaps in the shoring to release the pressure. Western Belgium is not completely flat, containing some ridges 50–100 meters (164.04–328.08 feet) high, especially around the northern, eastern, and southern sides of the city of Ypres. In this region the Ypres and Passchendaele Ridge formed the limit of German army occupation to create the famous Ypres Salient, bulging toward the east. Streams broke the ridge into sections, each of which was known by different names, such as Messines, Pilckem, Westhoek, and Broodseinde; all of these became famous in the history of the Great War. The German occupation of these ridge sections offered them great advantages on the surface in terms of defense and observation, but it was a disadvantage underground. The sand layer was much thicker here, extending up to eighty-one feet (24.68 meters), while under the British line to the west it was much thinner. British tunnelers could get through the sand on their end by tubbing and enter the inviting clay layer much more easily than their opponents.[31]

South of the Flanders clay region, from the area around Vimy Ridge and into French Artois and Picardy, is the chalk region. The chalk consists of "a soft, pure-white limestone overlain by a complex of soils created during the last ice age." Those surface soils—termed "limon"—proved easy to dig through, but the chalk beneath was "strong and dry." It was very noisy digging, too, but needed very little shoring except on the roof of galleries, where slabs of chalk tended to break off over time. It was difficult to hide the spoil because the whiteness was very visible. Even in the galleries the whiteness reflected lights and led to conditions something like snow blindness. Very porous, water percolates down through the chalk layer. But this was the least problem of mining in this area. The chalk tended to become badly fractured by mine explosions for a long distance from a chamber. At the center of the blow, according to Captain Reynolds, it became the consistency of cheese. Rather than debris settling back into the crater after a blow in chalk, it tended to spread out into a larger pattern across the ground.[32]

Farther south, the French and Germans engaged in digging galleries through quartzite at the northern end of the Vosges Mountains near Saint-Dié. It is "highly indurated sandstone, and conglomerates," as described by Brooks. "Both sides used air drills supplied from power stations located near the front line." The topography helped shield those operations to a degree. Sharp contours, deep valleys, and heavy tree cover served to hide the surface evidence of mining from visual observation, but the sound of the powered drills gave them away. "In general," concluded Brooks, "the plan [to mine in this region] was ill conceived and led to no decisive results."[33]

While military geologists engaged in work beyond mining during the Great War, especially in finding potable water sources for the troops, their greatest contribution lay in "determining the physical conditions that affect underground warfare," as Books put it. Their value can be illustrated by the water levels found along the line. In the chalk country that level could vary as much as thirty feet (9.14 meters) between summer and winter, rising in the latter season. David drew up tables of the water-level variations in this area by September 1916 that gave mining officers a guide as to how low they could dig their galleries, scrapping some ongoing projects and redigging them ten to fifteen feet (3.04 to 4.57 meters) higher. When the water level rose by January 1917, it just barely reached the floor of those shallower galleries whereas it would have completely flooded them if at the previous level. In late 1916 near Loos, the water level rose eighteen feet (5.48 meters) in only fifteen days, demonstrating how dangerous and frustrating this geological influence could be on mining operations.[34]

In places where it was not possible to adjust the depth of the gallery to avoid water, the only solution was to set up a system for extracting it from the tunnels. Pumps powered by gasoline motors appeared all along the line, but they came with many complications, not the least of which was the noise, the need to conceal the power stations from enemy view, the logistical problem of supplying them with fuel, and many other smaller issues. Experience showed, for example, that the threads of the jute used to make sandbags tended to float in water and would choke up a pump. Sump pumps located at the bottom of shafts in Flanders also tended to choke up because bits of material inevitably dropped down the shaft when hauling out spoil, leading crews to stop now and then to clean the pumps. Some machines could handle sludge of this kind, but most could not. Because of this, the British experimented with different types of commercially available pumps until they found one they liked. By the end of 1916, they had gathered enough usable pumps to support their mining operations reasonably well. Depending on the circumstances of each gallery, they might need to operate the unit for only one or two hours a day to keep the water down. But at other locations, massive pumping operations were required. At one place Australians set up ten generating sets to power an array of pumps that extracted 200,000–400,000 gallons (757,082.36–1,514,164.71 liters) of water per day.[35]

Water-level problems and how to solve them remained a major component of mining along most of the Western Front, which lay on relatively level ground. In the mountainous sectors near the southern end of the line, however, miners could use the varied elevations to their advantage. French engineers cleverly devised self-draining galleries near Saint-Dié-des-Vosges through tough sandstone and conglomerate.[36]

The Western Front witnessed a much greater utilization of advanced, commercially available equipment for digging mines and countermines than ever before seen in mining history. But when it came to charging those mines and countermines with explosives and setting them off, Great War miners went even further in their effort to use the latest components of commercial mining. The generation of explosives succeeding gunpowder had been developing for decades before 1914, but except for a handful of cases such as Port Arthur, those new high explosives had not been tested in the field. Electrical detonation had also been used at only a handful of locations in military mining before the Great War. The Western Front therefore became the greatest testing ground for advances in mine warfare in history.

Canadian mining officer L. B. Reynolds has well described the effect of any charge exploded underground. At the instant of explosion, the force creates a chamber of compression followed by the release of gas, which ruptures the surrounding ground and breaks through to the surface to form a crater. "In the detonation of high explosives all the damage underground is done during that first period of the formation of the chamber of compression. The shock of this first explosion rocks the galleries and ruptures the formation. The escape of the gases, expanding by heat in the chamber after its formation, gives the rending effect and the slow mighty upheaval that does the damage to the surface and within the area of the inverted cone from the charge to the surface."[37]

The type of explosive used, of course, had a decisive effect on this process, with high explosives spurring it more effectively than gunpowder. But it is interesting that in the very early days of British mining on the Western Front, gunpowder was still used. The British soon expanded their arsenal, however, and eventually tried thirty-six different kinds of low and high explosives. The latter's faster combustion rate meant that more of the entire amount burned at the same time than to be found among low explosives. The amount planted in each charge tended to be very high in comparison with historic examples of military mining. In chalk the British always used at least 2,000 pounds (907.18 kilograms), with an average of 5,000 pounds (2,267.96 kilograms) and up to 15,000 pounds (6,803.88 kilograms) in many cases.[38]

While trying many different types of explosives, British miners came to rely heavily on ammonal, which had been patented by a Viennese chemist in 1900. They varied the composition for military use and developed what Norton-Griffiths called "our excellent old pal." It filled a niche between the most volatile of the high explosives such as dynamite and the slower-burning gunpowder. Harry Trounce described ammonal as a "lustrous dark grey powder" consisting of TNT (15 percent), ammonium nitrate (65 percent), charcoal (3 percent), fine aluminum (1 percent), and coarse aluminum (16 percent). The British set it off electrically with a standard army plunger tipped with one-ounce dry guncotton primers. Ammonal was very sensitive to moisture and therefore was transported in "hermetically sealed tins" that weighed ten or fifty pounds (4.53 or 22.67 kilograms) each. Miners often transferred it at the front into forty-pound (18.14-kilogram) waterproof bags for easier stacking in the chamber. Three times more powerful than gunpowder, ammonal was far safer to move and use because shock did not disturb it. One could fire a rifle into a pile of ammonal or stick a flame into it and the mixture remained "stable and safe." It combusted at a slightly slower rate than guncotton and thus was ideal for lifting up huge masses of earth. In contrast, TNT was considered to have a combustion rate too fast for mining operations. Initially developed in Germany in 1863, its explosive potential was not realized until 1891, and the Germans began applying it to artillery shells. By 1916, the British were using ten times more ammonal than any other explosive in their mines.[39]

But miners soon learned the pitfalls of handling ammonal. Initially, they transferred it from the shipping tins to waterproof bags at the mine site on the assumption that they could more tightly pack them in the chamber and obtain a more effective explosion. Transferring the powder, however, was problematic. As a New Zealand miner explained, "the ammonal dust stain[ed] everything it came in contact with, including arms and faces, a violent and lasting yellow." They soon realized that packing the tins close together would give them an effective blast. According to the Royal Engineers official history, the shipping tins were not "absolutely watertight," so many miners transferred ammonal from them into empty petrol tins for long-term storage at the front. Each canister held forty pounds (18.14 kilograms) of the powder.[40]

Trials of other explosives produced results less satisfying than ammonal. Guncotton came in wet or dry condition. It could absorb 30 percent of its weight in water and when wet was safe to store and transport, even if it absorbed only 2 percent of its weight. It arrived in slabs of fifteen ounces (0.42 kilograms) each and had to be detonated by using a primer made of dry guncotton. At 1.2 times more powerful than TNT, it soon proved unsuitable for significant mine charges. When tried, guncotton did not produce good craters and left behind a great deal of carbon monoxide to endanger friendly forces working in the area of the blast. It came to be used mostly for demolition work and in small charges such as camouflets. Guncotton was the second-most commonly used high explosive in the British mining arsenal.[41]

The British gave several other high explosives a try. Blastine was a "light brown powder" consisting of a mixture of TNT, ammonium percholate, sodium nitrate, and paraffin wax. To a degree it was used in the chalk country but did not prove superior to ammonal. Neither did any of the other high explosives. Sabulite, also known as Belgian permite, produced only 85 percent of the lifting power of ammonal, while Amatol managed only 65 percent. The latter used less aluminum and thus seemed a promising alternative to ammonal because of the growing scarcity of the metal toward the end of the war. But its blasting effect in the field did not justify widespread use. Alumatol also was similar to ammonal but not as effective. Other explosives tried and found wanting included dynamite, blasting gelatin, and British Gelignite.[42]

Each of the other two main armies on the Western Front experimented with their own set of high explosives. The French used Melinite, which arrived at the mining site in a powder and proved to be two and a half times more powerful than gunpowder. They also used Cheddite, which had been developed by a French firm in 1897 and manufactured at the town of Chedde, Haute-Savoie, mainly for use in quarries. The Germans manufactured their own high explosives, using Gluckauf, Westfalite, and Donarit against their British and French adversaries.[43]

As historian Simon Jones has pointed out, the array of new high explosives was one of the most impressive aspects of military mining in World War I. They "had a major impact on the effectiveness of the miners and contributed to elevating military mining in 1916–17 to a scale and devastating power not reached before or since." Harry Trounce of the 181st Tunnelling Company exaggerated only a little when he described those formidable new explosives: "In this war, enormous charges of the highest and deadliest explosives known to man are used. Instant annihilation follows the slightest mistake or carelessness in handling such frightful compounds."[44]

While miners in all three armies generally used electricity to set off charges, the British, at least, also supplemented it with older methods. Improvements on older fuzes brought those devices up to date. Trounce has described a basic fuze of the era that was safely used in commercial mining and quite applicable to military purposes: "A core of black powder wrapped with hemp, cotton threads, or tape, with various water-proofing compounds between each, or on the outside, to provide a uniform burning speed of the powder core. Safety fuse usually burns at the rate of about 2 feet [0.60 meters] per minute, and is made in various grades, depending on the amount of water it is required to withstand."[45]

The Bickford safety fuze arrived on-site in rolls of 500–6,000 feet (152.40–1,828.80 meters). It became brittle in cold weather and needed to be warmed before unrolling it. The Dupont Company manufactured slightly different Bickford fuzes, labeling them damp, wet, and very wet. An instantaneous fuze burned underwater at the rate of 120 feet (36.57 meters) per second, and miners could attach blasting caps on the end. It had a tendency to jump when firing and so had to be weighted or fastened down. Cordeau fuzes contained "pure crystallized T. N. T. or other" high explosives. They had the advantage of touching off widely separated masses of high explosives due to their "sympathetic detonation" effect.[46]

Fuzes needed some sort of blasting cap, or detonator, on the end to communicate action to the mass of explosives. Usually made of copper wiring, different types were used for electrical systems of detonation. The British Army used the No. 8 detonator for safety fuzes and No. 13 for electrical detonation; both contained fulminate of mercury. No. 13 consisted of a wire inside a pencil-size copper tube filled with fulminate of mercury. The electrical current raised the wire to "a white heat," which ignited the fulminate, which in turn set off the charge.[47]

For fuzes, one unrolled the coil, laid it along the gallery, and lit the end. For electrical devices, a system of long, insulated copper wires called leads had to be laid out. The standard British Army issue actually had six copper and one steel strand inside a vulcanized india-rubber lead with compound coating. It arrived on-site wrapped around wooden drums. To cover the connection between the end of the wire and the blasting machine, operatives applied tape, liquid-rubber solution, or small rubber bags. The leads had to be laid carefully along the gallery, coiled a little to avoid the possibility of tugging it so tightly it affected the detonator, or were suspended from the roof of the gallery. Waterproof test boxes and connection boxes allowed for daily testing of the strength of the circuit if the system was meant to stay in place for some time before use. Daily testing consisted of sending a small current through the system large enough to deflect the needle of the galvanometer but not strong enough to set off the explosives.[48]

Series wiring was found to be more effective than parallel wiring. It was difficult to balance circuits in parallel, and leakage (or earthing) was more prone to happen in parallel wiring. Series wiring could overcome great resistance, and miners came to prefer three circuits in series "for the sake of safety." All electrical current must overcome some degree of resistance to its flow. That level of resistance, measured in ohms (named after Georg Simon Ohm), indicates how easily a current flows. Rubber resists electrical flow more than some types of metals, and a long, thin wire causes higher resistance than a short, thick one. In large and deep mining galleries, the electric current had to overcome a resistance of up to 100 ohms.[49]

Blasting machines, also known as exploders, were improvements on the basic design developed by Henry Julius Smith decades before the war, which was itself an improvement on the electromagnetic device of the early and mid-nineteenth century. They converted a mechanical movement into a surge of electrical power. Small ones were pocket size and generated the charge by the operator twisting the handle. Larger ones operated with a plunger. The British Army Mark V exploder was housed in a wooden box thirteen by eight by six inches (33.02 by 20.32 by 15.24 centimeters) in dimension. The operator pulled the handle up, connected the leads, and pushed the handle swiftly and smoothly down. He could lock the box so no one else could use it. The largest exploders were capable of firing 150 detonators at one push. A rheostat measured whether the machine was at full strength.[50]

Tunnelers went to great lengths to ensure that the explosion affected as much of the explosive material as possible and that the blast occurred

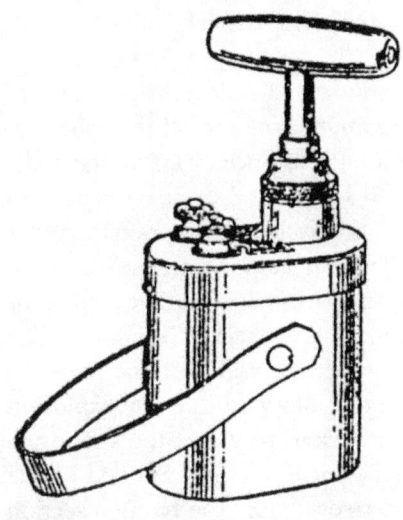

FIGURE 95.—Ten-cap exploder or blasting machine

Hand-Held Exploder, Western Front. The development of exploders had reached a stage by the time of the Great War that they could be so small as to fit into a satchel or pocket and activated by twisting the handle. This one was capable of exploding up to ten detonators with one twist. *Engineer Field Manual*, vol. 2, *Military Engineering*, pt. 2, *Defensive Measures*, 185.

sharply with no delays. Redundancy was the key. According to Trounce, British tunnelers set up two complete sets of firing circuits for each blast in case one failed. They also placed, as a rule of thumb, two detonators for charges up to 2,000 pounds (907.18 kilograms), three for up to 5,000 pounds (2,267.96 kilograms), and four or five for 10,000 pounds (4,535.92 kilograms) or more. But they did not employ more than five detonators for any one set of circuits. According to Coulthard, the Canadians placed a detonator in every tenth tin of high explosive, more if it was a particularly large mine, and spread unattached detonators loosely among the mass. Canadian engineer Captain Reynolds, however, argued that fewer detonators worked more effectively on large amounts of explosives than many strewn about the mass. Miners often mixed safety fuzes and instantaneous fuzes with electrical detonation systems for added peace of mind. Trounce reported that two blasting machines were set up for each electrical system and were pushed "at the same moment."[51]

This level of redundancy was a testament to the stress of underground warfare along the Western Front. After digging furiously for several weeks, the miners often gained only a tenuous and uncertain lead over their enemy, creating a tense atmosphere. After all that effort, no one wanted to lose the slim advantage by delays emanating from faulty wiring, too few detonators, or a blasting machine that did not work at the culmination of offensive miners' efforts.

When it came to tamping the gallery, little had changed over the centuries. Tunnelers primarily used sandbags and planned to fill the gallery at least equal to the distance of the line of least resistance or up to one and a half times that length. They always placed sections of sandbags separated by air space to better cushion the shock of the explosion. Captain Reynolds spelled out a typical scheme that involved fifteen feet (4.57 meters) of tamping, followed by ten feet (3.04 meters) of air space, with another ten feet of tamping and ten feet of air space, ending with ten feet of sandbags. The tampers filled up any holes in the sandbag sections with dirt and removed sets of shoring in the untamped section of gallery so that the explosion would not push the connected sets so far back as to wreck the shoring in the undamaged section of the gallery. Of course, the forward end of the gallery, near the powder chamber, was destroyed, and the tamped section itself was useless until cleared out. Alfred Brooks reported that miners often made a dogleg in the gallery near the chamber to help protect as much of their own tunnel as possible. This was an old concept that dated to the very beginning of the gunpowder era. He noted that this jog should be located at least twenty-five feet (7.62 meters) from the chamber and be offset from the main orientation of the gallery six to ten feet (1.82 to 3.04 meters). There is no doubt that at least some mines contained an offset like this, but how often it was employed is unclear.[52]

With a mixture of old and new techniques and employing the latest in high explosives, miners on the Western Front pushed their work forward with remarkable energy. They also improved on some older concepts that stretched their capabilities on the tactical level. For centuries, military miners had practiced using tunnels to blow out instant trenches on the surface, and at times Great War tunnelers did the same thing. Reynolds described the process as digging a gallery seven feet (2.13 meters) below the surface and drilling small bores forward through the face of the tunnel. Then the miner inserted explosives contained in a metal tube called a torpedo. These were five and a half inches (13.97 centimeters) in diameter and filled with ten pounds (4.53 kilograms) of high explosive for each running foot. The force of the explosion could blow out a trench to the surface. Rough and crude, it nevertheless accomplished most of the earth moving in a few seconds and could be improved by the infantrymen who occupied it. Brooks believed that this method could be effective under as much as fourteen feet (4.26 meters) of earth. It could also open a passage for attacking infantrymen through barbed wire or through a German parapet, but it is unclear how commonly this concept was employed.[53]

Another similar concept, commonly called the Russian Sap, was used sparingly. It involved digging a short gallery very near the surface so the miners could dig up and gain access to the surface at the end of the gallery. It was a modern version of the age-old concept of gallery mining to enter the enclosure of an enemy fortification rather than to bring down a section of the defending wall. Once broken to the surface, the gallery could be used as a communication tunnel. The French used Russian Saps in a major operation at Carency–Neuville-Saint-Vaast in August and September 1915. They dug surface saps as far as possible and then began Russian Saps from fifty to eighty meters (164.04 to 262.46 feet) from the German line. The tunnels were only fifty centimeters (19.68 inches) deep, and the plan was to emerge twenty-five meters (82.02 feet) from the target. The ploy failed to play a significant role in the French attack of September 25, as only a few of the Russian Saps were opened, and German fire stalled all progress in the assault. The British also used Russian Saps at Loos in an attack that same day—September 25, 1915. They opened the heads of their tunnels the night previously and then dug a new trench line connecting them. The infantry assault launched from that new trench the next day was only partially successful, but at least the troops had been able to reach a point closer to the target through the tunnels rather than charging across the entire surface of no-man's-land.[54]

Russian Saps failed to prove their worth as the launching point for an infantry attack. They had constricted room for rushing combat troops forward and usually doubled as two-way communication tunnels, with wounded men, couriers, and equipment being funneled through them at the same time. But they were very useful in the support of surface action near the point where the miners had opened up the tunnel. During a battle, the tunnel opening could serve as a machine-gun post, a first-aid station, or an observation post. The British effectively used Russian Saps to support a large trench raid by two hundred infantrymen just south of the Béthune–La Bassés Road on June 10, 1917. Three tunnels were dug forward, and a small mine was set off nearby to draw German attention just before the troops left the end of the tunnels. They easily broke into the first German line, which allowed British miners to closely examine and map the enemy galleries within reach and place charges to damage them. Russian Saps also served as good communications tunnels. Deep galleries would have been better protected from surface shelling but far more difficult for troops to travel through and for the movement of equipment and material.[55]

Camouflets, undercharged mines designed to disrupt or destroy enemy galleries, were exploded by the hundreds along the Western Front. According to Canadian tunneler Angus W. Davis, they started small early in the war, with fifty to one hundred pounds (22.67 to 45.35 kilograms) of ammonal "usually placed in bore holes." As the war lengthened, miners tended to plant larger charges, up to "several thousand pounds of ammonal," in camouflets. Many as a result broke through the surface to create small and large craters. Smaller camouflet charges were prearranged and stored on-site for quick use in case of emergency. The explosive was poured into cylinders made of galvanized sheet iron four to eight inches (10.16 to 20.32 centimeters) in diameter and a fuze and leads attached. The best way to collapse an enemy gallery was to avoid the face, top, or bottom and blow in the side, which was much more vulnerable than the other areas.[56]

At times miners tried not to create craters on the surface whenever possible. Craters altered the topography of the battlefield and served as obstacles to the movement of infantrymen and vehicles. But sometimes it was advantageous to create holes in the ground—they could serve as machine-gun posts, be used to start a short mine gallery, or provide shelter for soldiers caught in no-man's-land. They could also be deliberately created as part of an obstruction field for enemy tanks. A number of experiments took place along the front after the introduction of tanks halfway through the conflict. Tunnelers near St. Eloi blew craters of varied sizes and depths, staggering them in relation to each other, but found it to be unsuccessful in stopping enemy tank movement. Australians near Saint-Pol tried a different pattern of fourteen craters, which bogged down one large German tank. Modern-day archaeologists have discovered a group of forty-three mine craters near Messines, each one eight to twenty meters (26.24 to 65.61 feet) in diameter, most located on roads and thus meant to serve as obstacles to vehicle movement.[57]

When miners confined themselves to their traditional role of digging offensive galleries, as Great War tunnelers mostly did, they encountered a far more sophisticated array of listening practices by their opponents. In addition to advancements in explosives and the means of firing charges, the other major innovation to be seen on the Western Front was the employment of new devices for listening to enemy digging. Gone were old expedients such as placing pebbles on a taut drumhead. Civilian advances in science, technology, and commercial application infused countermining with new materials and methods to detect signs of underground warfare.

The transformation of listening underground did not take place quickly or smoothly. Initially, miners experimented with many devices and methods that failed to work. Norton-Griffiths borrowed stethoscopes from the London Water Board that had been used to listen for water leaks. Others used water bottles issued to French soldiers and attached medical stethoscope tubes to them. LCpl. Robert Leonard used a stick, pushing one end into the ground inside a listening shaft and holding the other end with his teeth. The explosions of German shells interfered with his work; he could not differentiate them from other vibrations.[58]

British cinematographer Geoffrey H. Malins, who filmed the preparation for the great offensive on the Somme in the summer of 1916, experienced listening for the enemy firsthand. Invited into a listening gallery by a mining officer, he followed him through an incline into the gallery. Even with the naked ear, Malins could hear "a queer, muffled tap-tap-tap coming through the earth on the left." The officer then gave him a device "very similar to a doctor's stethoscope. I put it to my ear and rested the other end upon a ledge of mud. The effect was like someone speaking through a telephone. I could distinctly hear the impact of the pickaxe wielded by the Bosche upon the clay and chalk, and the falling of the débris." As Malins discovered, even a simple device like a stethoscope could be very effective in detecting enemy digging. But one must keep in mind that even without it, he could hear German activity.[59]

By the spring of 1916, a number of new and more sophisticated listening devices began to appear in all three armies along the Western Front. The French developed the geophone, while the Americans provided the Allies a device made by the Western Electric Company, which the British considered too complicated for use in the field. The German company Siemans Halske provided its own version of a listening device. By far the geophone became the favorite on the Allied side of the front, although Brooks contended that the British used this French device more than the army that introduced it. The American version of the geophone, developed by the U.S. Bureau of Mines, was more sensitive than the original but was introduced too late in the Great War to replace it.[60]

The French geophone consisted of two mica disks held between vulcanized rings. Mercury and a thin space of air lay between the discs. When placed on the floor of a listening gallery, it picked up a slight vibration moving through the earth. "Due to the inertia of its large mass and the lasticity of the mica diaphragms," according to Harry Trounce, "the mercury does not follow the movements of the case, so there is a relative displacement between the two, and alternate condensations and rarefactions are set up in the small air spaces between the diaphragm and the case. These pressure changes are transmitted through rubber tubes to the ears, which recognize them as a sound."[61]

Trounce also expressed this process in a less technical way. The French geophone "transforms the earth wave into an air wave, which is heard by the ear as a sound, and at the same time amplifies it so that the sound is louder than if the ears were placed directly in contact with the earth." By

French Geophone in Action, Western Front. This photograph gives credit to the most successful listening device developed in World War I. In this posed image, a French engineer officer demonstrates the way in which the geophone was used. Imperial War Museum, Q069984.

placing a receiver on both ears, one could roughly determine the direction from which the sound was coming by measuring the time it took to receive it. For example, with both receivers on, if the sound seemed to come from the right, the listener could slowly turn right until he heard it in both ears at exactly the same time. The geophone allowed users to hear sounds of digging a least as far as one hundred feet (30.48 meters), and they could distinguish between different types of digging tools at half that distance.[62]

The French geophone had to be used by a single operator at the head of a listening gallery, which demanded shifts of listeners for twenty-four-hour coverage in each of hundreds of listening posts. French engineers soon developed an alternate system of remote-listening devices to cut down on the demand for manpower. One was a seismo-microphone that could detect sounds but not direction. It could be placed anywhere without constant supervision and wired to a central listening station that controlled microphones in up to fifty galleries. The wires were connected to a switchboard that was manned twenty-four hours a day. A listener with a geophone could be dispatched to any of the microphone locations if anything suspicious showed up. According to the British assistant inspector of mines, this system could detect sounds of picking in chalk up to only 175 feet (53.34 meters) compared to 250 feet (76.20 meters) with the geophone. Nevertheless, the advantage lay with the seismo-microphone because of its reduced need for manpower.[63]

While the French had developed the seismo-microphone system, the British used it more extensively. They saw its value in conjunction with geophone use and also sent men to patrol the system, visiting each listening station three times per day. They also tested the system by having a man make clicking noises at the microphone to see if the listening post could pick it up. But Canadian tunneler Robert R. Murray considered this system not very reliable "due to the distortion of sounds and the noise of the instrument itself." He contended that it was used only in charged mines that had to be held in abeyance for a while before exploding because it was too dangerous to place a valuable listener near the explosives. Perhaps that was the Canadian perspective, but all other sources indicate the system worked well and was widely used. Its defects could be mitigated by regular patrols and the dispatch of a geophone operator to investigate any suspicious sounds. The system was a splendid way to keep watch over a large number of listening galleries.[64]

Murray also provided details on a German listening device without naming the manufacturer. It consisted of a "telephone receiver and a very sensitive spring vibrator, so arranged in a steel case that the vibrator would tap the disk of the receiver." The Germans attached the steel case to a long rod that they drove into the earth. It could detect digging sounds within thirty feet (9.14 meters). While ingenious in its own way, this device paled in comparison to the sophistication and range of the French geophone and even the seismo-microphone.[65]

Geology played a role in listening. Solid formations such as clay transmitted sound better than loose formations such as gravel. Sound resonated in solid rock and chalk but less so in sand. British geologists estimated that sound carried two and a half times less well in a sandy clay than in a chalk formation.[66]

Whether using the geophone, the seismo-microphone, or the German devices, the quality of listening depended ultimately on the skill of the listener. Advanced technology could not entirely replace the human dimension. Capt. Herbert W. Graham of the 185th Tunnelling Company found that few men were qualified for the task. "A good listener must have equal hearing in both ears, a lack of imagination and the right temperament," he explained. After serving as his company's listening expert for six months, Graham found that most members of the rank and file tended to lack one of the three attributes. When tried, most of them lasted no more than six to eight weeks because "the work got on one's nerves." The British established listening schools to train listeners and weed out those unsuited to the task.[67]

With such a high level of listening capability, at least in terms of technology, miners digging offensive galleries had to be extra careful of making noise. They often padded their feet by wrapping empty sandbags around their boots or even went barefooted while underground. When very close to the Germans, Canadian miners gave up clay-kicking and instead loosened chunks of earth from the face of the gallery with a bayonet, catching the spoil before it hit the floor. Such precautions were especially useful when digging through chalk, which amplified the carrying distance of sound tremendously. But carefully picking one chunk of chalk at a time from the gallery face slowed progress to no more than a couple of feet every twenty-four hours.[68]

An alternate strategy was to make noise but hope to fool the enemy at the same time. It was not uncommon to rig a dummy picking apparatus that could be controlled by ropes. Trounce mentioned an apparatus like this that had been set up by his comrades in the 181st Tunnelling Company in Flanders when the Germans were estimated to be ten to fifteen feet (3.04 to 4.57 meters) away.[69]

At times opposing galleries came so close together that one side could take advantage of the situation to spy on their opponent. This opportunity came to a British tunneling company in the spring of 1916. Its members established a listening post and recorded conversations held in a nearby German gallery from May 31 to June 3. "That will soon be deep enough will it not?" "Yes—I do not know." "I will ask the Sergeant-Major"—so went the conversation on June 1. The next day it was of a more personal nature: "Karl, have you heard from your wife lately?" "No, she has not written for two weeks." "We shall not get any more leave now." "Why?" "I have heard it said."[70]

This example demonstrates how intimately opposing sides could come together without one side knowing about it. But the attempt to gain valuable information by eavesdropping on conversations obviously bore little fruit, and the spying in this instance continued only for four days, revealing a good human-interest story but little else.

Even more impressive than the new ways of listening for enemy mining was the creation of elaborate, preplanned methods of rescuing miners caught in collapsed galleries or suffering from the effect of gas. In the past this had always been done on an ad hoc basis, but like nearly everything else associated with military mining, it was standardized and prepared beforehand during the Great War. The threat of gas was greater than that of cave-ins and began to exact a toll early on, prompting all three belligerents to create rescue systems and study the proper way to treat gas poisoning in miners.

The primary problem of gas lay in the production of carbon monoxide during the explosion process. The more complete the combustion, the lower the rate of carbon monoxide given off. Ammonal produced less of that harmful byproduct, while guncotton gave off a high amount of it. One could not see, smell, or taste carbon monoxide, and the underground environment concentrated rather than dispersed it. The danger is that carbon monoxide displaces oxygen in the blood and thus starves body tissues of vital oxygen. The gas dissipated only slowly underground, meanwhile infiltrating every crack in the zone of rupture around the charge. It infiltrated less so in clay than in chalk, the latter having many cracks and cavities. An underground area could be poisoned with carbon monoxide for months, forcing miners who dug through it to wear protective gear. At times they hit large pockets of carbon monoxide, which made a swooshing sound as the gas escaped. In some cases it seeped all the way back to the shaft and up it, affecting nearby infantrymen on the surface. High explosives produced other gases, too, such as hydrogen and methane. Some could detonate inside the gallery and cause much harm to miners and damage to excavations.[71]

More miners were killed and injured by gas poisoning than by enemy action. One tunneling company suffered sixteen killed, forty-eight hospitalized, and eighty-six "minor cases" of gas poisoning (the latter treated in the field) during a six-week period. In one month another company lost twelve killed, twenty-eight hospitalized, and sixty minor cases to gas. In June 1916 a German camouflet explosion created only moderate damage to a British gallery but released a great deal of carbon monoxide, hydrogen, and methane, probably because it had been planted a long while before and had deteriorated. Those conditions led to an incomplete combustion with a consequent high level of gas. The effects were devastating. Four British officers and several enlisted men were in the gallery and tried to leave, but the sudden onrush of gas from the charge raced through the gallery and into a branch and transversal. Candles held by the men ignited the gas, causing death, severe burns, and shock. This explosion in the transversal created "a partial vacuum" that led to "another influx of gas into the galleries from the camouflet," killing two more officers and sixteen men, mostly through carbon monoxide poisoning.[72]

Danger of this magnitude demanded action, and all three armies sought help from the civilian mining industry. The British could choose from among several different types of respiratory devices, the two most popular ones being Proto and Salvus. The latter worked well for half an hour underground but was "heavy and complicated," much like another device called the Pulmotor. As a result Proto tended to be viewed as the "simplest and most efficient" respiratory aid. It was lighter than Salvus and worked for two hours at a time. Moreover, it allowed for more freedom of movement about the head and arms. The Novita device also was used to a limited degree. On the German side the Draeger device was preferred. All respiratory apparatuses, no matter who made them, worked on the principle of injecting oxygen from a cylinder into the gas victim to get rid of carbon monoxide. They were stored ready for use at rescue stations established within two hundred meters (656.16 feet) of every shaft or incline. The British also established mine rescue schools to train personnel in how to use the devices and the best method of dealing generally with gas infiltration and cave-ins. They probably attacked the problem of rescue earlier and more thoroughly than did the French or Germans.[73]

Rescuers took along extra Proto devices so as to give oxygen to injured men before moving them to the surface. The device also was used in other ways; for example, miners wore them to do the initial clearance work after an enemy camouflet because of the continuing presence of gas. In one case a German blow so saturated the zone of rupture that it was impossible to pump out the gas for two weeks, but the miners were desperate to stop any further German progress in that area. Seventeen men with Proto devices thus fixed a British mine in that gas-impregnated area, with four men standing ready with stretchers and Novita devices to rescue them if needed. They put in 5,000 pounds (2,267.96 kilograms) of ammonal in sandbags and erected sixty-five feet (19.81 meters) of tamping in forty-three hours, the whole time wearing Proto devices. This effort beat the Germans to the punch.[74]

Rescue men also had to deal with fires that broke out underground, although that was a rare occurrence in the galleries. Fires were more common in the underground shelters that infantrymen occupied. For such jobs, rescue personnel kept on hand asbestos hoods and coverings.[75]

The organization of rescue work involved peculiar problems. For example, the right kind of man was essential. Based on the effects of carbon monoxide poisoning on the human body, studies found that anyone inclined to obesity had to be excluded from mine rescue work because they were more susceptible to serious effects than skinny men. Those who did not readily faint were preferred, in the age range of twenty-six to thirty-eight, as well as those who had no tendency to catch colds or other ailments. Men with low blood pressure also were rejected, as were those with an oxygen deficiency. The army recruited physicians from civilian life who had experience in treating miners and put them through a course in gas poisoning and rescue work at the Army Mine Rescue Schools before assignment in the field.[76]

Some specialized equipment was developed for rescue efforts. Affected men could be carried out of a gallery if it had an inclined entrance, but shafts presented enormous difficulties in terms of lifting unconscious men. A Lieutenant Penman of the Royal Engineers devised a simple stretcher for this purpose. It consisted of flexible netting with three wooden skids on the bottom to allow an unconscious man to be dragged along the floor of the gallery. At the shaft he could be tied up in the netting and then lifted up to the surface.[77]

The British started training personnel in mine rescue during the summer of 1915 in their Army Mine Rescue Schools system. The training syllabus was rigorous. It included making every student wear respiratory apparatus and work for a time in practice galleries to accustom them to the gear. They had to recover a dummy from under debris and carry it to the surface in the stretcher. Students learned artificial respiration using the Schäfer Method: with the distressed man lying face down, the rescuer applied pressure rhythmically by hands to his lower thorax. The army disseminated information on rescue through the circulation of pamphlets such as *Memorandum on Gas Poisoning in Mines*, which was directed at all tunnelers and medical personnel. Discussing the details of carbon monoxide poisoning, it went through two revisions before the end of the war.[78]

This training richly paid off in a viable rescue procedure for mines. In one case study six British miners were working in a gallery when the Germans blew a camouflet that killed two of them outright. Rescue men left their station within 250 yards (228.60 meters) of the shaft, dodging shells while moving through the trench toward the mine. Within "a few minutes," they reached the four survivors; all were unconscious, and one was wounded by flying timber. They moved all four to the surface, administering artificial respiration and oxygen to two of them while the other two returned to consciousness upon breathing fresh air. The rescue men's work was not yet done, however, for they had to play a significant role in cleaning up the mess left by the camouflet. After waiting fifteen minutes, they returned to the gallery with Proto apparatus to look at the damage, taking along a warning mouse, which died a short distance in. The team cleared up the gallery, set up a ventilation system, and a few hours later miners resumed working in the area. It was not entirely safe; the next day a surge of gas went through the gallery, probably because of "settling of the ground" in the zone of rupture. Three miners lost consciousness because of this and had to be rescued, but all recovered.[79]

Great War miners learned the hard way how gas affected their health. In the early days it was assumed that walking off gas poisoning was best, but it wound up making the damage much worse. The best course was to apply oxygen, warm the victim, and allow him to rest. Exposure to cold air, however, often led a gassed man to regain consciousness. Gas poisoning could creep up on an unsuspecting miner if it lay in affected underground areas in small amounts, or it could hit one suddenly if in large amounts. As L. B. Reynolds explained, the former could be more dangerous than the latter. Creeping gas was more insidious because it did not alarm the victim, and its slower accumulation in the body saturated the tissues, causing more long-term health problems than a sudden blast. Speaking in 1919, Reynolds was sure that most miners who worked in the chalk regions, where gas saturation was especially bad, still suffered from the effects of slow poisoning.[80]

In general, the effects of gas poisoning were much less severe when explosions produced craters, for the hole served as an outlet for the dispersal of gas into the atmosphere. But to be certain, when planning the famous mine attack at Messines Ridge in June 1917, orders went out to evacuate all trenches and shallow dugouts within three hundred yards (274.32 meters) and all deep dugouts within four hundred yards (365.76 meters) of every expected mine crater. Infantrymen operating on the surface after the blasts were ordered not to go within ten feet (3.04 meters) of the lip of any crater. These may seem to be excessive precautions, but if a charge poorly detonated or the weather was dull and winds calm, gas tended to be a bigger problem within the zone of rupture underground and in craters on the surface. On at least one occasion, gas dispersing from a crater ignited, and the flame shot up to twelve feet (3.65 meters) in height, burning "for hours." A sight like that could easily justify extreme caution in anticipating what might happen after the explosion of a large mine.[81]

Military miners employed animals for gas detection, as did commercial miners, for the first time in history. Never before had they worked for so long in what became semipermanent mine galleries and thus had no prior need to use animal sentries to alert them to gas poisoning. Canaries and mice were preferred since they reacted much more quickly to the presence of gas than did humans. If, for example, a man was affected by a certain level of gas in the closed atmosphere of a gallery within thirty minutes, canaries and mice were affected by the same conditions in only two minutes. Canaries tended to be more reliable than mice, according to the Royal Engineer official history, because they were more sensitive to gas.[82]

Their service to the miners cost numerous canaries their lives. At first the bird would "get a mild 'jag' . . . and become very lively," recalled Reynolds. Then "as the gas became denser, he would drop over, and turn his legs up. It was time to get out quickly when this occurred." Engineer Lieutenant Mc-Cormack conducted tests on canaries to more accurately gauge the effect of carbon monoxide poisoning. He found that gas poisoning could be divided into three stages. In the first the "bird rubs its beak on the wires of its cage or against its perch, shakes its head vigorously and very often brings up seed as though slightly sick." In stage two the "bird pants" and spreads its legs more widely apart to better balance on its perch. In stage three the canary sways on the perch, then "suddenly makes a wild flight from the perch and falls" to the cage floor. Miners learned to clip the birds' claws so they could not steady themselves on the perch when in stage two—it was easier to notice their distress that way.[83]

Canaries were on duty in the galleries for six days at a time, with the next six days off in fresh air. Not surprisingly, their casualty rate was very high. Exposing the birds deliberately to gas and allowing them to suffer for it as an element of miner safety, Reynolds estimated that many hundreds perished in British galleries. Some miners became attached to them. After the conflict the Scottish National War Memorial at Edinburgh recognized animals of all kinds as unwilling participants of the Great War. Opened in July 1927, Phillis Bone designed a medallion along this line. It includes a special tribute to "The Tunneller's Friends," which consists of a frieze showing a cage with canaries and mice depicted under it. John Norton-Griffiths thought it "beautifully carved."[84]

Thus did tens of thousands of military miners fight the underground war along the Western Front, utilizing an astonishing array of new and old technologies, employing new and old methods. The old included the use of animals as early warning devices for gas poisoning, the employment of civilian miners in large numbers rather than relying exclusively on military engineers, and the utilization of techniques of tamping that had not changed at all for centuries. The essential doctrine of underground combat did not change very much, either. The need to aggressively push an offensive mine to stay ahead of countermeasures by the enemy remained the highest priority. Digging deeper to gain an advantage in the contest between opposing galleries also remained a high priority. Short-term tactical objectives intermingled with long-term aims. Miners not only sought to break open enemy trench lines to allow for massive infantry assaults but also aimed to blow up intermediate targets or to create craters in no-man's-land for various immediate purposes.

But on the Western Front, new developments in mining and countermining predominated in underground warfare. It was conducted on a scale that absolutely dwarfed any previous conflict in history. During the American Civil War, for example, probably no more than 500 men on both sides combined had been used to dig shafts and galleries, while during World War I, the British alone employed 25,000 men at one time during peak strength. If the French and Germans used roughly similar numbers, 75,000 personnel contributed to mining and countermining on the Western Front—and this is without counting those on other fronts of the Great War.

The new materials and methods used in the Great War tended to be centered on the employment of advanced technology, virtually all of it borrowed from or developed by the civilian mining industry in much the same way that the famous British mole was borrowed from Norton-Griffiths's mining company. This connection was especially important in creating powered devices to ventilate large gallery systems and pumping arrangements to keep them free of excess water. It also was important in the creation of new listening systems for defensive mining. The rescue system created to extricate miners in danger was completely new in military mining. In fact, the system of schools established for in-theater training in rescue and all other aspects of military mining was a new idea.

While much remained the same in terms of the doctrine of military mining, the Great War spurred more complicated application of that doctrine along with higher levels of tension in prosecuting underground conflict. Miners dug galleries longer and deeper than ever before, as much as two hundred feet (60.96 meters) deep. They went through several layers of earth while constructing multiple levels of galleries simultaneously as part of their effort to fool the enemy and give themselves options in pushing forward the underground offensive. They worked in a variety of mediums, from the sand dunes of the North Sea coast to the clay fields of Flanders to the chalk country of northeastern France to the mounts of the alpine region. Never before had military miners been so numerous, so busy, and so concerned with solving problems with new technology and methods as in the Great War.

12

1916

By 1916, mining forces in all three armies along the Western Front had worked out improved systems of operation that went far beyond most of the old ideas and materials used in previous underground warfare. This conflict was on such a vast scale, and with the widespread adoption of advanced equipment, explosives, and detonation devices, that a dramatically new period of mining history now appeared.

The enhanced characteristic of underground siege mining to be seen in 1916 included many large-sale mining fights at locations like the Hohenzollern Redoubt and Vimy Ridge. But another new development was to enlarge big projects like Vimy Ridge to another level, with multiple mines linked tightly into a large tactical-assault plan. This was first done by the British in preparation for their Somme offensive and would be greatly enhanced at Messines Ridge in 1917. The difficulties of coordinating the underground blow with the above-ground assault became even more pertinent with these big offensives, and the British did not quite work them out on the Somme. Even so, Britain emerged as the major mining power on the Western Front during 1916 and through 1917, although French and German activity continued during those two years. Ironically, 1916 saw an increased use of mines for discrete, small-scale objectives within no-man's-land as well. These local projects, typically for the limited gain of creating a crater that could be used for various purposes, actually produced far more mine explosions than those designed to break open enemy lines. This led to an increased number of small engagements designed to secure and utilize the resulting craters. The increased mining activity also led to opposing diggers more frequently meeting each other in underground combat, and much of that action, unlike previous examples of similar action, was well documented by its survivors.

Major mining projects developed at several sectors of the Western Front in 1916. One of the earliest aimed at the Hohenzollern Redoubt at Auchy-les-Mines near Loos-en-Gohelle. Underground approaches to the redoubt began on December 14, 1915, as tunnelers dug six shafts to the clay layer 20 feet (6.09 meters) deep and began galleries. They also dug even deeper and into a layer of chalk to start three additional galleries. By February 29, 1916, the miners charged the chalk-layer mines with 10,055 pounds (4,762.71 kilograms) of ammonal in one and 7,000 pounds (3,175.14 kilograms) in two, all three at about 30 feet (9.14 meters) deep and from 160 to 220 feet (48.76 to 67.05 meters) from the shafts. When detonated on March 2, they created craters from 100 to 130 feet (30.48 to 39.62 meters) in diameter but failed to provide the venue for a completely successful infantry assault. On March 18 the Germans exploded five mines as part of an effort to push back the British. They failed in this objective, and both sides settled into a stalemate, separated by little more than the width of mine craters. A total of fifty mines were blown at Hohenzollern Redoubt during March, April, and May, with little to show for the effort.[1]

At about the same time, British miners replaced their French counterparts at Vimy Ridge. They found that the Germans were gaining the upper hand here, with some galleries nearly under the Allied line, and a spirited contest soon developed. A total of seventy mines were exploded in the area during the next two months, with the Germans detonating most of them and the British barely holding them back. German operatives exploded thirteen mines on April 28 that wrecked sections of the British infantry line and underground dugouts. A local attack in May pushed British troops back seven hundred yards (640.08 meters) and allowed German miners to enter several Allied mine shafts. The pressure eased when the Germans scaled back their efforts in June, sending miners back to civilian jobs or to work on digging underground shelters on the new Hindenburg Line farther east. The British began to plan an offensive in August, but by October, the Canadian Corps took over the northern sector of Vimy Ridge and in February 1917 began to prepare its own major offensive in the area.[2]

But the dominating mining project of 1916 developed not at the Hohenzollern Redoubt or at Vimy Ridge but as part of the massive offensive in the Somme sector. When British forces replaced French troops in the northern part of this region, they found most shafts 20–30 feet (6.09–9.14 meters) deep and located at the first line. They concluded that these shafts and galleries were too small and too heavily timbered and began to change and enlarge the system. By early 1916, British countermines were generally about 40 feet (12.19 meters) deep with transversals and the offensive mines a bit less than 100 feet (30.48 meters) deep, just above the water level. They geared this system to support the big push scheduled for July 1.[3]

With five tunneling companies at work, the British placed nineteen mines to support the infantry attack. Some galleries were only 50 feet (15.24 meters) long, while one stretched 595 feet (181.35 meters) and was dug in only twenty-eight days. Many of the mines were grouped into minisystems and charged with ammonal to maximize effect. At Carnoy the minisystem consisted of two mines with 2,000 and 5,000 pounds (907.18 and 2,267.96 kilograms) of explosives; at Tambour Du Clos three mines with 9,000, 15,000, and 25,000 pounds (4,082.33, 6,803.88, and 11,339.80 kilograms); and at La Boiselle another three mines with 24,000, 36,000, and 40,000 pounds (10,886.21, 16,329.32, and 18,143.69 kilograms). The ninth mine, at Hawthorn Ridge, contained 40,000 pounds. Beyond these, the rest of the mines contained 500 pounds (226.79 kilograms) or less.[4]

The Hawthorn Ridge Redoubt mine became one of the most famous of World War I because it was the first photographed mine explosion in history. The gallery was 80 feet (24.38 meters) deep and 1,055 feet (321.56 meters) long. When it went up early on the morning of July 1, two men recorded images of the dirt plume. Ernest Brooks, an official military photographer, took several still photographs as the plume neared its apogee, while Geoffrey Malins, a filmmaker serving in the army, recorded its rise with a movie camera.[5]

Hawthorn Ridge was not the first mine explosion that Malins captured. A few days before July 1, he found out about a planned blow and secured a good filming spot in the forward British line eighty yards (73.15 meters) from the target. Malins carefully set up his position, moving some sandbags without drawing German attention. He then plotted his spot in comparison with the target on a trench map and determined that he would have to angle the camera 124 degrees to record the explosion. Malins then took an empty sandbag and cut a hole through the middle for the lens to poke through, providing some camouflage for the front of the camera and himself. Before dawn, Malins slowly raised the camera above the parapet and waited. One minute before the expected explosion, he began to crank the camera as sniper bullets started to land dangerously close. Malins managed to film the dirt plume rising up as a machine-gun opened fire on his position, and when he finished was nearly killed as a bullet sliced through his service cap.

Hawthorn Ridge Redoubt Mine Explosion (Malins), Western Front, 1916. At 7:20 A.M., July 1, 1916, Geoffrey Malins shot dramatic footage of the dirt plume from this mine explosion, one of nineteen designed to start the Somme offensive that day. The filmmaker incorporated twelve seconds of this footage into a documentary film released the following month throughout Great Britain. It became the first mine in history to be cinematically recorded and distributed for viewing by a wide audience. Still from *The Battle of the Somme*, directed by Geoffrey Malins (1916).

Unfortunately, the intrepid cinematographer never explained what he did with the film; it remains lost.[6]

Malins had less trouble filming the Hawthorn Ridge Redoubt mine on July 1. A division commander told him the best place to set up his camera—a spot called Jacob's Ladder, 150 yards (137.16 meters) from the redoubt—and a staff officer informed him of the time for the explosion. Malins had no trouble setting up his equipment less than twenty minutes beforehand and waited. He started to crank at 7:19 A.M., only half a minute before the detonation, because he wanted to catch the entire sequence of events from the moment the dirt plume began to rise. The charge went up exactly on time. "The ground where I stood gave a mighty convulsion," Malins remembered. "It rocked and swayed. I gripped hold of my tripod to steady myself. Then for all the world like a gigantic sponge, the earth rose in the air to the height of hundreds of feet. Higher and higher it rose, and with a horrible, grinding roar the earth fell back upon itself, leaving in its place a mountain of smoke."[7]

Thus, Hawthorn Ridge Redoubt became the first mine in history to be documented by either still or motion-picture photography as well as the first to be known to the public. Brooks's photograph became an icon of World War I trench warfare. For his part, Malins became famous for his July 1 cinematography, shooting footage of a lot of action before and during that day, which was quickly edited into a seventy-five-minute film called *The Battle of the Somme*. That film created a sensation when released in August 1916, eventually viewed by an estimated 20 million people throughout Great Britain. Only about twelve seconds of that film depicts the rise and fall of the dirt plume at Hawthorn Ridge Redoubt. Still, it was the first time in history that civilians had the opportunity to see what such a thing looked like. In the field, however, the mine became controversial because the follow-up infantry attack was deliberately delayed ten minutes after the explosion to allow debris to settle. That gave the Germans enough time to recover from the shock and prepare to meet the assault. As a result it produced little in the way of tactical gain and served as an important object lesson for the British never to delay a follow-up attack again.[8]

Five miles (8.04 kilometers) south of Hawthorn Ridge, the area around the village of La Boisselle had seen much mine warfare since early 1915. Here, in preparation for the Somme offensive, the British enlarged the established French system and dug galleries at depths of 30, 60, 90, and 120 feet (9.14, 18.28, 27.43, and 36.57 meters). German miners responded accordingly with an enlarged and deepened mine system of their own. The longest gallery on the British side began in a communication trench 300 yards (274.32 meters) to the rear of their first line and over 900 yards (822.96 meters) from the German first line at a depth of 50 feet (15.24 meters). The moles dug very quickly at first, but when nearing the target they took off their boots and walked on empty sandbags layered on the gallery floor. The men now talked in whispers and used bayonets with wood handles to pry out chunks of chalk, catching them before they fell on the floor. The target of this gallery was the Schwaben Höhe, a German position south of La Boisselle that dominated what the British called Sausage Valley. By mid-April, the gallery had reached a point 135 feet (41.14 meters) from the edge of the German fortification, but here the miners dug two branches, hoping to bury the German position with debris from the crater rather than continue digging so close to listening devices. The two branches were 75 and 45 feet (22.86 and 13.71 meters) long; the left was charged with 36,000 pounds (16,329.32 kilograms) of ammonal and the right with 24,000 pounds (10,886.21 kilograms), but the combined effect was designed to create one crater. When it blew, however, the effect was limited, and La Boisselle was not captured until after several days of costly fighting.[9]

Hawthorn Ridge Redoubt Mine Explosion (Brooks), Western Front, 1916. Photographer Ernest Brooks also caught the dirt plume of this mine explosion with a still camera, producing an iconic image of mine warfare on the Western Front. The official photographer to the royal family, Brooks was given an officer's commission so he could move freely along British lines, taking more than 4,400 photographs from 1915 to 1918. Imperial War Museum, Q000754.

At Tambour du Close the British did not plan a direct infantry attack. The three mines were designed instead to impede German flanking fire on troops advancing farther north. The explosions did help shield the first wave of troops but not the subsequent attacks. In fact, the southernmost of the three charges failed to explode at all due to dampness, and the mines played only a marginal role in support of the capture of Fricourt.[10]

In addition to nineteen offensive mines in the Somme offensive, the British deployed seventeen Russian Saps to aid the attack on July 1. Eleven of them were meant, after blowing out the forward end of the gallery, to give access to the surface and serve as positions for Stokes mortars and machine guns. The other six were designed to serve as communications conduits for the movement of troops forward and casualties backward. All of them wound up between forty and two hundred feet (12.19 and 60.96 meters) from the first German line at their forward end. British miners created an instant trench at High Wood to consolidate gains by using twenty-one small charges in a pipe. When blown, it created a communication trench fifty yards (45.72 meters) long, five feet (1.52 meters) deep, and twelve to fifteen feet (3.65 to 4.57 meters) wide at the top. The experiment with Russian Saps proved to be of limited success. Entrances tended to become clogged with wounded because it was not easy to evacuate them back through the narrow gallery. Additionally, in sectors where the infantry attack was successful, British troops quickly moved forward beyond the range of the mortars and machine guns positioned at the entrances to those tunnels.[11]

Most of the mining to support the Somme offensive centered on the first day of the massive operation, but some of it took place later in the four-month-long campaign. At High Wood miners sank a shaft 25 feet (7.62 meters) deep, then extended a gallery toward a German strongpoint 320 feet (97.53 meters) away. When detonated, 3,000 pounds (1,360.77 kilograms) of ammonal destroyed the strongpoint on September 3, but a quick German counterattack seized the resulting crater. Tunnelers cleared the wrecked gallery and blew a second mine on September 9 to destroy the German garrison at the lip of the crater, after which the British advance continued. The combined crater created by those two blows measured 85 by 135 feet (25.90 by 41.14 meters) and was 35 feet (10.66 meters) deep.[12]

The British also restarted efforts against the Hawthorn Ridge Redoubt before the end of the Somme offensive. They began to reopen the exploded gallery on July 5, but it was November before they were ready to blow a charge as part of a more tightly integrated assault plan. Once again planting a massive charge—30,000 pounds (13,607.77 kilograms) of ammonal—the British set it off on November 13. The troops advanced beyond their barbed-wire belt just before the blow and went in immediately, risking casualties from falling debris in order to maximize the advantage gained by shocking the German defenders. This time it worked, and a severe lesson had been learned. The emotional effect of mine explosions was great but very temporary, and the Germans had honed their ability to react. The only way to counteract their skill at seizing craters was to move as quickly as possible after the blow. Inspector of Mines Robert Harvey had tried to create this tighter coordination of underground and surface attack for the July 1 operation but had been overruled by staff and senor officers on many levels of the army units scheduled to take part in that assault.[13]

In addition to big mining projects such as these, miners participated in jobs designed to gain small increments of advantage over the enemy along many sectors of the front. Much of that activity took place in close cooperation with surface infantry operations, resulting in a kind of warfare that centered on creating and controlling mine craters as forward positions. This was not a new idea in the doctrine of mine warfare, but it became more prominent and widespread during World War I than in any previous conflict.

Brigade Maj. Alexander Johnston of the Seventh Brigade, Twenty-Fifth Division was a keen observer of crater warfare during April and May 1916. His sector included several British mine projects, and he recorded activity by the Germans and his own side as it affected surface operations. On April 25 the Germans detonated a mine fifty yards (45.72 meters) in front of the extreme left of the brigade. "Their object being not to blow any of our trenches up but to form a crater on top of which they could form a sniper's or machine gun post." Fifteen minutes after the blow, Johnston could see thirty German soldiers "working like ants and bringing up loophole plates." He was not much worried though, believing that this was at best a small advantage for the enemy. They could now more effectively fire along a British trench called Central Avenue, but the advantage was "nothing out of the usual."[14]

Much has been made of the German superiority in crater consolidation and exploitation, but Johnston felt that it really depended on the quality of officers in any given unit. He personally witnessed several cases where British infantry managed to hold their own or better the Germans in taking advantage of craters as forward posts in no-man's-land. At both Broadmarsh Crater and Duffield Crater, British and German infantry dug in on opposite sides of the hole, protected by the lip and whatever sandbags and metal plates they could erect in a standoff separated only by one hundred feet (30.48 meters) of upturned earth. At Broadmarsh on the night of April 30, a British party counterattacked to reestablish control of the crater after losing it to an earlier German assault. They went in under the cover of darkness and gained the lip of the crater before the Germans noticed them. The men dug fifteen yards (13.71 meters) "of good trench round the lip" before dawn, with a barricade at each end to serve as a traverse against enfilade fire. Nearly three weeks later the British again lost this post, but Johnston helped the soldiers regain it. He suggested they set up a Lewis gun only thirty yards (27.43 meters) from the enemy and fire to draw their attention as a party advanced in the dim light just before dawn. When close, the infantry pelted the enemy position with hand grenades as Johnston personally used a rifle to shoot any Germans who raised their heads above the lip. This enabled the infantry to retake the position.[15]

Tenacious struggle characterized crater fighting just as it did the underground war. Blowing mines for small tactical purposes in no-man's-land was not as important as trying to tie in mine attacks with major offensives such as the Somme, but it was easier to do and had a better chance of success. In that sense there is little wonder that small plans predominated by 1916; they served a useful, although marginal, purpose in the overall scheme of things.

Mining activity of all kinds, however, eased off during the latter months

DEFENCE OF CRATERS.

(a) Near Lip Defended.

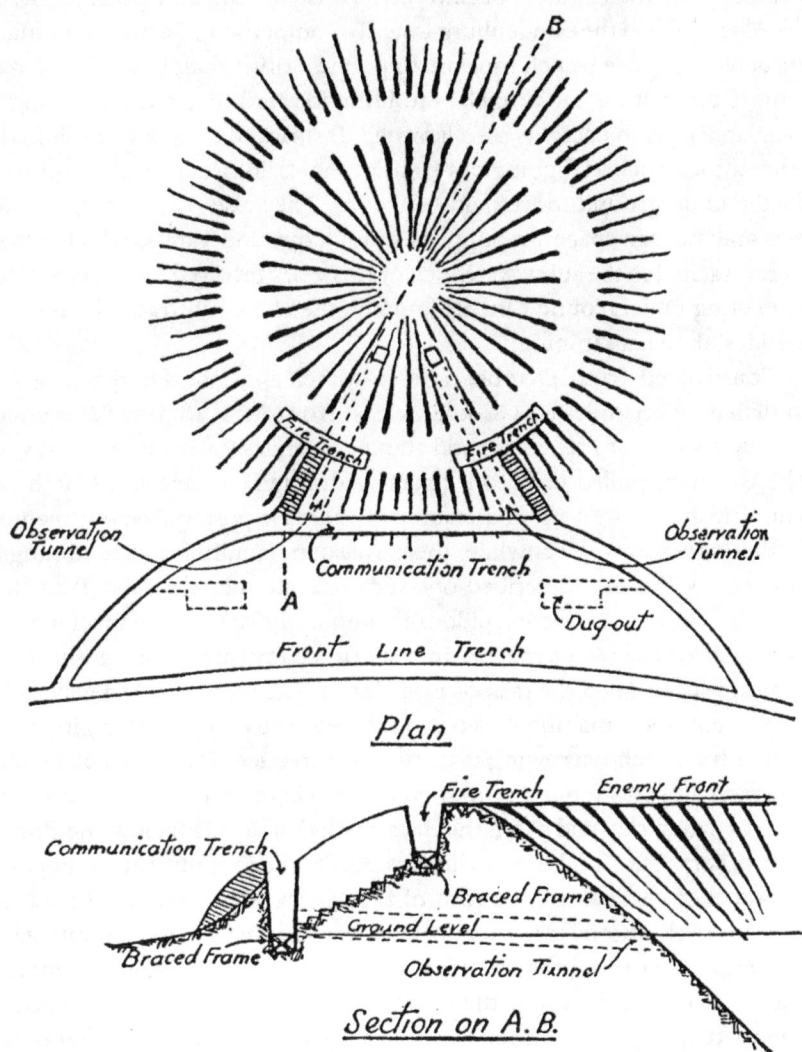

Plan

Section on A.B.

Near-Lip Defense of a Crater, Western Front. This diagram illustrates the best way to defend a mine crater in no-man's-land by fortifying the near rim. Elliott, *Trench Fighting*, 157.

of 1916 along the British sector because of a German plan to contract their lines and more heavily fortify them in order to reduce troop strength. They sent many of their miners back to help construct the new position, which the Allies dubbed the Hindenburg Line. To compensate, German commanders deployed more trench mortars to pound British shaft heads. This made it more difficult for the tunnelers to bring forward material for mining, so their underground activity also lessened. Instead, tunnelers were shifted to other duties, such as digging out transit tunnels, informally called subways, for the underground movement of troops. They also placed hospital facilities and storage space for supplies in underground vaults, which usually were attached to the subways. Much of this work involved the enhancement of existing underground tunnels found in the chalk substrata, the result of decades of civilian mining.[16]

Constructed from October 1916 to March 1917, the Hindenburg Line stretched ninety miles (144.84 kilometers) from Arras to near Soissons and gave up a swath of French ground about nine miles (14.48 kilometers) wide. The Germans pulled back to it in stages during February and March 1917. The British followed up and examined the new position before planning further operations. Meanwhile, there was opportunity for local mining action. Harry Trounce described one such operation by the 181st Tunnelling Company against a concrete pillbox that housed two German machine guns two hundred feet (60.96 meters) in front of the forward British position. His section was assigned the task of reducing it. The men started a gallery in a dugout entrance, making it two by four feet (0.60 by 1.21 meters) in dimension at first, then narrowing it to two by three feet (0.60 by 0.91 meters). The men dug only ten feet (3.04 meters) deep in order to stay within the top layer of sandy clay and avoid the hard chalk that lay just below the floor of their gallery. They timbered as little as possible to save time but kept two to three feet of earth above the roof of the gallery. Along the way they broke into two shell craters, using sandbags smeared with mud to camouflage the openings. After two days and nights of heavy digging, Trounce's men hit the concrete foundation of the pillbox. They planted five hundred pounds (226.79 kilograms) of ammonal; set up a double set of primers, detonators, and leads; and tamped the gallery with sandbags for thirty feet (9.14 meters) back. When they set off the charge just before dawn, it destroyed the pillbox, and the follow-up infantry assault not only secured the position but also captured two hundred yards (182.88 meters) of the nearest German line. A second attack cleared another three hundred yards (274.32 meters) of the line. In examining the captured ground, Trounce found a German electrical listening device in a shaft located in the chalk layer but none in the sandy clay layer that his men were using to approach the target. His section had

DEFENCE OF CRATERS.

Far Lip Defended.

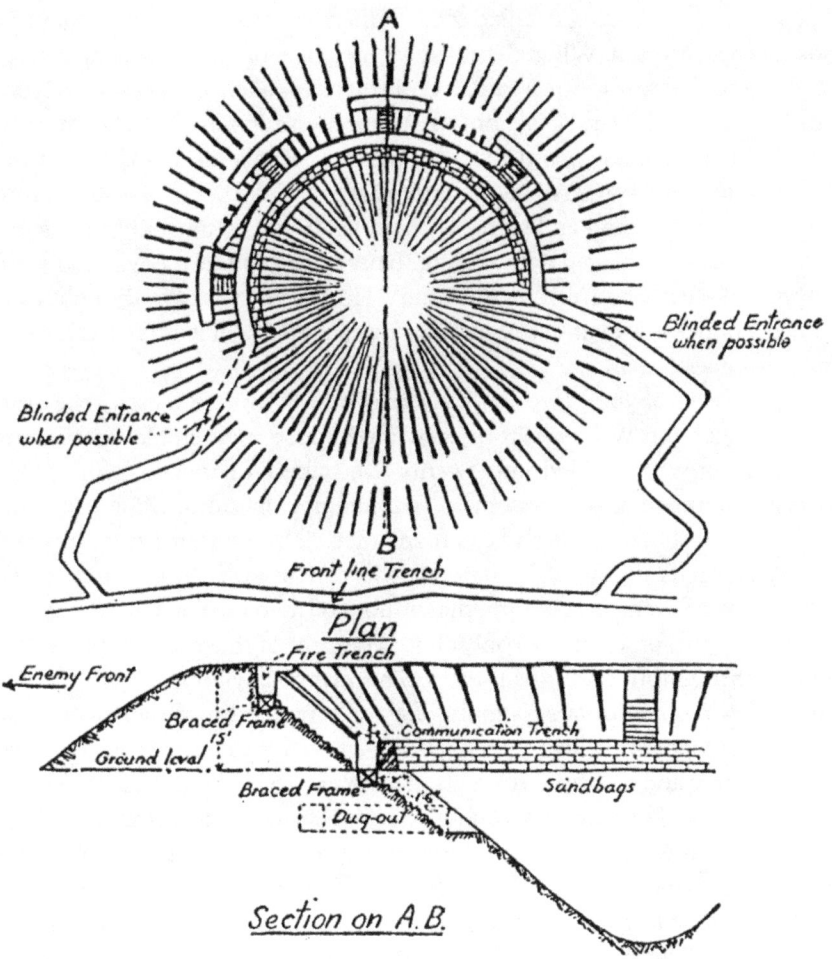

Far-Lip Defense of a Crater, Western Front. This diagram illustrates how the British fortified the far rim of a crater in no-man's-land. Elliott, *Trench Fighting* (1917), 158.

demonstrated how effective offensive mining could be to achieve limited, local results.[17]

By 1916, the size of British mining units increased. Starting with 349 men of all ranks, the 2nd Canadian Tunnelling Company expanded to 675 by the fall of that year. Adding detailed infantrymen, the strength of the company soared to 2,000 men. When the 185th Tunnelling Company fell dangerously short of manpower in April 1916, its commander called for volunteers from the Fifty-First Highland Division. About 150 men answered the call, most of them miners from Scotland, and two officers came with them. At first the Scots volunteers were assigned temporarily to the 185th, but later they were permanently attached and some of them entered the Royal Engineers. All of the volunteers served with the company to the end of the war and were happy to be excused from infantry duty. Herbert W. Graham thought they added a noticeable element of military bearing to the unit as well as civilian mining expertise.[18]

By the end of June 1916, on the eve of the Somme offensive, the British sector of the Western Front contained 25,000 men employed in mine work. They were divided into twenty-five imperial companies and seven overseas companies. Equivalent to two infantry divisions, Mine Inspector Harvey later thought this was too many men. "No modern army can stand this drain on its strength," he stated in a lecture in 1929. Yet as Trounce had pointed out eleven years before, the number of men involved in mining and countermining amounted to only about 1 percent of the total manpower deployed throughout the British sector. This historically unprecedented force of miners required a large supply of varied material to do its work. The First Australian Tunnelling Company alone used 135,000 sandbags, sixty-two pumps, three and a half tons of nails, 180 meters (590.55 feet) of suction hose, and two tons of corrugated iron in January, February, and March 1917.[19]

With such a large force working underground, the number of blows increased dramatically in 1916. While the Germans opposing the British sector exploded 700 mines and countermines that year, the British set off 750 charges. That amounted to a total of 1,450 underground explosions along nearly half the front in one year. This by itself would constitute the largest number of military mine explosions in world history, even if one does not count the hundreds of charges set off by the Germans and the French on the rest of the Western Front before, during, and after 1916.[20]

The Royal Engineers maintained a record of exactly when those charges were blown, numbering 1,025 mines and countermines (499 British and 526 German) over seven months of 1916. They found that the Germans preferred to set off their charges between 5 P.M. and 9 P.M. The British also preferred evening and early nighttime for their explosions, but the period from 7 A.M. to 3 P.M. was relatively quiet.[21] The preferred timings of these blowups strongly imply that the overwhelming majority of them were not tied to above-ground assaults.

With so much digging on an unprecedented scale and so many men busy underground, it is not surprising that the number of incidents in which foe met foe in a personal way under the surface of the Western Front should increase dramatically. Throughout mining history, underground confrontations between individuals had always been assumed but rarely documented. Other than the vivid evidence of an underground combat discovered by French and American archaeologists at Dura-Europos, we have mostly textual references to this subject, most of which are so vague as to cause one to hesitate before accepting them as reliable.

But that does not apply to World War I. The literature is filled with reliable evidence that opposing miners frequently broke into each other's galleries. The result was a wide range of underground contact between opponents that varied from a handful of miners breaking into an enemy gallery and doing mischief to vicious gunfights in darkened galleries. In a few cases miners actually stole the explosives from charges set by their opponents before the enemy had a chance to spring them. In the area of underground combat as in most others, the Western Front ushered in a new phase in the history of military mining.

At times the subterranean encounter was a somber affair. The First Canadian Tunnelling Company dug deep and got behind the head of German galleries at the Bluff near Ypres. The men blew a charge and cut off an upper German gallery that was seven hundred feet (213.36 meters) long. "Our tunnellers afterwards broke into this system and found it occupied with the bodies of German miners who had died at their work," recalled Angus W. Davis. "Some of them lay close to the wrecked ground and had evidently been engaged in trying to dig through the wrecked area when overcome. Gas from the explosion probably killed the entire party."[22]

Near Cambrin on the Béthune sector, the 251st Tunnelling Company worked at a place called Mad Point. The men unknowingly dug a branch just under a damaged German gallery. They only realized their proximity when a small hole developed "in the top left hand corner of the face" and air flowed in. The moles plugged the hole with empty sandbags and stationed an armed listener at the spot, who soon reported hearing Germans at work trying to clear the damaged gallery. The British tried to maintain their secrecy as long as possible, planting a camouflet but leaving some space for the listener to crawl through and keep up his work. Just when they had almost finished the tamping, another hole developed "some distance back from the face." The moles quickly installed detonators and leads, but a German clumsily broke through the second hole. His booted foot came crashing down and hit a Lieutenant Hansen in the back as the British officer was lying down. Fortunately, the German did not realize what was happening, and Hansen was able to finish the job and leave before the British exploded the camouflet.[23]

There were several instances of British miners stealing German charges. Capt. Clay Hepburn and two lieutenants of his 172nd Tunnelling Company accomplished this feat near St. Eloi by breaking into a German gallery and nabbing 1,350 pounds (612.34 kilograms) of explosives over two days. They then used the pilfered munitions as part of a mine blow soon after. "On several occasions I have unloaded enemy mines," reported Capt. L. B. Reynolds without offering details.[24]

The worst underground encounters, however, remained those in which shots were fired. Bernard Newman, coauthor of an early history of the British tunnelers and an infantry officer on the Somme sector, claimed that "dozens of men actually faced this startling experience" during the war. Sappers Garfield Morgan and Albert Rees were among those who knew this firsthand. While digging at a face, they broke into a German tunnel at Hill 60 near Ypres. Calling Lt. Thomas Black, the three turned on a flashlight when a German shot at them, tearing Black's tunic. The trio fell back, regrouped, and came back to the scene. They could detect no enemy but saw a small explosive. Cutting the leads and laying their own charge, 250 pounds (113.39 kilograms) of guncotton, they left it to be exploded later. On another occasion, at the Bluff Lt. Richard Brisco used a bayonet to enlarge the hole his tunnelers had accidentally made in the side of a German gallery. When large enough to admit a man, he boldly explored the enemy space and fired at a group of Germans to drive them away. Brisco then planted a charge that sealed the German gallery when it exploded. Sometime later at a different spot, Brisco again explored a German excavation. This time the side of a British-held crater subsided because of heavy rains and exposed the enemy gallery. The daring lieutenant entered and initiated a firefight with a German party. Brisco used some abandoned sandbags to block the hole and later blew a camouflet to seal it properly.[25]

One of the most thoroughly documented cases of underground combat involved Trounce and his comrades of the 181st Tunnelling Company. On March 8, 1916, in a gallery at Cordonneire north of Neuve Chapelle, a clay-kicker broke into the top of a German gallery. He plugged it and reported his find, which led to two officers planting fifteen pounds (6.80 kilograms) of guncotton near the hole and tamping the area even as they could hear the Germans trying to enlarge the hole. Right after the blow Trounce and Sapper Doherty went in with Proto respirators and a canary to explore. The bird died quickly, but the men were able to evaluate the damage. They found the bodies of three German miners but realized that the heavy timbering of the gallery had limited the damage. Trounce looked for and cut any wires in case the Germans had installed a charge and then lay an air hose he had brought along. After evacuating the damaged enemy tunnel, his men pumped in fresh air for an hour, after which another party entered but found no evidence of German occupation for some distance.[26]

The British now decided to form three parties and see just how far this tunnel remained empty. Each party consisted of one officer, one noncommissioned officer, and two sappers armed with revolvers and grenades. Trounce's group acted as a reserve while the other two explored this underground no-man's-land. Lieutenant Wright's party moved along the gallery toward the shaft, cutting the lines feeding the electric lights to maintain darkness. Before long the lights came on in a section ahead that lay around a bend, and Wright's men saw two Germans round that corner walking toward them. The British opened fire and felled one of the men before falling back. Wright dropped a fifteen-pound (6.80 kilograms) box of guncotton and arranged to spring it as another man fired over his head at the remaining German.[27]

Wright had indeed found the forward edge of no-man's-land and had temporarily stopped enemy penetration toward the hole in the gallery. After a while another party of nine men carrying cases of guncotton explored through that hole once more. They encountered Germans once again, proving that Wright's charge had failed to block the gallery. After exchanging shots, the British party retired to the hole, where the men set up the biggest charge of this incident and fired it just before dawn of March 9, sealing the hole.[28]

A number of lessons were learned from this harrowing incident of accidental contact underground. The British found that the Germans used "small sandbags made from all kinds of coloured materials, similar to diapers' remnants," a reflection of the crunching British blockade of German seaports that already was beginning to strangle the kaiser's war effort. They found no listening devices or tools and only one mobile charge. The British concluded not only that the Germans heavily timbered their galleries but also had a less sophisticated listening system than their own. Most of all, the incident demonstrated that "small untamped mobile charges" failed to close galleries but simply enlarged preexisting holes. Larger and properly tamped charges were necessary to seal a discovered enemy tunnel.[29]

Underground contact happened often enough so that patterns developed in how it played out. Rather than cut and run, both sides tried to see if they could gain any advantage from the encounter. They pushed forward as soon as possible to explore enemy diggings, hoping to find maps, documents, equipment, or anything else that could enlighten them about how the other side operated. Examining the tunnels also was a high priority, hoping to learn something from the physical evidence or to use enemy diggings for their own purposes. When fighting broke out, it always was spontaneous, the result of chance encounters, and usually lasted only a few minutes, serving to draw battle lines between opposing sides. At that point it was time to set charges to seal off tunnels and pull back. Most of all, underground combat was best conducted as a group endeavor, ordered and superintended by higher-level officers, rather than by plucky individuals acting on their own as had Brisco.[30]

While most underground firefights were brief affairs, Lt. John Westacott of the Second Canadian Tunnelling Company endured much worse subterranean fighting than was typical. A twenty-five-year-old civil engineer, he engaged in such combat at Mount Sorrel near Ypres in the spring of 1916, then again in late June. In this last encounter the Canadians blew in a German gallery and sent armed parties with Proto apparatus to explore. By luck, two parties managed to trap between them a German party that also was exploring no-man's-land. A three-minute firefight ensued, with losses on both sides, but the Canadians captured the German survivors. They then blew two small charges to destroy all they could of the enemy gallery.[31]

On September 16, 1916, Westacott endured his third underground combat, which was an unusually long and bitter battle. He was working with eighty men in the mines at Mount Sorrel when the Germans launched a local attack on the surface that captured part of the British line. Only when Westacott emerged into the trench system at the end of his shift did he realize that all of the shafts were now in German control. Chased by enemy infantrymen, the lieutenant escaped back into the mine system and organized his men for battle. They were able to kill individual German soldiers who descended the shafts before Westacott ordered nearly all entrances blocked, leaving only one open that he hoped the enemy would not discover. But the Germans were determined and began to dig out several of the shafts, soon infiltrating Westacott's mine system, which was half a mile (0.80 kilometers) wide. The result was a full twenty-four hours of intense combat. The ventilation system was off, and there was no electricity, so the battle took place in cramped galleries and in total darkness because it was too dangerous to use candles or torches. Most of the fighting took place at arm's length. The German uniform included epaulettes unlike the Canadian dress, so it became necessary for Westacott's men, armed with knuckle knives strapped to their wrists, to feel the shoulders of anyone they encountered before deciding whether to kill or disable him.[32]

In waves the Germans came down and tried to advance through the gallery system, only to be met with fierce resistance by the Canadians. The free use of hand grenades damaged the galleries and created a pall of carbon monoxide that mingled with the smell of blood to create an absolutely hellish environment. "I'd never seen anything like it," Westacott recalled in 1960. "As fast as we got them down, there would be somebody on top of you again. The corpses blocked the place right up, we had to drag them out before we could do anything. But we knew our mines, knew our own workings—they didn't." Westacott was wounded by a grenade blast, and eventually sixty of his eighty men fell in this horrible nightmare. He remained proud of them for the rest of his life. "It was 'no surrender' with them, and they would have fought till the last man dropped. We could not have lasted more than a few hours longer before we should all have been gone as we had such a small number of men left, and these totally exhausted." The miners managed to hold off the Germans until a British counterattack on the surface pushed the enemy back and retrieved the shaft heads. "It was murder down there," Westacott admitted, "I've had some shocking nightmares since."[33]

Another extraordinary story of endurance arising from the tensions of underground contact involved a rescue from an almost certain deathtrap. At Petit Bois the Germans blew two countermines at 6:30 A.M. on June 10, 1916. The one to the north contained 4,000 pounds (1,814.36 kilograms) of explosives and was located forty feet (12.19 meters) above British Gallery SP13, which was ninety feet (27.43 meters) below the surface. This explosion collapsed a section of SP13 for about three hundred feet (91.44 meters). Thirteen moles were trapped at the forward face of the gallery. Most of them naturally wanted to begin digging their way out, but one advised not to attempt it. Sapper William Bedson, a prewar civilian miner, told his mates that a rescue attempt would obviously be launched. Trying to dig out would consume precious oxygen in the gallery. He urged the others to lie down and wait, but they refused.[34]

While a dozen of the trapped men tried to dig their way out, Bedson made a matting of empty sandbags at the face of the gallery, which happened to be at least twenty feet (6.09 meters) higher than the rest of the excavation. Here he waited patiently, consuming minimum oxygen. Bedson had no food and precious little water, so he merely rinsed his mouth now and then and put the water back into the bottle. He had undergone a similar experience before in his civilian job; the only thing he could not anticipate was how long he would have to wait. His mates died one by one because they used up oxygen in the lower part of the isolated gallery. It took rescuers some time to clear the debris—reports vary from six days to ten days. At any rate, it was a very long time to wait in the dark with no nourishment and the knowledge that colleagues were dead and dying just a few yards away. But Bedson's patience, foresight, and the fact that his raised section of the gallery offered a bit more breathable air than elsewhere saved his life. After rescuers extracted him, Bedson spent a week in hospital, another week to get back on his feet, and then was sent home.[35]

Chilling stories like those of Westacott and Bedson remind us of the human cost of military mining on the Western Front. Underground combat dramatically increased both in terms of frequency and intensity because of the hugely increased mining and countermining activity along the front. The nature of those confrontations was every bit as dangerous and traumatic as any that took place on the surface. But they took place below ground, where only the participants could truly know how grueling they were, even though comrades who worked feverishly to offer help had an inkling of what was taking place.

Hellish underground fights were a part of what the military miner had to endure in World War I as the pace and intensity of mining and countermining accelerated throughout 1916. Historically, mine warfare reached its highest state of scope, sophistication, and intensity that year on the Western Front. It resulted in multiple underground attacks in support of the Somme offensive; created far more smaller, localized attacks between the opposing lines than ever before; and resulted in far more and better documented examples of underground combat as opposing miners met each other in the dark. While the British gained a reputation as the best mining power of the war by 1916, it was not enough to give them a decisive advantage in above-ground operations. The age-old difficulty of linking siege mining with infantry attacks persisted despite the extraordinary nature of mine warfare by the midpoint of World War I.

I3

1917

The year 1917 saw the apogee of mining not only on the Western Front but also for all of world history. The offensive mining system that supported the British attack at Messines Ridge on June 7 dwarfed all previous mine projects in the global past. Even so, it was by no means the sum of military mining activity on the Western Front that year. Extensive work took place at Vimy Ridge by British forces and at Les Éparges and the Butte de Vauquois by both French and German miners. But after the enormous expenditure of energy and resources that year, mining activity dropped off dramatically during 1918 and would hardly be noticed in subsequent military operations for the next hundred years.

Perhaps it is ironic that at the start of this important year in the history of military mining, there already were signs of decreasing reliance on the ancient art. The need to dig deep underground shelters began to take root in all three armies on the Western Front, and miners were the most capable personnel to dig them. The Germans started this trend in 1915, making good use of them to protect their frontline troops in the Vimy Ridge area and during the Somme offensive. The French learned this lesson after capturing some deep German shelters in May and June 1915 and began digging their own variety, usually with about twenty-two feet (6.70 meters) of cover overhead—anything shallower, typically called cut-and-cover shelters and established just below the surface, had proven inadequate to withstanding heavy artillery fire. While the Germans and French placed their deep shelters at the front or main firing line, the British preferred to place theirs on reserve lines, believing that it was too difficult to rush large numbers of troops out of them in time to meet enemy attacks. By 1917, miners in all three armies were increasingly digging deep shelters rather than working on offensive or defensive mines.[1]

But all armies still had enough in the way of mining personnel and resources to support some major campaigns in 1917. For an attack at Arras on April 9, the British developed a unique approach to marshaling troops and equipment to support the assault. They discovered a marvelous system of caves east and southeast of the town that had been dug beginning in the seventeenth century as civilians mined chalk blocks in the area. These tunnels were forty to ninety feet (12.19 to 27.43 meters) below the surface and marked irregular paths because the miners had searched for the best quality chalk to be found, "just as a miner follows pay ore." British miners explored and developed this system for military use. The improved caves could accommodate thousands of men, with staging areas for attacks and room for support services. New galleries were dug from this system toward the German position to provide underground passages for assaults, which the British called subways to distinguish them from offensive mine galleries.[2]

A total of twelve subways were constructed at Arras from October 1916 to support the Canadian Corps assault on Vimy Ridge, which took place on April 9, 1917, but most of the work occurred during the last three and a half months prior to the assault. One of them, Goodman Subway, was 1,883 yards (1,721.81 meters) long, a little more than a mile. Another, Grange Subway, was 1,343 yards (1,228.03 meters) long. The total length of all twelve subways was six miles (9.65 kilometers). Relatively spacious, most were six feet, six inches (2.01 meters) high and wide enough for two men to pass each other. No shoring was necessary because the chalk held up well by itself. The moles could make good progress in this type of digging, even though the size of the tunnel was much bigger than usual. Sixteen feet (4.87 meters) of gallery in twenty-four hours was a good day, but some sections managed to dig forty-eight feet (14.63 meters) in that time; one section achieved the incredible record of sixty feet (18.28 meters) in twenty-four hours, although no one could dig that fast for long. Five tunneling companies were involved in the massive project at Arras. Digging these subways was "a great relief to us," admitted Capt. Herbert W. Graham of the 185th Tunnelling Company, who was a South African mining engineer before the war. It was "more in keeping with our professional work—it was clean, straightforward, less dangerous and compact."[3]

The subway complex at Arras consisted of branches from the main dozen tunnels to allow wider access to a broad front. One main stem alone offered access to two miles (3.21 kilometers) of front and involved more than 3,600 yards (3,291.84 meters) of galleries, including branches. Each branch ended with an incline to bring personnel close to the surface near the target. Electric lights illuminated most of them, and trams were set up in some. The subways became home to dressing stations to care for the wounded and storerooms for ammunition and equipment. Traffic officers, signboards, maps, and even some traffic lights were placed to control movement at the time of the attack. These were wise precautions because from April 5 to April 11, a total of 9,700 troops moved through one subway alone.[4]

The subways linked the old chalk caves with access to the front line, and those caves, enlarged and improved by the miners, became the forward support areas for the attack. A total of twenty-five were used and linked together by newly dug galleries. Two reservoirs with twenty-five feet (7.62 meters) of overhead cover were constructed of wood lined with tarpaulins; each contained 50,000 gallons (189,270.59 liters) of water. Hospital accommodations were established as well. Thompson's Cave alone had room for seven hundred wounded men, and its operating rooms, kitchens, and latrines allowed for injured soldiers to stay as long as needed. When ready for evacuation to a general hospital, ambulances could motor into the cave system for easy transport.[5]

The British also found that the modern sewer system of Arras was spacious enough for military purposes. One sewer was lined with brick and had a pathway on each side. Miners connected this system with the chalk caves by digging galleries to extend their already extensive and complicated network of underground movement, storage, and support even more.[6]

Mining played another role in the coming attack at Vimy Ridge. Officers who studied the terrain of the battlefield identified nineteen groups of existing mine craters along four miles (6.43 kilometers) of front. They recommended the infantry avoid four of those groups because their clay content had caused them to fill with water over time. The other fifteen crater groups would not pose much of a hindrance to an assault. Planners initially conceived of twenty-six galleries to target the total of twenty-six salient points in the German defense line, but in the end the majority of those galleries were never completed or used. This was largely due to concerns about creating too many fresh, large craters that could impede infantry movement on the surface. "As a result," writes historian Michael Boire, "of 26 large mines of various types initially planned to support the assault, four were abandoned due to technical difficulties, two were detonated before 9 April for defensive purposes, five were fired as planned, eight were prepared but not used and seven others, though mentioned in the earliest stages of planning," were not recorded as having been dug. The attack was eminently successful at pushing the Allied line farther east, and miners played a large role in that victory.[7]

The combined cave-subway-sewer system of underground movement and support at Arras was unique in the Great War and in mining history generally, dependent as it was on the fortuitous placement of the civilian quarries in the right place to support a major attack in April 1917. Unique in its own way was another major operation heavily supported by mining that took place only two months later near Ypres. The British attack at Messines Ridge on June 7, 1917, has become an icon of military mining, taking its place as the largest and most successful mine assault in world history. It was the greatest achievement of John Norton-Griffiths's moles and represents the best that was possible in the concept of underground warfare.

Norton-Griffiths claimed responsibility for the idea of using a massive series of mine explosions to prepare the way for the infantry attack at Messines Ridge, with a decided twist to the normal plan. In July 1915, when he initially proposed his idea, it was rejected. Pitching it again in January 1916, he managed to convince British generals. The key to his idea was that Messines Ridge was a series of connected sand hills resting on "blue London clay." He explained that "the intention was not to 'blow out' but to shake the whole of the Ridge" to create what he called "a little earthquake" preceding the assault to shock the Germans. While that idea made some headway with the generals, Norton-Griffiths felt that they were finally convinced only when he asserted that the plan could save 10,000 lives and that "the Army could walk to the top, smoking their pipes."[8]

A total of nineteen charges eventually were incorporated into the attack plan for the assault at Messines Ridge. They were planted at the end of eleven offensive galleries, six of which held more than one charge. Five of the galleries had been started back in 1915, five in 1916, and only one in 1917, but all were adjusted to accommodate the 1917 attack plan. The galleries were spread out along the outer edge of a bulge in the Germans' forward line that stretched for over eight miles (12.87 kilometers) of front.[9]

The northern-most gallery was located at Hill 60, the site of numerous underground conflicts for the past two years. It had been started on August 22, 1915, and was 870 feet (265.17 meters) long. Two branches extended from that main gallery. Branch B veered to the right for another 510 feet (155.44 meters) to end under a terrain feature of the German line known to the British as the Caterpillar. Here a charge of 70,000 pounds (31,751.46 kilograms) of ammonal was planted 100 feet (30.48 meters) below

Location of Gallery and Mine (North to South)	Start Date	Completion Date	Depth of Charge in Feet	Weight of Charge in Pounds	Length in Feet
Hill 60-A	August 22, 1915	August 1, 1916	90	53,500	Gallery 870 and Branch 240
Hill 60-B (The Caterpillar)	August 22, 1915	October 18, 1916	100	70,000	Gallery 870 and Branch 510
St. Eloi	August 16, 1916	May 28, 1917	125	95,600	Gallery 1,640
Hollandscheschuur Farm-1	December 18, 1915	June 20, 1916	60	34,200	Gallery 825
Hollandscheschuur Farm-2	December 18, 1915	July 11, 1916	55	14,900	Gallery 405 and Branch 45
Hollandscheschuur Farm-3	December 18, 1915	August 20, 1916	55	17,500	Gallery 405 and Branch 395
Petit Bois-1	December 16, 1915	August 15, 1916	57	30,000	Gallery 1,810 and Branch 210
Petit Bois-2	December 16, 1915	July 30, 1916	70	30,000	Gallery 1,810 and Branch 260
Maedelstede Farm	September 3, 1916	June 2, 1917	100	94,000	Gallery 1,610
Peckham Farm	December 20, 1915	July 19, 1916	70	87,000	Gallery 1,145
Spanbroekmolen	January 1, 1916	June 6, 1917	88	91,000	Gallery 1,710
Kruisstraat-1 (unused)	January 2, 1916	July 5, 1916	57	30,000	Gallery 1,615
Kruisstraat-2 (unused)	January 2, 1916	July 12, 1916	62	30,000	Gallery 1,310 and Branch 170
Kruisstraat-3 (unused)	January 2, 1916	August 23, 1916	50	30,000	Gallery 2,160
Kruisstraat-4 (unused)	January 2, 1916	April 11, 1917	57	19,500	Gallery 1,615
Ontario Farm	January 28, 1917	June 6, 1917	103	60,000	Gallery 1,290
Trench 127-1	December 28, 1915	April 20, 1916	75	36,000	Gallery 770 and Branch 250
Trench 127-2	December 28, 1915	May 9, 1916	76	50,000	Gallery 1,355
Trench 122-1	February 15, 1916	May 14, 1916	60	20,000	Gallery 350 and Branch 440
Trench 122-2	February 15, 1916	June 11, 1916	75	40,000	Gallery 350 and Branch 620

Table of the Messines Mines, Western Front, 1917. Adapted from Ball, "Work of the Miner," 222–223.

the surface on October 18, 1916, and maintained for nearly eight months until needed. Branch gallery A extended 240 feet (73.15 meters) to the left and held a charge of 45,700 pounds (20,729.17 kilograms) of ammonal planted 90 feet (27.43 meters) deep on August 1, 1916. This left-branch charge also contained 7,800 pounds (3,538.02 kilograms) of guncotton in slabs, packed in the empty spaces between ammonal tins. Three sets of leads were set up for each of the two charges, with sixty detonators sprinkled about. The main gallery, sometimes called the Berlin Tunnel, was 4.25 feet by 2.25 feet (1.29 by 0.68 meters) in diameter and was tamped for 240 feet (73.15 meters) to contain the blasts. While the 175th Tunnelling Company had constructed the gallery leading to both targets, the 3rd Canadian Tunnelling Company had planted the explosives. The 1st Australian Tunnelling Company then took control of the area in November 1916 soon after the plantings.[10]

Even before the Messines operation began, the Hill 60 area had witnessed heavy fighting and mining. Located 60 meters (196.85 feet) above sea level, it is an artificial mound of sand from the nineteenth-century construction of a railroad cut crossing the ridge. After two years of mining, the British tunnel system here consisted of four levels at 15, 45, and 90 feet (4.57, 13.71, and 27.43 meters) deep, with the beginning of another level at 100 feet (30.48 meters). Fifty listening posts were in operation. During the previous fifteen months before the Messines attack, a total of eighty-four mine charges had been blown in this compact area.[11]

Two miles (3.21 kilometers) south of Hill 60 along the Messines Ridge, the next gallery approached the area of St. Eloi. Begun on August 16, 1916, it stretched for 1,640 feet (499.87 meters), but a charge of 95,600 pounds (43,363.43 kilograms) of ammonal was planted 300 feet (91.44 meters) short of the face of the gallery and at a depth of 125 feet (38.10 meters). This was the deepest of the Messines mine charges. The ammonal was packed in 1,912 tins, each of which held 50 pounds (22.67 kilograms) of the powdery explosive. Six hundred feet (182.88 meters) of tamping with 1,800 sandbags completed the charge. The 1st Canadian Tunnelling Company did most of the digging of this gallery, hauling out 1,700 cubic yards (1,299.74 cubic meters) of spoil in 75,000 sandbags. They incorporated part of a gallery previously dug by the 172nd Tunnelling Company more than a year before and finished everything by May 28, 1917, a little more than a week before the big show.[12]

The next site, one and a half miles (2.41 kilometers) south of St. Eloi, held three charges connected to one gallery in the vicinity of Hollandscheschuur Farm. The gallery had been started on December 18, 1915, and was pushed to completion by the 250th Tunnelling Company. The men extended it only 405 feet (123.44 meters) before reaching a point where two branches diverged. The main gallery continued from there another 420 feet (128.01

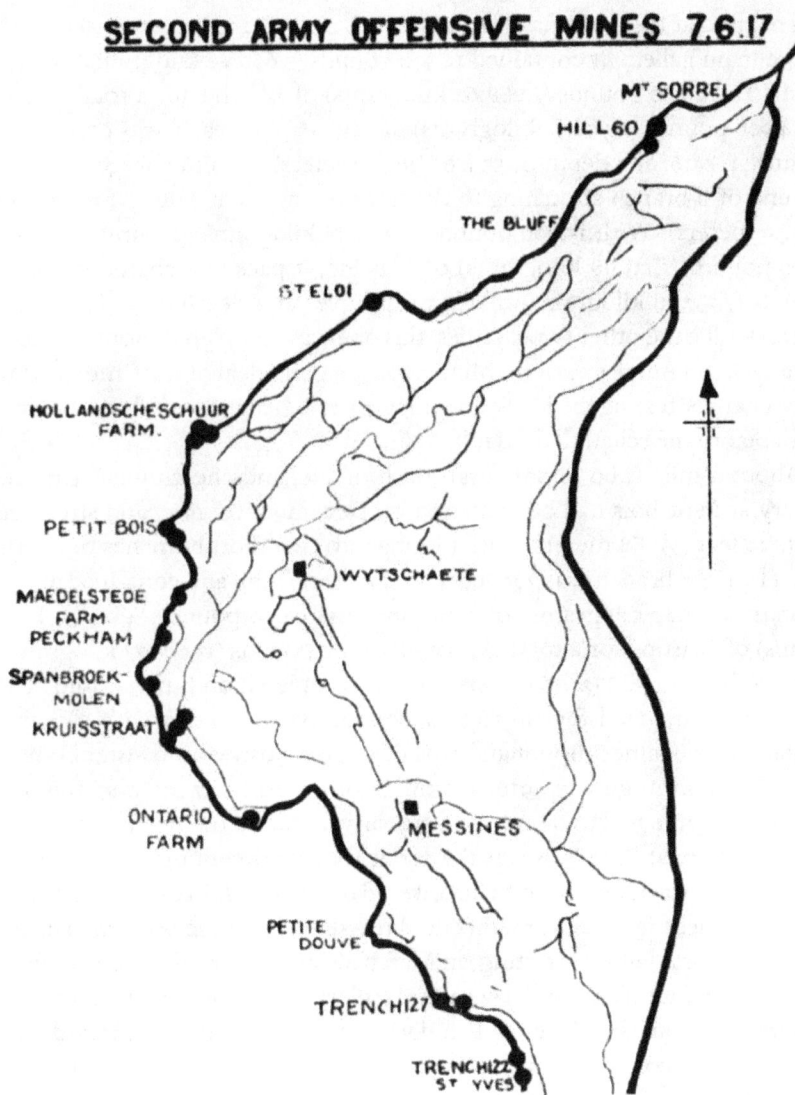

SECOND ARMY OFFENSIVE MINES 7.6.17

M⁺ SORREL
HILL 60
THE BLUFF
ST ELOI
HOLLANDSCHESCHUUR FARM.
PETIT BOIS
WYTSCHAETE
MAEDELSTEDE FARM
PECKHAM
SPANBROEK-MOLEN
KRUISSTRAAT
ONTARIO FARM
MESSINES
PETITE DOUVE
TRENCH 127
TRENCH 122
ST YVES

Map of the Messines Mines, Western Front, 1917. From Ball, "Work of the Miner," 224.

meters) to make the entire length 825 feet (251.46 meters). A charge was planted at the end of the main gallery consisting of 30,000 pounds (13,607.77 kilograms) of ammonal and 4,200 pounds (1,905.08 kilograms) of Blastine, a total charge of 34,200 pounds (15,512.85 kilograms) of high explosives, at a depth of 60 feet (18.28 meters). This charge was completed on June 20, 1916, and was maintained for nearly a full year before being used. Another charge was planted at the end of a short branch of 45 feet (13.71 meters) to the left of the main gallery. It contained 12,500 pounds (5,660.90 kilograms) of ammonal and 2,400 pounds (1,088.62 kilograms) of Blastine for a total charge of 14,900 pounds (6,758.52 kilograms) of high explosives. It was completed on July 11, 1916, at a depth of 55 feet (16.76 meters). A third charge rested at the end of a branch stretching to the right of the main gallery for 395 feet (120.39 meters). With 15,000 pounds (6,803.88 kilograms) of ammonal and 2,500 pounds (1,133.98 kilograms) of Blastine, it packed a charge of 17,500 pounds (7,937.86 kilograms) of high explosives at a depth of 55 feet (16.76 meters). Like the other two charges, this one was completed months before the attack, on August 20, 1916. Miners spent a great deal of time monitoring these charges, testing the leads, and hoping the Germans would not discover them before the scheduled assault.[13]

About a mile (1.60 kilometers) south of Hollandscheschuur Farm, the gallery at Petit Bois had been started on December 16, 1915, and stretched for 1,810 feet (551.68 meters) until forking into two short branches right and left. The right branch was 260 feet (79.24 meters) long and contained 21,000 pounds (9,525.43 kilograms) of ammonal and 9,000 pounds (4,082.33 kilograms) of Blastine for a total charge of 30,000 pounds (13,607.77 kilograms) at a depth of 70 feet (21.33 meters). It was completed on June 30, 1916. The left branch stretched for 210 feet (64.00 meters) and ended with a charge containing the same amount and type of high explosives as its sister branch. Completed on August 15, 1916, at a depth of 57 feet (17.37 meters), this and the other charge had to be nursed for nearly a year before needed.[14]

The gallery at Petit Bois was the venue for experimenting with the Stanley Heading Machine. It had to be lowered down a shaft in dissembled parts for eighty-eight feet (26.82 meters) and reassembled at the bottom. The experiment was a failure. Starting on March 4, 1916, the machine began well, cutting a six-foot-diameter (1.82-meter) gallery at the rate of two feet (0.60 meters) per hour, but it tended to dive down while moving forward, and the operators could not find a way to stop that tendency. Then another problem developed. As it continued forward and down, the machine encountered such heavy clay that, when it was turned off, the clay swelled and engulfed its edges. Operatives had to dig it out several times before finally giving up entirely on the experiment, leaving the machine to rust forever, and clay-kicking their way around it.[15]

Three-quarters of a mile (1.20 kilometers) south of Petit Bois, the gallery at Maedelstede Farm supported only one charge. Started on September 3, 1916, it ran for 1,610 feet (490.72 meters) and ended with 90,000 pounds (40,823.31 kilograms) of ammonal and 4,000 pounds (1,814.36 kilograms) of guncotton, a total weight of 94,000 pounds (42,637.68 kilograms) of high explosives, at a depth of 100 feet (30.48 meters). It was the second-largest charge among the Messines mines and was completed on June 2, 1917, only five days before the offensive.[16]

A little more than half a mile (0.80 kilometers) south of Maedelstede Farm, another single-charge mine gallery was dug toward the area of Peckham Farm. Begun on December 20, 1915, the gallery was 1,145 feet (348.99 meters) long. Miners charged it with 87,000 pounds (39,462.53 kilograms) of high explosives—65,000 pounds (29,483.50 kilograms) of ammonal, 15,000 pounds (6,803.88 kilograms) of Blastine, and 7,000 pounds (3,175.14 kilograms) of guncotton. At a depth of 70 feet (21.33 meters), the charge, tamping, and installation of leads was completed by July 19, 1916, and maintained for almost a year before use.[17]

The next gallery, half a mile (0.80 kilometers) south of Peckham Farm, began early but very nearly was not used in the Messines attack. Men of the 250th Tunnelling Company started it with a shaft 60 feet (18.28 meters) deep on January 1, 1916. By the third week of that month, after driving the gallery 90 feet (27.43 meters), they handed it over to the 3rd Canadian Tunnelling Company, which pushed it to 790 feet (240.79 meters) in three months. Then the 175th Tunnelling Company took over and two weeks later was replaced by the 171st Tunnelling Company. This unit finished the gallery at 1,710 feet (521.20 meters). It also planted 91,000 pounds (41,276.90 kilograms) of ammonal 88 feet (26.82 meters) deep by June 28 and then maintained the charge indefinitely.[18]

The Germans probably would not have disturbed the gallery at Spanbroekmolen, the next mine south of Peckham Farm, if the British had not become ambitious about using it for other purposes. In early 1917 they dug a branch from it toward a spot in the enemy line called Rag Point. This branch came to the attention of the defenders, and they blew several camouflets that not only damaged the branch but a section of the Spanbroekmolen gallery as well. This sparked a desperate effort by the moles to restore control of their huge charge of ammonal at the end of the gallery. They dug a bypass about 450 feet (137.16 meters) long parallel to the main gallery to go around the damaged junction of the Rag Point branch. But they hit a pocket of earth disturbed by one of the German camouflets that still contained gas, which killed three men before rescuers could reach them. After this tragedy, the moles edged past the damaged earth to continue their bypass, racing against time to finish the job before the planned attack. They barely managed to do so, securing and checking the ammonal charge on June 6, literally a few hours before the start of the offensive.[19]

Half a mile (0.80 kilometers) south of Spanbroekmolen, at Kruisstraat, the 171st Tunnelling Company planted four charges at the forward end of one gallery but encountered so many problems that none of them were blown as part of the Messines attack. They dug 2,160 feet (658.36 meters) but planted two charges 545 feet (166.11 meters) back from the face of the gallery, approximately 1,615 feet (492.25 meters) from the shaft. The miners packed one with 30,000 pounds (13,607.77 kilograms) of ammonal at a depth of 57 feet (17.37 meters) by July 5, 1916, and the other with 18,500 pounds (8,391.45 kilograms) of ammonal and 1,000 pounds (453.59 kilograms) of guncotton at the same depth by April 11, 1917. After setting those two charges, the miners placed two more. A third chamber was made at the end of a branch gallery that veered off to the right and was 170 feet (51.81 meters) long. It was packed with 30,000 pounds of ammonal at a depth of 62 feet (18.89 meters) on July 12, 1916. The fourth chamber, with 30,000 pounds of ammonal 50 feet (15.24 meters) below the surface, was completed at the face of the main gallery by August 23, 1916.[20]

As at Spanbroekmolen, German activity endangered the use of this gallery network at Kruisstraat. A camouflet so damaged the tunnel that the British, unlike at Spanbroekmolen, decided it was not worth restoring control of the four charges. One of them detonated when lightning struck the ground nearby during a thunderstorm in 1955, but the other three still lie there, apparently ready to be set off by the right kind of stimulus.[21]

The 171st Tunnelling Company also was responsible for the gallery at Ontario Farm three-quarters of a mile (1.20 kilometers) south of Kruisstraat. Beginning on January 28, 1917, the moles dug the tunnel 1,290 feet (393.19 meters) and placed one charge of 60,000 pounds (27,215.54 kilograms) of ammonal 103 feet (31.39 meters) down, the second-deepest of the Messines mines next to the one at St. Eloi. As with Spanbroekmolen, the miners at Ontario Farm finished the work on June 6, 1917, barely in time for the offensive.[22]

The next-to-last set of mines was attached to a gallery that extended from a section of the British line identified as Trench 127 about one and a half miles (2.41 kilometers) south of Ontario Farm. The 3rd Canadian Tunnelling Company started the gallery on December 28, 1915, and extended it 1,355 feet (413.00 meters), planting a charge of 50,000 pounds (22,679.61 kilograms) of ammonal 76 feet (23.16 meters) deep by May 9, 1916. A second charge was placed at the end of a branch 250 feet (76.20 meters) long. Containing 36,000 pounds (16,329.32 kilograms) of ammonal at a depth of 75 feet (22.86 meters), it was completed on April 20, 1916.[23]

A little less than half a mile (0.80 kilometers) south of Trench 127, a gallery at Trench 122 constituted the southernmost set of mines for the Messines offensive. The location was known as Factory Farm. Its gallery, which was started on February 15, 1916, extended 350 feet (106.68 meters) before forking into two branches. The left fork was 440 feet (134.11 meters) long and ended in a charge of 20,000 pounds (9,071.84 kilograms) of ammonal 60 feet (18.28 meters) below the surface. It was completed by May 14, 1916. The right fork was 620 feet (188.97 meters) long, supported a charge of 40,000 pounds (18,143.69 kilograms) of ammonal 75 feet (22.86 meters) deep, and was completed by June 11, 1916.[24]

From August 22, 1915, to June 6, 1917, the preparatory phase of the underground attack at Messines Ridge was by far the longest of any mine operation in history. The twenty charges contained 933,200 pounds (423,292.40 kilograms), or 446.6 tons, of high explosives. The combined length of galleries and branches equaled 23,620 feet, or 4.47 miles (7.19 kilometers). Messines Ridge dwarfed any mine project in history except the British mine preparation to support the Somme offensive of July 1916, when nineteen mines were placed. The charges of the Messines mines, however, were much larger than those used at the Somme and were far more intricately tied together as a unit. Whereas the Messines charges were deliberately planned to act together to shock the Germans and quake the ground of their defenses, the Somme charges acted individually on selected target areas, and all of them were not exploded at exactly the same time. British officers planned the ultimate mine offensive at Messines and managed to pull it off in spectacular fashion.

But Messines Ridge was the most difficult British mine project of the war because of its size, complexity, and duration. It was highly unusual that sixteen of the twenty charges were finished months before they were needed, necessitating a system of daily monitoring and maintenance of those dangerous accumulations of high explosives. Ammonal, which constituted the bulk of the charges, was highly vulnerable to moisture, but the sealed tins worked well to keep it dry in deep underground chambers surrounded by wet chalk for months at a time. The leads had to be tested daily to make sure they still worked. But the biggest threat to the project lay in the possibility of German detection, and in that regard the British were fortunate. Despite the many mines and the long duration of the project, the enemy failed to discover their plan. It is true the Germans damaged two galleries—the British abandoned one while putting the other back into use—but the fact that this is all they managed to do reflected poorly on the effectiveness of their listening systems.[25]

The Messines offensive was designed to gain a limited objective—only the ridge itself. The British Fifth Army deployed an attack force of 80,000 troops supported by three hundred airplanes and seventy-two tanks. From May 26 onward, 2,400 artillery pieces bombarded the German position with 3.5 million shells. In consultation with the mining officers, commanders set guidelines for how the troops should behave when the charges went off. No one was permitted to be within two hundred yards (182.88 meters) of a mine, and troops were cautioned not to stand near a tree, a wall, or a parapet for fear the shock waves might make those items fall. All deep shelters within four hundred yards (365.76 meters) of a charge were to be evacuated just before the blow, and advancing men were to avoid the bottom of the newly formed craters.[26]

Finally, after many months of tedious preparation, eleven firing posts were ready to touch off nineteen mines at 3:10 A.M., June 7, 1917. They all went up within forty-five seconds of each other, and many men witnessed the sight through the dimness of early dawn. They felt the earth shiver as nineteen mounds of earth and debris rose into the sky and then fell. At Hill 60 one could hear gases hiss out of the raw craters and see tongues of flame. Mine Inspector Robert Harvey was watching from the top of Mount Kemmel but could barely take in the experience. He had a good view of Spanbroekmolen, but the other mines, more distant, springing up one after another until he "found it difficult to concentrate on looking. . . . [T]here was so much going on and the scene, which baffles description, developed so quickly." Reports later circulated that it had been possible to hear the mine explosions in London and even in Dublin.[27]

The craters formed by those explosions varied in dimension. Their diameters at ground level were always smaller than what the Royal Engineers termed the "diameter of complete obliteration," by which they meant the sum of a crater's diameter at ground level added to the width of the rim. The depth of a crater always was less than the depth of the charge. That latter fact developed because much of the displaced earth and debris settled directly down after being lifted some distance, filling in much of the crater. The rest of the displaced earth formed a pronounced rim around the circumference of the hole and also was scattered around the area. At every crater the rim was much wider than it was tall. Spanbroekmolen is a good example of all these statistics. Here the diameter at ground level was 250 feet (76.20 meters) while the zone of destruction was 430 feet (131.06 meters) in diameter. Crater depth was 40 feet (12.19 meters), less than half the depth of the charge, which was 88 feet (26.82 meters). The Spanbroekmolen rim was 90 feet (27.43 meters) wide but only 13 feet (3.96 meters) tall. The Ontario Farm charge was an outlier among the Messines Ridge mines. For some reason, it did not form a good crater. The diameter at ground level was 200 feet (60.96 meters), but the zone of destruction extended to only 220 feet (67.05 meters) in diameter. Crater depth was described by the Royal Engineers as "practically nil," even though the charge had exploded at a depth of 103 feet (31.39 meters). The rim was only 10 feet (3.04 meters) wide and 4 feet (1.21

meters) tall. In fact, reports indicate that Ontario Farm left "a vast circular, pulpy-looking patch that bubbled slowly for days like porridge coming gently to the boil."[28]

The mining work at Messines Ridge was an unparalleled technical success, but even more importantly it played a key role in the tactical success of the Fifth Army offensive. The Germans were taken completely by surprise. Many of the troops in their first line, just on the British side of the ridge slope, were killed. The rest retreated in fear. Advancing British troops easily occupied the abandoned first line and fought for control of the villages of Messines and Wytschaete, which the Germans had fortified since they shielded their second line on top of the ridge. After they fell, the second line was abandoned. British troops began to dig in on top of the ridge to consolidate gains that afternoon, and a week later the Germans evacuated their third line, located on the eastern slope of the ridge. As a result, the offensive was unusually successful at biting off a large salient along the Western Front. At comparatively light loss, Fifth Army captured a chunk of French soil more than twelve miles (19.31 kilometers) wide and three miles (4.82 kilometers) deep, which included the only high ground in the region.[29]

No wonder that Messines Ridge became an icon of military mining and its potential for promoting success in surface operations. "Messines is to the Tunneller what Waterloo was to Wellington," declare W. Grant Grieve and Bernard Newman, veterans of the Western Front and early historians of the British moles. Historian Alexander Barrie agrees: "It had certainly been the tunnellers 'finest hour.'" There can be no doubt that the miners played the key role in the overwhelming success of this impressive attack. Grieve and Newman do not exaggerate when they conclude that "the Battle of Messines was won in the bowels of the earth."[30]

In the mind of John Norton-Griffiths, who had inspired the Messines Ridge work, the effect of his moles was unmistakable. In an after-dinner speech given at a meeting of the tunnelers veteran organization ten years later, he called Messines Ridge "the greatest mining effort in the history of the world." He was certain that the infantry attack "would have been an absolute failure" without the explosions. "This stupendous artificial earthquake shook the ridge from end to end . . . and enabled the army—as we had promised—to walk to the top of the ridge in comparative safety." Grieve and Newman, writing almost twenty years after the battle, conclude that Messines Ridge proved beyond a doubt that the British tunnelers were superior to the Germans "in purely offensive mining."[31]

The characteristics of the mine project at Messines set it apart. The nineteen charges covered a total area of more than eighteen acres (7.28 hectares), which was equal to the 236 mines that had been exploded around the Ypres

Salient combined since the start of the Great War. The galleries tended to be longer than usual, and most of them were dug long before need. Once the offensive was approved, mining officers tried to incorporate as many previously dug galleries into the plan as possible. Given the complexity and size of the mining operation, it became necessary to develop plans for maintaining charges over many months until needed. The size of the charges tended to be much bigger than usual, which is why the total came to nearly one million pounds of high explosives. The depth at which those charges were planted tended to be deeper than normal as well. Packing so much high-explosive power in so comparatively small an area, the resulting shock waves reportedly were felt at Lille, France, nearly twelve miles (19.31 kilometers) away.[32]

The British had time afterward to examine the captured ridge. They found extensive damage caused by the nineteen mines. Some Germans had been killed instantaneously in their shelters, still in poses they had assumed just before the moment of death, and some had smiles on their faces. A few German trenches on the surface had collapsed, burying occupants in standing positions. A number of captured Germans were emotionally shocked by the charges and had broken down, especially by the fact that the explosions took place one right after another instead of all at the same instant. While it could not be ascertained how many of the 20,000 German casualties of June 7 were caused by the mines, it was possible to verify that 679 were killed at Hill 60 and the Caterpillar and another 400 at St. Eloi by those explosions.[33]

On examining the German shafts and galleries within the captured ground, British mining officers came away with a reinforced view that their own work surpassed that of their opponent. It was obvious from the progress marks on shoring that the Germans worked much slower than the moles—about six feet (1.82 meters) a day compared to twenty-six feet (7.92 meters) daily by the British. The heavy shoring and shafts lined with concrete denoted thoroughness in the German approach to mining that the British disdained as unnecessary and time consuming.[34]

Messines Ridge was the swansong of the British military miner. Underground projects dropped off quickly from the spring of 1917, not just on the British sector but everywhere on the Western Front. There is no clear reason for this except that mining had been pushed to its extreme in 1916 and 1917 so that the bubble burst, so to speak, and all three belligerents began to realize that the tactical results generally did not justify the enormous investment of manpower, time, and material resources. This was a potent reason why Messines tended to shine so brilliantly in the minds of British advocates of mining—it was such an unusual event in the global history of military mining as well as the last major mining project of the moles.

English-language sources naturally tend to give more prominence to British mining operations on the Western Front, leaving the work of French and German miners in the shadows. While there are plenty of sources in the French and German languages that document this part of the war, little of it has been translated into English. But those two armies engaged in extensive, deadly, and dramatic mining operations. None of them were as decisive as Messines Ridge, but all of them constituted the heart and soul of their underground-war efforts.

A hill located northeast of Reims in Champagne, Mont Cornillet, be-came the focus of a gruesome episode in May 1917. The German lines rested on top of the hill, and their miners drove galleries under it, starting from the eastern side. These were not designed for attack purposes but to shelter the troops. Miners found chalk quarries and incorporated them into three mil-itary galleries that were 10 feet (3.04 meters) wide, 7.5 feet (2.28 meters) tall, and from 393 to 469 feet (119.78 to 142.95 meters) long. A transverse gallery linked them for ventilation. Soldiers sheltering in those galleries emerged to repel French attacks on April 17 and May 4, 1917.[35]

Rather than countermine to solve this tactical problem, the French de-vised a plan to use artillery fire against the German gallery entrances. Be-ginning on May 18, artillery spotters in airplanes directed massive fire across Mont Cornillet to land on the area of the entrances and partially close them. The next day French gunners fired phosphine gas shells at the ventilation ports and entrances. On May 20, the climactic day of the operation, artillery fire essentially closed all entrances. Carbon monoxide produced by explod-ing shells seeped into galleries while the sides and ceilings of the tunnels also began to crumble. The shelters had become deathtraps. Those Germans able to move tried to escape, and while some managed to do so, hundreds remained trapped. On the afternoon of May 20, an infantry attack captured the hill; in consolidating their gains, the French simply left the German bod-ies inside the galleries. An effort to recover the remains in 1933 failed due to the continuing presence of poison gas that endangered anyone venturing underground. It was not until 1973 that a joint effort by French and German army personnel finally entered the galleries and recovered the bones of 321 men for burial in a nearby cemetery.[36]

Operations at Mont Cornillet demonstrated the danger of relying on deep shelters to protect infantrymen if the enemy could reach the entrances with accurate artillery fire. Even the German tendency to thoroughly tim-ber the sides and roofs of galleries failed to protect their occupants from this murderous heavy shelling.

A major focus of mine warfare developed at a ridge known as Les Éparges, located fourteen miles (22.53 kilometers) west of Verdun. The lines settled near the western end of this feature, which was 1,050 feet (320.04 meters) tall and half a mile (0.80 kilometers) long, and hardly moved for four years. The French began to sap and mine on the north and west slopes in the middle of November 1914. They blew mines at the end of four saps, in true Vaubanian fashion, and attacked on February 17, 1915, to capture the western end of the ridge. But afterward the opposing forces remained locked in this static position for two and a half years of digging and blowing.[37]

Mine activity at Les Éparges tended to come in spurts. By May and June 1915, it picked up as huge craters dotted the top of the ridge between the lines. The Germans blew several in September and October that year; one of them, a huge charge, obliterated five infantry sections of the French 303rd Regiment on October 13. Three days later two large French mines created craters that were seized by friendly troops. The pace continued, with an average of four charges blown by one side or the other every week during late 1915 and early 1916, until there was "an almost continuous furrow of mine craters" on top of the ridge. Many of those charges landed up at the end of galleries that had been started in the slope of the ridge rather than from shafts or inclines. Mining continued into 1916 but began to lessen the next year. Even so, a few were charged and sprung as late as August 1918 until a combined French and American attack captured the ridge on September 12–13 without any mining to help the effort.[38]

The intensive underground effort by both sides at the western edge of Les Éparges was by no means decisive in influencing surface operations. It represented the application of a field tactic that suited the terrain and tactical situation. While the many craters certainly altered the terrain, creating a crazy moonlike surface of denuded ground, mining could not alter the tactical stalemate.

The same was true of another, similar location along the French-German lines. The Butte de Vauquois rises 950 feet (289.56 meters) high nineteen miles (30.57 kilometers) west of Verdun. With steep sides and measuring 520 yards (475.48 meters) long by 371 yards (339.24 meters) wide, it supported the medieval village of Vauquois, which no longer exists. The Germans settled their line on its top soon after their first offensive into France stalled. French troops then gained a foothold on top and controlled part of the village by February 1915. The next month the Germans began a nearly continuous mine contest that lasted almost for the rest of the war. In response, French miners dug nine shafts by mid-April but were unable to bring any of them into play to support a failed ground attack in early May. On the might of May 13, German miners broke through the roof of one French gallery and used pistols and grenades to clear it before exploding charges to collapse the tunnel. But other French galleries came close enough to the German surface position to damage trenches when blown.[39]

This first phase of the mine war at Vauquois was followed by a new, more concentrated phase in June 1915. The French dug shafts every fifty to sixty-five feet (15.24 to 19.81 meters) along a 437-yard (399.59 meters) front to create a dense countermine system. The Germans also shifted into a new phase of their offensive efforts by digging a new set of galleries from their second line. As they had previously done, they started the new galleries from the cellars of houses in the village but now used electric drills for faster progress through the sandstone that constituted most of the butte. The fact that no shoring was needed also added to the speed of digging. The Germans also constructed transverse galleries for ventilation. When both sides struck a large void in the sandstone, intense underground combat ensued before both blew camouflets to deny its use to their opponent. These second-phase galleries tended to be fifty to sixty-five feet (15.24 to 19.81 meters) deep in contrast to the shallower galleries dug in the first phase, which were thirteen to nineteen feet (3.96 to 5.79 meters) deep. By the middle of 1915, the French had blown seventy-seven charges, and the Germans had set off fifty-one mines. Two-thirds of those explosions broke through the surface, but rather than battle for control of the craters, both sides tended to occupy the lips on their side of the hole and incorporate the position into their respective trench system.[40]

By the end of 1915, both sides entered the third phase in the mine warfare at Vauquois. This period was characterized by bigger charges, with the French setting off 29,762 pounds (13,499.81 kilograms) of high explosives to create a crater 65 feet (19.81 meters) in diameter. The Germans also began to dig galleries starting in the steep northern slope of the butte, again avoiding the need for shafts. This also allowed them to dig galleries even deeper than before, starting about 82 feet (24.99 meters) down from the top of the butte and declining the tunnel to a point where it was 131 feet (39.92 meters) below the surface. They installed a small-gauge railroad to remove spoil and dug deep shelters in the sandstone for troop protection.[41]

All of this led to a vicious mine war at the butte during 1916. A German charge of 36,376 pounds (16,499.87 kilograms) of high explosives placed 114 feet (34.74 meters) deep devastated part of the French line on March 3. The French retaliated with six tons (5,443.10 kilograms) of explosives that similarly devastated sections of the first and second German lines on March 23. This phase culminated in the springing of the largest single mine charge of World War I. The Germans packed sixty tons (54,431.08 kilograms) of high explosives at the end of a gallery 282 feet (85.95 meters) long, distributed five hundred detonators, and laid down three sets of leads. Thirty men tamped this charge with eighty-seven yards (79.55 meters) of material, including 25,000 filled sandbags and 19 feet (5.79 meters) of concrete, nearly closing the entire length of the gallery. When this huge mine exploded on May 14, more than a hundred French infantrymen in the first, second, and third lines went missing. A sector of the defenses 114 feet (34.74 meters) wide was destroyed, and three French shafts and galleries collapsed, with nine miners killed. A crater 196 feet (59.74 meters) in diameter was left.[42]

This terrible explosion spurred the French into starting six new galleries from shafts located 98 feet (29.87 meters) behind their first line. The galleries descended at an angle of forty-five degrees until reaching a depth of 98 feet below the first French line, then continued toward the enemy on a horizontal plane. The Germans blew a large mine to destroy one of the galleries on December 10, 1916, killing twenty-one French soldiers, but fifty others were rescued by their comrades. While all this was going on, the Germans had been working on another system of their own. Started in April 1916, their new galleries began in the north slope of the butte 137 feet (41.75 meters) below the top. By late July, they used the galleries to blow several mines, some of them to enlarge the already big crater produced by the May 14 blow and another to enlarge a crater created on March 3.[43]

The intensity of mine warfare on top of the butte wore on everyone involved. The three French engineer companies primarily responsible had lost 215 men killed by August 1916. They were replaced by three fresh units, which in September started yet another system of three galleries, this time beginning in the southern slope of the butte 124 feet (37.79 meters) from the top. They set up an electrical power system to light the galleries and used electric winches to raise and lower material, in addition to using compressed air to run pneumatic drills. A small-gauge railroad removed spoil. Despite this increased level of activity, the Germans already were deeper than these galleries. A new set of French galleries, one leading from the southern slope and the others from shafts on the top, were started near the end of 1916. Still, the French were not gaining on their opponents in this deadly contest.[44]

Overall, while the Germans never dominated the mine war at the Butte de Vauquois, they did manage to counter French offensive efforts rather consistently. But in 1917 both sides began to wind down their intense digging. The Germans blew their last big mine on March 9, and the last French mine that broke the surface exploded on June 30. From that time on, both sides concentrated on camouflets to counter their opponent rather than on offensive digging. When German miners broke into a gallery and captured a French listener in late July, the French retaliated, which in turn led to another German strike. While the mine contest at the Butte had descended to harassment of countermines and listening posts, it also tended to become pretty intense. From February to August, 1917, the Germans set off thirty-eight countermines, and the French exploded forty-seven camouflets. Still, the perceived need to remain at a deeper level than the French spurred German engineers to plan their last, deepest system at Vauquois. In the summer of 1917, they began three new galleries from the north slope, inclined sharply to reach a depth of 196 feet (59.74 meters), and then angled in a more shallow way to continue the descent to a depth of 311 feet (94.79 meters) by March 1918. The French were unaware of this deep penetration.[45]

The project proved to be the last new digging at Vauquois. The French blew their last camouflet on March 21, 1918, and the Germans did the same a few days later. Italian troops replaced French infantry in this sector on May 16, and French engineers evacuated their entire mine system the next month. By this time, the Germans considered the Butte de Vauquois less important to their operations. They moved their main line one and a half miles (2.41 kilometers) away in January 1918 but held a forward observation post on top. That post fell on September 26 during the early phases of the American Meuse-Argonne offensive, as troops of the Thirty-Fifth Division bypassed the butte and rendered it unimportant. All that remained now was a shattered surface, scarred by huge craters, as the village that once was the home of 170 people had been obliterated. Eleven miles (17.70 kilometers) of combined French and German galleries lay empty.[46]

For two and a half years, the opponents had been locked in a strange fixation on this small corner of the Western Front, a fixation that led each to invest an enormous amount of manpower, resources, and time in attempting to blast the other out of their lodgment on top of the butte. During that time, the French exploded a total of 320 mines, while the Germans set off 199 charges—a grand total of 519 mines blown in an area 520 yards (475.48 meters) long by 371 yards (339.24 meters) wide, easily making it the most mine-saturated terrain in global history. New craters merged into old craters, and no-man's-land became a bizarre, angular landscape that was virtually unredeemable after the war. Neither the village nor the ground was reconstructed; the butte stands as one of the most vivid memorials to the tragedy of the Great War.[47]

What can one make of the mine warfare at the Butte de Vauquois? It was the most intense underground contest in world history in terms of scale and duration, but it failed to win anything for either side. Each opponent managed to counter the offensive efforts of the other, with the Germans usually one step ahead of the French. They had an advantage in that the northern slope was steeper and lent itself better for the start of galleries than the southern slope. In fact, starting galleries in slopes was relatively uncommon throughout the history of mining and countermining, so the butte provided the venue for more such activity than was usual. The sandstone made for relatively easy digging, which also facilitated the drive toward a more intense effort by both sides. Terrain and geology offered the military engineer a tempting environment at the butte, and neither side could resist it. But at the same time, neither side could find a solution to the age-old problem of the military miner—how to make his effort pay off by helping surface troops win their objective.[48]

During four years of draining operations along a front that crossed international boundaries, the British, French, and Germans planted and exploded an astounding number of mine and countermine charges. There is no truly comprehensive data for the entire Western Front during the whole war that is readily available, but two large chunks of it are accounted for in reliable terms. In 1916 the British blew 750 mines along their sector, while the Germans opposing them exploded 700 charges. The number drastically reduced in 1917, with 117 British charges compared to 106 German explosions. The most active month for both sides was June 1916, with 101 British and 126 German mines blown. Combined with the accurate data on French and German blows at the Butte de Vauquois (a total of 519 charges), we know for certain that at least 2,192 mines and countermines were exploded on the Western Front. But that total only accounts for the British sector and one intense but small part of the French sector. Extrapolating this reliable data, one can reasonably conclude that 5,000 mines and countermines were set off by the British, French, and Germans along the entire Western Front during World War I.[49]

While it is difficult to estimate how many mine charges had been exploded during the course of history since the advent of gunpowder three hundred years before the Great War, there is no doubt that it could not be anywhere near five thousand. In all previous mining projects, typically no more than half a dozen or so mines were set off at each operation. At Sebastopol, where the largest number of mine blasts had occurred before 1914, the French set off eighty-seven and the Russians eighty-five charges. During the entire American Civil War, only half a dozen mines were blown. It is quite possible that World War I witnessed more than double the number of mine and countermine explosions than all previous conflicts combined. In mining history World War I was an astounding departure from previous siege mining practices.

The British succeeded more consistently in mine warfare on the Western Front than did the French or Germans. The primary reason for this was the infusion of civilian miners into their engineer forces. It was not only

Butte de Vauquois, 2018. The shattered top of the butte, remaining largely as French and German miners left it as far as their craters are concerned, provides a stark image of mine warfare on one small spot of the Western Front, its effect on the environment, and its legacy. Still from Scott, *Butte de Vauquois Drone 4k* (June 2018), YouTube, December 24, 2018, https://www.youtube.com/watch?v=Y54AjzGsyNo.

the drive and vision of John Norton-Griffiths that inspired his colleagues and the rank and file but also the expertise and experience brought in by dozens of mining engineers from around the empire and commonwealth. The moles brought with them a hardy and confident attitude toward their work as they strove to beat the Germans in, first defensive, then especially offensive mining. They readily understood the needs of siege mining and adapted their digging skills to the task.

When it suited them, the British also tended to deviate from the standard practice of military mining. Their drive led to impressively increased rates of digging not envisioned by the military manuals. The chief advantage in this was in the clay-kicking technique, but the rate of digging also was spurred by an aggressive spirit no matter what technique was used. The British were quicker to reenter a gallery after a blow, whether exploded by themselves or the Germans, and set out to bypass an affected area as fast as possible to continue work on an offensive gallery. They also avoided wasteful effort whenever possible. While the French and Germans tended to shore their galleries pretty thoroughly (especially the Germans), the British adjusted the degree to which they shored according to the nature of the soil so as not to waste time and material. They also found that less shoring allowed them to better judge how much sagging was taking place along the walls and ceilings of galleries over time. All of those factors led to a pretty consistent domination over their opponents. Given time, material, and freedom to do as they thought best, British miners usually succeeded.[50]

The French and German mining efforts were massive in their own right but lacked many of the elements that led to British success. Both belligerents infused civilian miners into their military forces, although to a much lesser degree than did the British. The French and Germans tended to be more strictly guided by prewar mining doctrine and technical details as well. And their miners could become fixated on intense mining activity, like that at Les Éparges and the Butte de Vauquois, but without either side gaining a decisive advantage over the other comparable to what the British achieved at Messines Ridge. Neither the French nor the Germans truly gained an upper hand over the other in any sector of the Western Front, expending their time, energy, and resources in an underground stalemate that mirrored their stalled operations on the surface. The Germans especially were satisfied with a comparatively leisurely pace of digging and focused a great deal on heavy shoring. They were good miners, in their own way, if one were digging for ore in peacetime. But they failed to adjust as readily to the peculiar demands of military mining as did the civilian miners of Britain.[51]

In 1917 Messines Ridge witnessed the most impressive underground attack combined with surface attack in the history of mining. That year also saw the culmination of French-German mining obsession at the Butte de Vauquois. In between, the British enjoyed the luxury of extensive underground shelters carved out of chalk near Arras. By the end of the war, while probably 5,000 mining charges had been blown along the Western Front, the best mining was turned in by British moles.

Although the Western Front witnessed the greatest episode in the global history of military mining and countermining, it was not the only area of World War I in which mining took place. Wherever static operations ensued, some degree of underground warfare also occurred to a much greater degree than in any other conflict in history.

14

Other Fronts of World War I

The Western Front was a vivid culmination of trends evident in the history of military mining since the introduction of gunpowder. In the level of manpower and material employed as well as the number of mining projects, the number of shafts and galleries, and the huge number of charges planted and exploded, it far exceeded any other venue for underground warfare in history. That culmination also owed something to the venue provided by the great powers in the Great War. More troops were packed into static positions for longer periods of time and with more material of all kinds at their disposal than ever before in history. But other overlooked fronts of the Great War also witnessed mining. In operations on the Gallipoli Peninsula, the expansive eastern front, and the Italian front as well as to a very limited degree to some operations in the Middle East, miners conducted their work on a much smaller scale than on the Western Front. They used modern technology but had less opportunity to experiment with higher levels of mining management and fewer chances to learn from experience because the number of ignited charges and follow-up attacks was much more limited.

One lesson to be drawn from military mining on the subsidiary fronts of the Great War is that underground approaches were very adaptable to circumstances of topography, geology, force structure and strength on the surface, and the ebb and flow of attack and defense. Mining took place in small operations like that at Gallipoli; in alpine regions, boring through solid rock, on the Italian front; along a front in eastern Europe that was longer, less densely saturated with troops and artillery, and less static than the Western Front; and briefly in a highly mobile campaign in the Middle Eastern deserts. Previous wars had taken place in all of these kinds of situations before without any mining. But there was something unique about World War I in the global history of underground warfare to spur miners and their commanders to greater efforts wherever belligerent formations happened to be campaigning.

The joint British-Australian effort to knock Turkey out of the war by landing on the Gallipoli Peninsula early in 1915 has become justly famous, garnering a great deal of attention from historians and students of the Great War. But most overlook the fairly extensive tunneling that took place during this failed campaign. The Allies managed to attain nothing more than a toehold on the shore of the peninsula and were locked into static positions for months to come. Gallipoli's geology is conducive to digging. The soft sediments that underlay the dry surface make for good tunneling—in other words, easy digging and little need for shoring facilitated mining operations.[1]

Australian forces discovered signs of Turkish mining pretty early, on May 5, and quickly instituted countermeasures. Three days later they started three tunnels from the bluff slope so as to avoid shafts, removing the spoil in sandbags, and set up listening posts. Australian diggers continued the three tunnels, with only eighteen inches (45.72 centimeters) of earth between the roof of each and the surface, and joined them with a transverse gallery ninety feet (27.43 meters) forward of the front line. Ventilation and observation holes broke through the surface every forty-five feet (13.71 meters) along this transverse tunnel.[2]

The Turks exploded the first mine at Gallipoli on May 29 but without achieving much tactical benefit. Mining continued as the Australians planned a significant effort to support surface operations. Their plan was to spring four mines at the end of those three tunnels just before an attack on Turkish positions near the Valley of Despair on the night of July 31. Two of the four exploded as planned, a third delayed and killed both Australian and Turkish troops, while the fourth charge never went off at all. Nevertheless, the attack achieved its limited objectives. The three tunnels were opened near the end to help in consolidating the new ground, and the troops held their gains against Turkish counterattacks.[3]

As work progressed on the mine attack against the Valley of Despair, the Australians had also been working on a larger project to attack Lone Pine. Here they dug seven shallow tunnels linked by a transverse gallery forward of their own line. Two of the tunnels continued forward from the transverse toward the target. The plan was for infantry to leave those galleries after miners broke open to the surface only forty to sixty yards (36.57 to 54.86 meters) from the Turkish trench on August 6. When the tunnels were opened at 5:30 P.M., the attack went in and captured the position.[4]

By September, mining activity picked up so that underground confrontations began to take place. When Australian miners hit an enemy gallery on September 11, a firefight broke out, with Turkish miners firing rifles into the Australian tunnel. Both sides constructed barricades, and the Australians managed to plant and explode thirty pounds (13.60 kilograms) of guncotton. Then they did what their compatriots on the Western Front often did and constructed a bypass gallery under this point of conflict to reach the enemy gallery and collapse it with a charge. At another location the Turks broke through an Australian gallery on September 14, but the Australians had heard their digging and were ready, firing through the breach and forcing their opponents back. The Turks later advanced to bomb the Australians, but blocking the hole with sandbags stopped that tactic. Then the Australians placed a charge and fired it, destroying forty feet (12.19 meters) of their own gallery to permanently seal off Turkish access to it.[5]

On October 20 the Allies dramatically escalated their mining plans by starting several galleries sixty feet (18.28 meters) deep. By mid-November, the heads of these were near the Turkish trenches. This represented something akin to what had been happening on the Western Front—initial limited efforts failed to achieve a breakthrough in surface operations but led to increased activity underground. If the campaign had continued for many months, mining might have increased over time, but by late November, the Allies decided to evacuate the peninsula. Much discussion ensued as to what to do with the tunnels, eventually incorporating them into the evacuation plan. Sixteen mines were charged and readied to be set off if needed. On December 19 six of them were exploded at Cape Helles to distract the Turks from the evacuation of Anzac Cove and Suvla Bay. The next day three more were set off at the Neck. When the last troops evacuated Cape Helles on January 9, 1916, seven charges were left unfired (and still remain more than a century later). Of a total fourteen and a half tons (13.15 metric tons) of high explosives planted in the sixteen mines, the Allies left eleven tons (9.97 metric tons) unexploded.[6]

The Australians used about a hundred miners at Gallipoli. One of them, Sgt. Cyril Lawrence, wrote of the experience in a diary and two reports, providing much detailed information. The soils "ranged from black earth to clay and from a hard kind of sandstone to coarse loose sand." Little if any shoring was needed, so the miners expanded the dimensions of their galleries, averaging five feet, nine inches to six feet (1.75 to 1.82 meters) high and two feet, six inches to three feet (0.76 to 0.91 meters) wide. They found some ancient pottery "of a deep red colour and very fine and close in consistency." Initially, sappers dug while detailed infantrymen removed spoil, but later the infantrymen helped dig as well. As they progressed four to eight feet (1.21 to 2.43 meters) every eight-hour shift, the men made small ventilation holes up to the surface every fifteen feet (4.57 meters) by poking through the shallow earth above the gallery roof with bayonets attached to rifles.[7]

The tunnels at Gallipoli were generally shorter and used smaller charges than those on the Western Front. In the B group of galleries, with which Lawrence was intimately familiar, none apparently extended more than 180 feet (54.86 meters) in length, with laterals up to 60 feet (18.28 meters) long. The biggest charge he recorded was one hundred pounds (45.35 kilograms) of ammonal. Miners often heard sounds of Turkish picking and adjusted their own work accordingly, even bending one gallery to avoid an area where a Turkish soldier was buried. Inclined galleries were needed only if the approach crossed a ravine. In one such case Lawrence noted that the gallery declined one foot (0.30 meters) every three feet (0.91 meters), then came back up again, similar to the Union gallery that approached Stockade Redan at Vicksburg during the American Civil War. The Australians tamped their charges in the traditional way, in one example with 7 feet (2.13 meters) of tamping material, then an air gap of 5 feet (1.52 meters), and another 5 feet of tamping for a charge of thirty pounds (13.60 kilograms) of guncotton. All the charges early in the campaign were shallow, little more than 12 feet (3.65 meters) deep, thus there was no need for huge amounts of high explosives.[8]

Lawrence provided much human interest to our understanding of life in the Gallipoli mines. "It is jolly hard work picking and shoveling by candle-light and after working for a while you are wringing wet and then you spell and immediately get cold, because down here in these burrows it is both cold and damp." At one point Lawrence came down with "a chill or something. It's those damned tunnels that are giving me this and the toothache."[9]

Working in and near the tunnels could also be dangerous. Early one morning near the end of Lawrence's shift, a Turk shell happened to explode just at the mouth of the gallery in which the sergeant was working at the face. The bursting shell decapitated an Australian who was standing at the mouth, and its blast flowed unimpeded all the way into the tunnel, extinguishing the candles and tearing off the empty sandbags used to cover the air holes at night to prevent light from being seen on the surface. The blast even pushed Lawrence against the face 150 feet (45.72 meters) away from the mouth as it filled the tunnel with smoke. After restoring the sandbags and arriving at the mouth, he saw an awful sight. Blood and brains not only were scattered around but had sprayed five feet (1.52 meters) down the gallery as well.[10]

Encounters with the enemy could be sudden and startling. One night a Turk "shoved his hand down the air hole in the drive just opposite me," Lawrence wrote. "It scared the wits out of us all. We could hear them crawling around above us all night." Far more deadly were the camouflets, which exploded without warning. This led to frenzied efforts to rescue comrades caught in the zone of destruction and a combination of gas and smoke filling sections of galleries. The Australians apparently had not established a rescue system with trained personnel or specialized equipment. On one occasion Lawrence and a comrade were poisoned by gas from a blast and tried to leave the tunnel but collapsed halfway out. They had to be dragged the rest of the way. "We were all violently sick, splitting headache and am very groggy in the legs," he wrote in his diary. Officers assigned him to office duty until he fully recovered.[11]

By July 1915, the Australians had dug a total of one and a half miles (2.41 kilometers) of galleries at Gallipoli. While they failed to develop a good rescue system, their work in general was similar to that of British, French, and German miners on the Western Front but with a few interesting exceptions. In their early phase they dug tunnels at very shallow depths, similar to the Russian Saps of the Western Front, and used them differently than their comrades in Belgium and France. The Australians literally turned their transverse galleries into firing lines by digging out recesses in the forward wall of the gallery for infantrymen to occupy and then breaking open a firing port to the surface. In this way the troops could shoot at the Turkish line through a hole in the ground. The recesses were opened up enough to give the man good lateral aiming, but the transverse gallery remained intact to provide all-around protection. One such firing line was located ninety feet (27.43 meters) forward of the first Australian trench. On July 15 Lawrence's crew finished the basic outline of the recesses on one transverse and then handed it over to the infantrymen, "who are going to open them up and finish off their new abodes. They have just swarmed into them, digging with picks, shovels, bayonets and entrenching tools, and blocking the tunnels with earth so that you can only get along by crawling." There is no evidence that underground firing lines such as these were made on the Western Front or in any other venue of mining operations in history.[12]

After the small gains made in the attacks on the Valley of Despair and Lone Pine in early August, it was possible to examine Turkish galleries and compare them to the Australian diggings. Lawrence found the enemy works "not as high as ours but are wider and more of the sewer type. The faces of their tunnels are all very square and the miners apparently live in them." The Australians found dugouts and latrines "cut in the sides" of the galleries. Lawrence noted, however, that when the Turks dug a tunnel with the intention of exploding a charge at the end, they made the dimensions smaller. This rendered them a bit less vulnerable to countermeasures because the Australians would have had to explode a charge closer to such tunnels to damage them.[13]

Mining and countermining at Gallipoli thus not only followed some well-worn paths charted on the Western Front but also deviated in some ways. The deviations apparently came from individual ideas stemming from the Australian engineers and from the peculiar circumstances of the soil, the tactical situation, and the needs that prevailed on the peninsula. Neither side gained a decisive advantage over the other underground for any substantial period of time, although the Australian miners' role in promoting tactical success in the attacks on the Valley of Despair and Lone Pine stand out. If the campaign had continued, there is a real possibility that the Australians might have achieved a stunning success with their enlarged tunneling project that could have tipped the balance at Gallipoli in the Allies' favor.

The eastern front of World War I, which dwarfed the Western Front in terms of length and witnessed huge battles that swayed back and forth over a wide extent of ground, has been given second billing in the work of Great War historians. It represented the longest continuous line of combat thus far in global history. The trenches stretched from 932 to over 1,000 miles (1,499.91 to over 1,609.34 kilometers) in length across the extreme western extent of the Russian Empire and through parts of Austria-Hungary. In contrast, the Western Front was about 475 miles (764.43 kilometers) long for most of the war. With huge armies arrayed along both sides of no-man's-land, one would expect a high level of mining and countermining to have occurred on the eastern front.[14]

Yet that was not the case. Although the documentation is sparse, there is enough evidence to give a picture of mining history in the East, and it paled in comparison with that of the Western Front. A total of fifty-eight locations have been identified as sites of tunneling, mostly along the southern sector where Russians confronted Austro-Hungarians, but some were located along the northern sector opposite German units. The time period of this activity stretched from February 1915 to September 1917. Both the Germans and Austro-Hungarians used geologists and specialist mining units.[15]

A number of factors contributed to the comparatively low incidence of mining projects on the eastern front. These include the fact that the lines were not so locked in one place for extended periods of time as in the West and the much longer length of the eastern front, which saw a much lower density of manpower per linear mile. The mining resources likely were on a much smaller scale than that in the West, and it is quite possible that there were fewer good targets for offensive mining efforts in the East due to terrain and geology. All of those factors are in need of further study to illuminate this major theater of the war.

The fighting between Italian and Austro-Hungarian forces along the Italian front also has been overshadowed in Great War studies. But some intensive mining developed in one sector, the Dolomite region of the Alps in northeastern Italy. The terrain consists of a mountain range with eighteen peaks ranging from 9,833 feet (2,997.09 meters) to the highest, Marmolada, at 10,968 feet (3,343.04 meters) elevation. Many of them rise in nearly vertical slopes, creating deep, narrow valleys. All in all, the Dolomites constitute the most unusual venue for military mining in history. Engineers had to dig galleries through solid rock using chisels and handheld mechanical drills.[16]

From January 1, 1916, to March 13, 1918, miners on both sides of this strange war exploded a total of thirty-four mines—twenty by the Italians and fourteen by the Austro-Hungarians. They tended to be smaller than on the Western Front, ranging from 240 to 110,000 pounds (108.86 to 49,895.16 kilograms) of high explosives. The galleries were generally short, but at least one extended 1,320 feet (402.33 meters). They were almost evenly divided between offensive and defensive mines, but of the thirty-four dug, only two played any role in supporting a successful, although limited, surface attack. The largest number of charges exploded at one place occurred at Mount Pasubio, where ten mines were blown.[17]

The unusual topography and geology of the Dolomites forced alterations in the experience of military mining. Military miners had to learn lessons from civilian commercial mining in hard-rock regions and borrow specialized equipment. It proved virtually impossible to keep most of the mining projects a secret from the enemy. The noise of mechanical drills and the carrying tendency of soundwaves through rock proved to be good aids to any listener on the other side of no-man's-land. In many cases mine explosions led to rock slides ranging from dislodged large individual stones to entire areas of a mountain slope filling in sections of valleys. The third Austro-Hungarian mine to be touched off on Mount Lagazuoi, on May 22, 1917, stripped off a section of rock 220 by 150 yards (201.16 by 137.16 meters) in extent, resulting in 240,000 square yards (200,670.56 square meters) of material filling the nearby valley. In another incident fifty Italian soldiers were killed when one of the Austro-Hungarian mines at Pasubio broke off part of the rock face and sent it tumbling down the mountainside. Austro-Hungarian troops also were killed and injured in several Italian mine explosions. It is likely that more men were killed and injured from rockslides than directly from mine charges on the Italian front.[18]

The last and the largest mine explosion on the Italian front took place at Pasubio on March 13, 1918. The Austro-Hungarians planted 110,000 pounds (49,895.16 kilograms) of high explosives in two chambers and fired them at dawn, just before the Italians had planned to explode a mine of their own. The explosion wrecked the north face of the mountain and buried numerous Italian soldiers. But it obviously had been improperly tamped: back blast entered the Austro-Hungarian gallery and killed and injured friendly troops as well.[19]

The end result of mining on this front was not unlike that everywhere else in the Great War—a failure to play a decisive role in above-ground operations. From a technical perspective, both the Italians and the Austro-Hungarians mastered the topographical and geological conditions in the Dolomites to achieve success in digging tunnels, planting and exploding charges, and weakening enemy defenses. The thorny problem of integrating this technical achievement underground into successful operations on the surface were not solved any more on the Italian front than anywhere else in mining history.

It may seem impossible for mining to play a role in the mobile field operations that swirled across the deserts of the Middle East during the Great War, but so it did and to a very slight degree. Lt. Gen. Sir Frederick Stanley Maude conducted a major British offensive up both sides of the Tigris River beginning on December 16, 1916, capturing several fortified Turkish positions along the way. One of the major strongholds was located at Sannayait, taken in an assault on February 17, 1917. After the capture, Bvt. Maj. W. H. Lang of the Tenth Cavalry examined the defensive fieldworks and found that the Turks had dug several saps fifty yards (45.72 meters) forward of their first line. In them they constructed positions for snipers and for men to toss hand grenades. Lang also found at least one defensive tunnel running from the Tigris River parallel to and about twenty-five yards (22.86 meters) in front of the first line. This gallery was three feet, six inches (1.06 meters) square, with a roof only two feet (0.60 meters) below the surface. The Turks had shored it with planks and corrugated iron mainly to support the roof. Because this gallery ran across and under several saps, the Turks had made air holes for ventilation through the roof into their floors. Lang found "no signs of mines or electric wires," so he concluded the tunnel was "purely a defensive gallery against our mining." But Maude had no intention of engaging in such a time-consuming and laborious tactic on his campaign.[20]

Mining on all the other fronts of World War I paled in comparison with the Western Front, but the work at Gallipoli, on the eastern front, in the

Dolomites, and in the Middle East demonstrated how readily adaptable underground warfare could be. A willingness to dig was the key, and that willingness was at an all-time high during World War I compared to previous conflicts. That high level of interest in mining had been building for some time before 1914, spurred by technical advances that increased hope for success both in the process of exploding underground charges and in the possibilities of incorporating them into surface operations in an effective way.

In short, the Great War was the ultimate in siege mining not only because of the Western Front but also to a lesser degree because of mining on other fronts. From the introduction of gunpowder to the principles of Vauban and the experiments that documented the globe of compression (later called the zone of destruction) to the development of high explosives and electrical detonation systems, a trajectory can be seen aiming toward intense and extended use of mining by World War I. Important sieges from Schweidnitz to Sebastopol and Port Arthur demonstrated the utility of those new developments. Given the size of its armies, the professionalization of specialist units within them, and the long duration of static confrontation on several fronts, the Great War provided the perfect venue for the culmination of that trajectory.

The big question was would military mining continue to play such an elevated role in field operations after the Great War? Although it had reached apogee by 1917 with the impressive show at Messines Ridge, mining had still not proven itself a consistent support for success in above-ground assaults. Much would depend on how veterans of World War I interpreted the history of their efforts underground as well as on whether surface operations would remain the same after 1918 as they had been during the Great War.

15

After the Great War

The key characteristics that made the Western Front such a fertile field for mining also gave rise to the explosive growth of field fortifications. The armies were large enough to form a nearly continuous line along national frontiers, with no opportunities for flanking fortified positions. This led millions of soldiers to burrow deep into the surface of the earth for protection; adding heavy artillery to supplement field artillery encouraged them to dig even deeper. The trenches of World War I lodged massive numbers of men in the same locality for long periods of time. Many efforts to break this peculiar kind of positional warfare failed miserably until attrition, combined with innovative tactics and operations, restored some degree of mobility to the Western Front during the last few months of the conflict. The operations during that time were never so freewheeling as the early weeks of conflict in Belgium and France, but they certainly promised a true end to the positional warfare at some point in the near future. The sudden collapse of Germany in early November 1918 prevented observers from seeing how soon fully mobile operations might have resumed.

All of this led to a considerable amount of speculation during the years following the armistice as to what one might expect of the next major Continental war. As far as military mining was concerned, the conclusions divided into diametrically opposite viewpoints. On the one hand, there were those who were convinced that future conflicts would largely mimic the Western Front. "The next war . . . will be an underground war—air forces will see to that," wrote W. Grant Grieve and Bernard Newman. Authors of the first popular history of mining on the Western Front—this book appeared in 1936—their opinion was naturally flavored by their deep interest in the history of mining during the Great War. They believed that improvements in the air arm would overwhelm ground operations and force armies to continue digging for protection. Grieve and Newman also warned that the home front would suffer enormously because of air power. They advocated the digging of deep underground shelters even on home territory, employing out-of-work miners to make them before any war broke out. As far as the battlefield was concerned, Grieve and Newman assumed the same scenario would occur as had happened in 1914. An initial mobile phase would slow and stall, leading to a heavily fortified position that would once more require large numbers of tunnelers.[1]

Grieve and Newman were not alone. An American coast-artillery officer named Robert E. Cron Jr. was equally convinced that "mining operations, and tunneling under enemy positions, will become the ordinary method of fighting" in future wars. If this was so, then one needed to be prepared for it. In the later 1930s, as war clouds darkened over Europe, Harry Trounce wrote a multipart article for the American service journal *Military Engineer*, laying out the basics of mine warfare in simple and clear terms. The U.S. Army's *Engineer Field Manual*, published in 1918, had already devoted more than thirty of its pages to covering all aspects of mining.[2]

The viewpoint that mining would inevitably attend a future war in Europe was largely promulgated by men experienced in or specially interested in military mining. But ironically, those who held the opposite viewpoint were led by the men who superintended the British mining effort of World War I. After speaking with those British officers, American military geologist Alfred H. Brooks understood and sympathized with their viewpoint. "Those who directed the active period of mine warfare generally concede that the results achieved were not commensurate with the expenditure of personnel, matériel, and time, and it is improbable that military mining will ever again be employed on so large a scale as it was during this war."[3]

We do not have to rely solely on Brooks to bring out this point. In 1929 Robert N. Harvey, who had served as British inspector of mines during the war, spoke his mind in a lecture delivered to the Society of Military Engineers at Chatham. "The principal lesson to be learned is that any officer in a position of responsibility in the next big war should do everything he can to prevent mining being carried out except for strictly defensive purposes." He noted that the British devoted the equivalent of two divisions of personnel to military mining. In Harvey's view, that was too severe a drain on troop strength for the little gain achieved underground. Tunnelers, however, were very good at digging deep shelters for infantry. "This is useful mining," Harvey asserted. But "whole sale mining under the front trenches as was done on the British Front in France should be discouraged."[4]

In other words, the man who supervised the British mine war on the Western Front, which was not only the most impressive of that conflict but in world history, had thoroughly soured on the usefulness of large-scale mining in the field. Even the Americans, who had not been involved in any mining projects during the Great War, agreed with Harvey. An engineering textbook for entering freshmen at the U.S. Military Academy notes that "advance by mining is effective . . . but is very slow and costly." Modern listening devices and countermeasures could easily negate progress, rendering offensive mining "extremely dangerous and unsatisfactory." The U.S. Army's field manuals indicate the low importance placed on mining in American doctrine. In regulations issued in 1926 concerning the importance of underground water to field fortifications, the army also noted that mining was "tedious, disheartening work at best," and if a gallery was broken by an underground source of water, "the effect on morale is decidedly bad." Six years later the engineer field manual devoted much more information about how to dig deep underground shelters than how to conduct offensive or defensive mining. By 1940, the manual dropped the section on mining altogether.[5]

The message was very clear—avoid large-scale military mining at almost any cost. Such an endeavor was prohibitively expensive in terms of personnel, material, and time, even if a perfect venue for practicing the ancient art was presented as it had been in the Great War. Harvey allowed that one had to defend one's position if the enemy initiated offensive mining, but he urged future officers not to be lured into a grueling contest for supremacy under no-man's-land.[6]

War underground was not a contest that anyone could win easily enough to justify the effort. The most difficult impediment was the enemy's ability to counter offensive digging, while the second-most difficult problem lay in coordinating underground efforts with above-ground attacks. Sophisticated listening devices and quick countermine construction were formidable enemies of the offensive miner. And the problems of developing a truly integrated plan to make mining work for above-ground offensives had never truly been solved during the Great War despite the sterling but limited success of the Messines Ridge attack. In fact, one could point to that 1917 operation as proof that offensive mining was hardly worth it. The level of technical success had been unprecedented, to be sure, but it had been purchased at the cost of literally months of intensive digging and a good deal of favorable luck. Messines Ridge had been anything but typical of mining projects on the Western Front. It was emblematic of what could be accomplished if one invested huge resources of manpower, time, and material and if everything else worked out as hoped. But all those factors could not be expected to fall into place very often.

Even those commentators who argued that extensive mining should be avoided in the future admitted that tunnelers could still play a significant role when it came to constructing deep shelters. In the United States the *Engineer Field Manual* of 1940 continued to offer instruction on how to make cave shelters. Critics of extensive mining often admitted as well that small offensive mines aimed at specific strongpoints in an enemy position could be effective if there were no better ways to deal with the tactical problem and conditions were right for mining. That left the doctrinal door open for continued mining in future wars no matter how much mobility was added to military forces by the fuller development of tanks, mechanized infantry divisions, and tactical air support.[7]

Opportunities to dig shelters and to conduct the odd offensive mining project cropped up now and then throughout the extended period embracing the complicated events of World War II. Early in the Spanish Civil War (1936–1939), Republican forces besieged a small force of Nationalists in the Alcázar at Toledo. An "old Moorish fort" on a hill that was the home of the nation's military academy, the siege began on July 23, 1936, and lasted sixty-eight days. The Republicans enlisted the aid of civilian miners from Asturia to construct two galleries aimed at the southwestern tower and the west wall a bit north of the tower. The garrison heard sounds of digging on August 16, but rock inhibited countermining, and sorties could not locate the starting point of the two galleries. The Asturians had actually started their tunnels by digging shafts inside two houses eighty yards (73.15 meters) from the west wall and had finished one gallery by September 16. Two days later, at 6:21 A.M., charges in both galleries exploded, with newsreel crews and still photographers from Madrid covering the event. Five members of the garrison were killed in the blasts. But a follow-up attack by 2,500 men, supported by a tank and two armored cars, failed. The Asturians did not give up. They targeted a stable with another gallery and blew up part of the structure with TNT, but a subsequent attack failed to fully capture the site. Later they blew another charge at a water conduit, which did little harm. Any further effort was cut short by a Nationalist relief force that raised the siege on September 28.[8]

As the example of the Alcázar indicates, the only real need for mining would be if surface operations ground to a halt in efforts to attack or defend

a strong position for some time; it also would depend just as much on decisions by local commanders and access to trained personnel and appropriate
materials. In the expansive, intense, and deadly operations that occurred
around the globe during World War II, the nature of military operations
only rarely offered such conditions. Improvements in many weapons such
as the tank, mobile artillery, and tactical air support during the 1930s, led by a
resurgent Germany, largely negated the possibility of a major front stalling
in trench warfare similar to that in France and Belgium during the Great
War. Periodically, surface operations in various theaters of World War II
did begin to resemble static, positional warfare, but commanders in those
regions almost never called on the military miner for help. They preferred
to seek quicker solutions to their tactical problems and almost always found
them.

As a result, there was no significant military mining or countermining in
all of World War II, the biggest conflict ever to occur in global history, even
though it began within the lifetime of the generation that had conducted
the most extensive mine warfare in history during 1914–1918. Of course, engineering manuals had dropped instruction in gallery mining by the end of
the 1930s, a historic departure from centuries of traditional attention paid to
it in the military manuals of several nations. Even the possibility of digging
deep underground shelters, a role that the military miner could fulfill on
demand, seemed outdated in the context of mobile surface operations.[9]

But before completely dismissing gallery mining from military history,
we must acknowledge that at one small place on the huge Eastern Front of
World War II at least two traditional mines were constructed to good tactical effect. In the massive struggle for control of the Soviet industrial city of
Stalingrad in 1942–1943, operations bogged down in bitter house-to-house
combat. In both cases of mining, Soviet sappers targeted buildings used
by the Germans as strongpoints. Their first attack involved digging a shaft
16 feet (4.87 meters) deep and extending a gallery 141 feet (42.97 meters) to
the target in fourteen days of work. Then the sappers planted 6,600 pounds
(2,721.55 kilograms) of explosives to blow it up.[10]

In the second major effort, sappers began their gallery in the cellar of
one building and aimed at a German strongpoint in a building across the
street. On the first day they broke into a water pipe and continued forward.
One sapper broke off chunks of earth at the face while another scraped it
out of his way and others put it in sandbags for carrying back to the cellar,
where the bags were used to strengthen fighting positions. The sappers had
to set up shoring frames, consisting of two vertical pieces joined at the top
by a horizontal piece, close to each other along the way. Their gallery was
only five feet (1.52 meters) high and three feet (0.91 meters) wide, which

allowed for faster progress, but the men were forced to walk through it in a bent posture. They worked continuously in shifts of five to six hours, using pocket lanterns and lamps powered by storage batteries for illumination. Sappers also fashioned hand-operated bellows for air circulation, although some degree of natural ventilation entered the gallery both from the cellar and from the water pipe. By the end of the fourth day, the Soviets could hear the Germans above them, so they worked more quietly, wrapping cloth around their shoes. When it came time to plant the charge, they filled the chamber with TNT, stretched a lead, and tamped the gallery. The miners exploded the charge with a hand-held detonator; it so disrupted the German strongpoint that an infantry assault captured the building.[11]

If we can find but one small area of World War II in which mining took place, all it can indicate is that this method of dealing with a stubborn tactical problem in surface operations had not completely died out. And World War II was not by any means the last war to witness tunnel mining. Aggressive underground approaches to blow up a strongpoint did not appear during the Korean War (1950–1953), even though that conflict was largely stabilized in positional warfare after some nine months of highly mobile operations. Strong lines of field fortifications spanned the Korean Peninsula to provide a rough approximation (on a much smaller scale) of the Western Front for more than the last two years of the conflict. But Chinese forces did construct tunnel shelters to protect troops and supplies from aerial bombardment.[12]

Lt. Col. Robert S. Mayo of the Marine Corps Reserves provided an explanation as to why military mining never was attempted in the Korean War despite the apparently good venue for it. He noted that the geology of the peninsula presented daunting obstacles. It was not only hilly but largely solid bedrock underlying a thin layer of soil. Most of the tunneling would have had to bore through that bedrock with power equipment that would create a lot of noise and be difficult to hide from the enemy. In the valleys the upper layer consisted of loam and tended to be quite wet except for a couple of months during the year.[13]

As we have seen in World War I, military miners did not hesitate to tunnel through solid bedrock or deal with soggy upper layers on the Italian front and in Belgium. Mayo was not aware of their achievements, but he was fully aware of how civilian miners in the United States contended with such problems, briefly describing some of this in the pages of *Military Engineer*. To deal with quicksand or "running ground," civilian engineers cased it with steel and removed it with devices that employed pressure. They also could solidify running ground "by injecting Chemical Emulsion through pipes jetted into the face ahead of the tunnel. Temporarily solidified sand has the consistency of soft clay; permanently solidified is about the hardness of a soft sandstone and may be mined with a pneumatic pick without blasting." Mayo warned, however, that all civilian methods of dealing with problematic geological conditions were time consuming, labor intensive, and required equipment not normally issued to engineer units in the field. The gallery would need to be four feet (1.21 meters) wide and six feet (1.82 meters) high "to provide the necessary room for drilling and blasting" in bedrock.[14]

Mayo was aware of fundamental concepts used in ventilation, noting that a two-inch (5.08-centimeter) pipe and blower operated either by hand or by motor could be set up. He even wrote of creating a bellows in the field by using pieces of plywood and canvas. For getting water out of a gallery, a hand- or motor-driven pump would be needed. Lighting the gallery was easy. An automobile storage battery could power a system of wires running "along the roof with a small bulb hung at every station. Very little light is required except at the face of the tunnel." Mayo was aware of the Great War geophone, mentioning that it still was the primary tool of listeners even in 1953.[15]

Military mining as practiced in World War I had gone quite a bit farther than Mayo's rather limited conception of it in 1953, the only exception being in the area of preparing soft ground for mining by using modern chemicals and methods honed in civilian practice since 1918. No one serving in the army during the Korean War era would have had any personal experience at military mining. If theater commanders had been willing to authorize tunneling, someone would have had to do quite a bit of historical research to discover the best practices of thirty-five years in the past as well as to glean more information from midcentury civilian mining practices in the United States. But as in World War II, no one thought it was worth the effort to resort to tunneling during the Korean conflict.

That was not the case after the war on the peninsula ended in 1953. Without a firm peace settlement, tensions simmered for decades along the narrow demilitarized zone (DMZ) separating the armed forces after the armistice was signed on July 27. By the late 1960s, South Korean authorities began fortifying their side of the DMZ, which was 150 miles (241.40 kilometers) long and 2.5 miles (4.02 kilometers) wide. They constructed the concrete foundation for a fence; erected wire, sensors, and watch towers; and established concrete bunkers and antitank ditches. Unknown to them at that time, the North Koreans had already begun to tunnel their way under the DMZ in the early 1960s and accelerated their efforts when the field fortifications began to appear along its south side. Each of the ten North Korean divisions deployed along the north side started two tunnels. The first sign of this activity appeared in November 1973, and after that three of the tunnels were found by South Korean officials and neutralized. The first one was 148 feet (45.11 meters) deep, lined with concrete slabs, and had a small-gauge railroad running through it. The second was discovered in 1975 at Ch'orwon and was 160–500 feet (48.76–152.40 meters) deep, 6.5 feet (1.98 meters) wide and high, and a bit over 2 miles (3.21 kilometers) long. The third was located in 1978 and measured 240 feet (73.15 meters) deep and a bit more than a mile (1.60 kilometers) long. The reason for those incredible depths is the hilly nature of the landscape in this area. Galleries started at much shallower depths and pushed basically at a level under a mountainous terrain would automatically be much deeper in the middle sections than at either end.[16]

The purpose of the North Korean tunnels fit the long global pattern of military mining. They were an offensive tactic, not to undermine a strong-point but to infiltrate operatives into South Korea. The North Koreans had been doing this for years, before the South began fortifying their side of the DMZ, by pushing small parties along the surface of the zone at night. The purpose of those teams was to destabilize the South Korean regime, in-spire revolt, and punish the South Koreans for supporting the American war effort in Vietnam. The South Korean field fortifications had been started to hinder this surface infiltration, but they inadvertently increased the un-derground approach. Relatively little detail about digging technique and material is publicly available—even in the twenty-first century the DMZ in Korea remains one of the most sensitive places in the world—but it seems that North Korean engineers were fully aware of best practices for military mining and employed them to construct major tunnels for the transit and shelter of troops.[17]

North Korea's DMZ tunnels may well be the last example in world his-tory of gallery mining for the sake of offensive effort. But even before they were started, the second-to-last example of offensive mining had already taken place at the siege of Dien Bien Phu in 1954. While the North Korean galleries represented a historical link with the British subways of Arras, the North Vietnamese tunnels at Dien Bien Phu represented a throwback to Vauban's method of undermining a strongpoint in a fortified position. During the last major campaign of the French Indochina War of 1946–1954, a small French force was heavily besieged by North Vietnamese troops. The only slim advantage it had were field fortifications consisting of earthwork strongpoints and bunkers. Initially relying on heavy artillery bombardment, North Vietnamese commander Võ Nguyên Giáp decided during the siege to close in on the defenders because of a looming shell shortage. His plan was to dig saps toward key strongpoints to neutralize French artillery fire and allow his troops to more effectively use small-arms fire.[18]

The North Vietnamese conducted this new approach with skill. On reaching the minefield and wire belt protecting French strongpoints, they began digging short and shallow galleries from the saps, pushing under those obstructions and sometimes directly into French trenches. They carefully planned this effort to support above-ground strikes. On the night of April 22, the North Vietnamese opened three galleries inside a strongpoint named Huguette-1 and poured troops through, which helped capture the work with minimal losses. In another effort the Eighty-Third Engineer Company dug a gallery under Eliane-2, a heavy bunker that sheltered French troops and from which they had conducted counterattacks to fend off North Vietnamese strikes. The engineers veered slightly off course but managed, when exploding a ton of explosives on the night of May 6, to disrupt the structure enough to aid in its capture that night.[19]

The North Vietnamese effort at Dien Bien Phu demonstrated an awareness of the deep history of military mining, the successful application of Vaubanian practices in the mid-twentieth century, and an impressive last hurrah for the professional military miner in his chosen sphere. In contrast, the North Korean tunneling under the DMZ represented something that also fell within the realm of legitimate military mining but without the satisfaction of blowing something up at the end of the long, difficult job of digging tunnels under mountains.

While it may have lacked the high drama of exploding a massive charge under the enemy, the construction of infiltration tunnels and deep shelters for troop safety and supply or as refuge for civilian populations was actually the future of the military miner after the Great War. Beginning with the deep shelters of the last half of World War I and the development of those wonderful subways and shelters in the chalk under Arras, the benefits of using tunnelers to provide underground shelter and staging areas for troops inaugurated a long trend in the twentieth and twenty-first centuries. This is not to say there was a cause-and-effect relationship—there is no evidence that the people who began to use underground complexes for making war studied the deep shelters and subways of World War I. But the desire to hide and shelter underground is a universal one, especially for a weaker party in any armed confrontation. While mainstream formations got away from the trench stalemate and the underground war through increased mobility after 1918, more marginally powerful forces tended to seek strength through underground strategies. In order to do so, they needed to know how to dig shafts, galleries, and transit tunnels even if they did not plan to use them to blow up an enemy target.

The Chinese adopted a strategy of using underground shelters to protect civilians and as a base from which to harass occupying troops in surface operations during their long war with Japan from 1937 to 1945. Also, Japanese military forces employed miners to dig extensive cave shelters in coral-rock formations on many of the islands they held in the Pacific during World War II. These not only sheltered troops and material but often were used for defensive fighting purposes, with artillery emplacements and machine-gun positions placed at the mouth of the caves. South Vietnamese communist forces also used extensive underground shelters during the American Vietnam War of 1965 to 1973. And during the seemingly never-ending troubles in the Middle East, insurgent forces have increasingly adopted tunnel warfare for shelter, infiltration, and small-unit surface combat from the turn of the twenty-first century onward.

In China, tunnel complexes were dug by both the Nationalist forces and the Communist forces to fight the Japanese. They often were designed to shelter civilians as well and were equipped with traps to snare entering opponents. That trend continued during Japan's war with the United States in the 1940s, although in this case Japanese island defenders dug in to maximize their strength and firepower and to protect themselves from aerial and artillery bombardment. While U.S. Army and U.S. Marine authorities were, of course, highly interested in understanding those underground shelters and defenses, most of the information readily available about them now comes from the work of postwar archaeologists who have explored, mapped, and photographed those systems. They typically consisted of a combination of manmade tunnels linking natural caves. Storage areas were built in the form of tunnels, which also sported fighting positions at their entrances. Natural caves that happened to be located in a militarily advantageous position for overlooking approaches were enhanced by Japanese engineers for fighting purposes. Months of work on digging those systems out of coral or volcanic rock (without shoring) forced the Americans to fight bitterly in efforts to root out defenders. Many Pacific islands consisted primarily of karst, a term denoting any water-soluble bedrock such as limestone that readily develops sinkholes and natural caves. Archaeologists have documented at least twenty-seven Pacific islands with Japanese karst defenses. Noteworthy examples of these were on Watom Island off the northern end of New Britain in the Bismarck Archipelago, Biak Island off the coast of New Guinea, Saipan in the Marianna Islands, Peleliu in the Palaus, Iowa Jima in the Bonin Islands, and Okinawa in the Ryukyu Islands.[20]

Cave-and-tunnel shelters also played a role in the Korean War. Chinese forces constructed them at selected locations, usually strongpoints on high ground, to protect their troops from American bombing and artillery barrages. Viet Cong forces relied heavily on growingly sophisticated tunnel systems for shelter, concealment, and support bases in Vietnam, building on the legacy of such systems dating back to the French Indochina War. Tunnel complexes were especially evident in the A Shau valley and in Cambodia. Bunker complexes at many locations often included tunnels that linked underground shelters. Some estimate that about forty-five miles (72.42 kilometers) of tunneling were constructed during the conflict with France, while nearly two hundred miles (321.86 kilometers) were dug during the U.S. war in Vietnam. Built by a combination of forced labor and army troops, the systems often were large and complex, with numerous booby traps and some fighting positions. One of them, used as a communications center, included a total of six miles (9.65 kilometers) of galleries.[21]

It is not surprising, then, that several insurgencies during the past fifty years have invested considerable time and energy in constructing underground support complexes. This remains one of the chief ways to conceal movement and strategic buildup from a force that dominates the surface of contested territory. Afghans who battled occupying Soviet forces in the 1980s utilized existing water-irrigation and cave systems as readymade military bases. Years later Hamas, the Palestinian military organization, constructed an extensive system of tunnels in Gaza from which it operated against Israeli settlements. They also used those underground passageways to smuggle goods when either Israel or Egypt blockaded Gaza. The Israeli Defense Force has found that the Hamas tunnel complex was "the critical component of its ground operations" during brief wars with their opponents.[22]

The use of tunnel complexes to support military operations has a long history, but that use is quite different from traditional military mining and countermining. From the classical era to the twentieth century, miners had utilized underground approaches to reduce a fortified position or to counter underground attacks. Support tunnels of the twentieth and twenty-first centuries more fairly fall into the category of field fortifications than mine approaches, even though they share many technical and physical characteristics with attack and defense galleries.

Sheltering civilians and troops, storing material, and basing soldiers in tunnel complexes made sense from every perspective. They were relatively easy to dig with minimal equipment, provided less obvious targets for enemy forces, neutralized air superiority, and reduced the opponent's dominance in ground operations. It was a logical extension of military tunneling operations—so logical that its initiation during the last half of World War I was never questioned. The moles who were energized by an opportunity to blow up a German strongpoint tended to view deep-shelter digging as a chore, but they did it well at any rate. With a mixture of civilian and military help, the Chinese saw the advantage of adapting underground shelters as an important weapon of the weak when faced with a brutal occupying force. The Japanese saw them as an advantage for an outnumbered force isolated on a remote Pacific outpost, especially given the geology and terrain of many of their occupied islands. And various insurgent groups saw the utility of underground shelters in conducting their own operations, using modern methods and equipment to dig sophisticated multipurpose tunnel systems. In each instance the other side was forced to examine, understand, and develop tactics to deal with this new form of underground warfare.

War beneath the surface, in other words, did not end with 1918. Some examples of traditional siege mining continued in limited ways during the Spanish Civil War, on the Eastern Front of World War II, and in the French Indochina War. But more importantly, mining morphed into a new kind of underground warfare by and for the weak when faced with opponents who controlled air space and dominated battle grounds. Tunnel complexes provided a defensive structure that could also be used to base offensive action against that opponent and were not easy to detect or capture. The use of tunnels in these new ways is likely to continue, extending the life of underground conflict indefinitely.

Conclusion

From the ancient world through the twenty-first century, gallery mining has resonated throughout military history. It has played a significant role in military operations in all eras and in many areas of the world. Mining and countermining has been more heavily concentrated in the Western military tradition, stemming from Near East origins and extending through the Greek age and especially the Roman era to Medieval Europe. From the start to the fifteenth century, it was a relatively simple process of digging short and shallow tunnels, excavating a cavity under a fortress wall and propping it up, stuffing combustibles in the cavity, and setting them afire. This method changed very little over that great expanse of time, in part because the methods of civilian mining changed very little. Yet mining offered the commander a viable option that could and did work often enough to tempt him when other methods such as blockade seemed unattractive or the limitations of pregunpowder siege artillery became obvious.

The introduction of gunpowder in the fifteenth century altered the trajectory of military mining from the simple prop mine designed to collapse a wall to the explosive mine designed to blow it apart. Gunpowder improved the result of an underground attack, although it did not necessarily increase its ability to support above-ground assaults. It also complicated the method of siege mining and required new ways of constructing the gallery to reduce backfire from the explosion. Gunpowder made military mining more dangerous both for the victim and the perpetrator, increasing the deadly stakes involved in success or defeat in underground warfare. While gunpowder was the method of choice for those who wished to undermine a defensive position from the fifteenth century on, it was succeeded by a complex range of high explosives in the latter part of the nineteenth century that represented a huge increase in punching power. Along with the next generation of explosives came the application of electricity as a much safer, more reliable method of touching off charges. With new explosives, new detonating systems, and ample resources of material and manpower, the Western Front of World War I provided the perfect venue for what can only be termed an explosion of military mining and countermining from 1914 to 1918.

Gunpowder had come into play just when the Scientific Revolution also began to have an influence on Western culture. The seventeenth century witnessed a new turn in how practitioners approached the military craft of mining. They began to examine the process rationally with experiments, careful measurement of results, and analysis of data to find out exactly how gunpowder acted on soil and how it blew out craters. As a result they discovered the globe of compression, what a later generation called the zone of destruction. This discovery led military miners to realize how they could use that globe to destroy enemy galleries or open craters in no-man's-land as a way to continue the siege approach by using the craters as starting points for new, shorter galleries. Vauban's system of siege approaches tightly connected underground tunnels with the digging of parallels and saps on the surface. Starting an offensive mine from the end of a sap led to short and shallow galleries that did not need ventilation.

Arriving at the same time as gunpowder's introduction to Western warfare, the mechanical printing press played a role in spreading technical information about military mining. Prior to the fifteenth century, information about mining spread slowly and primarily by word of mouth, based solidly on experience in civilian ore mining in many different areas of the Near East and the Greek and Roman worlds. We know of relatively few books of that era that could be termed military manuals, and they do not contain much in the way of technical information on gallery mining and countermining.

But the mechanical printing press changed that. A proliferation of technical manuals on military matters, including details and diagrams about mining, began to be published in the sixteenth century. The trend started in Italy and spread to other areas of Europe, reinforcing the Western military tradition's role as the center of global mine warfare. The prevalence of wars, both large and small, throughout Europe during the period stretching from the Italian Wars of the fifteenth century to World War I provided another spur to the perpetuation of mine warfare in the West. This might explain why comparatively little of it took place in North and South America. Vauban's system of siege approaches, which included mining at the end of a surface sap, was not only the product of rational analysis but also a response to a strong need exerted by an ambitious monarch who wished to capture enemy strongholds in a never-ending hunger for land and glory. Europe, as the venue for intense and repeated conflict during this era, became the center of global military mining. England, because of its physical separation from yet close relation to this world, was part of the Western tradition of military mining but not at its center. And the North American colonies, which became the United States at an even greater remove from the center, also became a marginal part of this Western tradition. In the rest of the world, Middle Eastern cultures spawned effective traditions of mine warfare, but too little is known about military mining in Asia and Africa, at least in English-language sources, to gain a firm handle on it. How often it occurred, exactly how it was practiced, and how information about it was disseminated in those cultures await further inquiry.

The comparatively frequent application of mining to military operations in Europe, combined with improved technology and the publication of technical manuals, spurred the growth of this element of siege tactics from the fifteenth to the twentieth centuries. This trajectory led to the great splurge of mining and countermining activity on the Great War's Western Front. One historian of the Italian Wars, F. L. Taylor, has asserted that mining operations changed little from that era up to the Great War in "the essential principles of their design and in the exceptional amount of slow, painful, and continuous labour which they require."[1] Taylor was right in regard to the labor required but wrong in other ways, for the technology of military mining forced alterations in design, technique, and success rates by 1914.

Because of its sporadic rather than consistent use, military mining rarely caught up with advances in civilian mining, from the earliest origins of the technique into the early twentieth century. One saw many new developments in civilian mining, especially after the Industrial Revolution, precisely because it was a widespread and continuous human activity. By comparison, military mining occurred every now and then, skipping wars and often skipping many campaigns within those conflicts where it did make an appearance.

Nevertheless, several prominent sieges of the modern era highlight progress in the techniques and methods of military tunneling. Prussian miners employed the globe of compression in sophisticated ways at Schweidnitz in 1762 and pushed their efforts with a consistently aggressive manner. They dug deeper than usual to bypass countermeasures but still focused on only one underground approach against a small target. French miners at Sebastopol in 1854–1855 expanded their underground work to several simultaneous approaches along a line of earthworks only to be met with intense countermeasures. The Russians used electrical detonation for the first time in history during this campaign, which resulted in a more successful rate of explosions than in any previous siege. Japanese miners at Port Arthur in 1904 employed high explosives for the first time and pressed their efforts with intensity in multiple approaches against a long defending line of earth and concrete structures.

All this led to what happened on the Western Front. Now, for the first and only time, military mining mostly caught up with civilian mining in the full employment of new technology. The opposing armies remained locked in one place for four years, and officials in all armies supported massive mining efforts as necessary adjuncts to their tactical and operational plans to find a breakthrough. "The colossal scale of the mine attacks and defenses on the western front from 1915 to 1917 completely dwarf all underground operations of previous wars," concludes Alfred H. Brooks. "During this brief period the art of mine warfare made more progress than during its entire previous history of over 20 centuries." Harry Trounce, who participated in this apogee of mine warfare, believed the "scientific development" of Great War mining techniques was unprecedented in global history, with Brooks convinced that the "tactical application" of tunnel warfare was similarly unmatched.[2]

World War I also led to a sharp increase in the average length of offensive and defensive galleries. Before 1914, based on thirty-nine available and reliable reports, the average gallery length was 125 feet (38.10 meters). In the American Civil War, fifteen cases reveal an average length of only 111 feet (33.83 meters). But in all theaters of the Great War, sixteen cases demonstrated an average gallery length of 535 feet (163.06 meters). Along the Western Front, a dozen cases averaged 568 feet (173.12 meters). This fell significantly after 1918, with two cases averaging 190 feet (57.91 meters). The average Western Front gallery length itself exceeded the longest reliably reported mine gallery before 1914, which was the Pleasants mine, 510 feet (155.44 meters) long, at Petersburg during the American Civil War.

The much longer galleries of World War I came about because of several reasons. Much greater manpower was available for digging mines, and much more sophisticated ventilation systems were available to provide air for long tunnels. The tactical situation called for starting shafts in or behind friendly trench lines; it was too dangerous to start them in craters located in no-man's-land because heavy artillery and trench-mortar fire saturated the exposed terrain between the lines. A combination of operational circumstances, manpower-mobilization initiatives, and new technology allowed for the digging of longer galleries.

But neither the sophistication of technical knowledge nor of tactical use could compensate for major problems that shaped the nature of mine warfare on the Western Front. Robert N. Harvey has pointed out that, as always, coordinating underground effort with surface operations was the key to success, and that was even more difficult in World War I than before: "The blow from below required far too much time to stage, and can not be readily adapted to meet the ever-changing factors which control the occupation of trenches on the surface."[3] The depth of trench systems in World War I, which typically consisted of not one but three trenches several hundreds of yards from each other, severely limited the chances of a clear breakthrough of the line.

So, even with the impressive level that military mining reached by 1918, it still could not overcome a number of factors that limited its contribution to operations even on the Western Front, which was the most fruitful venue for its development. When Harvey, the officer who oversaw most of the successful British mining efforts of the war, bluntly told his fellow engineers to "do everything you can to prevent mining being done" in any future war, one has to take that advice seriously. Brooks, a keen observer of Western-Front mining, has pointed out that once engaged in a major mine war, the action takes on a life of its own. "It is generally much easier to initiate mine warfare than to terminate it." His warning was well taken; one could become enmeshed in a tactic that promised rewards incommensurate with the effort and not know how to stop it as well. Historians generally agree with Harvey and Brooks; they see the history of military mining as indicating that local successes were possible but usually did not translate into anything more important.[4] It is possible to speculate that they could have had more chances of translating into significant larger consequences in a smaller war of limited goals, such as many of the conflicts that occurred during the classical, medieval, and early modern eras, but less so in bigger, more complicated wars of the modern era.

Before World War I, something less than half of mining projects had produced even local successes. The percentage probably was higher on the Western Front but far from overwhelming. And there was no indication that even the local successes in France and Belgium had any real hope of translating into a true breakthrough of the opposing trench lines. The fact that the static nature of the front persisted so long was an important reason for the proliferation of underground activity—mining was one of several efforts by all armies to find a solution to the collapse of mobile operations. But like the others, it failed to provide the answer to that deadly problem.

Mining was an expensive effort, too, not just in terms of material, manpower, and time but also in terms of the emotional consequences of underground fighting. Anyone who might think that it was not more emotionally demanding than surface combat because it took place enclosed by the earth, and thus it was safe or clean, would be very much mistaken. Even with the gung-ho attitude of the typical British sapper, there were moments when the awfulness of what they were trying to do to other human beings came uppermost in their consciousness. When he watched the mine go up at St. Eloi for the famous Messines Ridge battle, Canadian George H. Morley could not help but wonder at the effect of his actions on the Germans. "It was not a sight that a man could see without being impressed, without feeling that even though it was glorious it was awful—awful to think that there were hundreds of men in that inferno you'd help let loose. There never was nor ever will be anything like it."[5]

Far worse was personal combat in the tunnels, which happened far more often on the Western Front than in any previous mining effort in history. In many ways tunnel fighting produced a more emotionally challenging environment than surface combat because of the tightly closed spaces and the utter darkness produced by small-diameter galleries. One did not have to worry about damaging artillery fire or machine-gun nests; combat was reduced to its bare essentials, and warriors often were brought very close together in ways that mimicked sword combat in the classical and medieval eras. Lt. John Westacott's experience of underground combat, as related in chapter 12, is one of the most gripping examples of desperate fighting on the surface or under it that can be found on the Western Front.[6]

Westacott's story reminds us of the human cost of mine warfare, and it brings back a natural desire to understand the effects of tunneling on the individuals who dug the galleries. When Alexander Barrie wrote a popular book on the British tunnelers of the Great War, published in 1962, he based it primarily on contributions from sixty-eight men who were willing to tell their stories to the public. Even after the tunneling generation had passed away, popular interest in those men continued. The engaging story of Sapper William Hackett of the 254th Tunnelling Company has moved not only thousands of readers but also numerous visitors to a monument erected in his memory at the site of his death. Hackett, a civilian coal miner with a wife and children, was among five tunnelers trapped in a gallery that was collapsed by the German explosion of a mine that created what later was named the Red Dragon Crater near Givenchy-lès-la-Bassée on June 22, 1916. Frantic rescue efforts reached the men two days later, but one of them was too badly injured to emerge through the small hole. Hackett was fully capable of escaping but refused, preferring to stay and take care of his injured mate. Unfortunately, soon after that, further digging caused a serious collapse of the damaged ground at the site. After another three days of digging, with no signs that the two men were still alive, rescuers gave up the effort. Hackett's dedication to his injured comrade led to the erection of the memorial to him at the site in 2010.[7]

In addition to recorded memories by the tunnelers and monuments to them, a small number of leftover mines have bequeathed a disturbing legacy to postwar Europe. Four of the mines at Messines were left intact and unexploded after the battle. British officials promised to dig them up after the war but failed to do so because they had lost the exact location after the German offensive of spring 1918 recaptured the ground. One of them exploded spontaneously on July 17, 1955, probably triggered by an electrical storm. Fortunately, no one was hurt in this bizarre reminder of a dangerous past. Three other mines at Vimy Ridge were removed in the 1990s, while a charge located in a field near Arras suddenly exploded in 2015. How dangerous leftover mine charges really are is something of a guessing game. Obviously, if one happens to be near one that goes off, it can be deadly; but what are the chances of unexpected explosions? Historians Peter Barton, Peter Doyle, and Johan Vandewalle speculate that charges with intact leads probably are the real danger, basing this on the fact that the 1955 blowup probably occurred because of an electrical storm. But they admit that no one really knows for sure what constitutes the danger, nor does anyone have a good idea how many charges were left intact when the war came to an end. It is possible that numerous small charges—camouflets—were prepared because of the close approach of enemy galleries and, since a reason to set them off never materialized, they were left behind and forgotten.[8]

Other than the threat of blowups from leftover mines along what was the Western Front, the other major legacy of underground warfare is what remains of craters and galleries. In this regard there was very little evidence prior to 1914. The Roman workings at Uxellodunum were discovered in the nineteenth century, but the most prominent mining site found and excavated dating to the long classical era remains Dura-Europos in modern Iraq. Excavations by a French-American team in the early 1930s uncovered a rich picture of life in that Roman outpost as well as the most astonishing physical evidence of mining, countermining, and underground combat from any pre–World War I military mining site. It was studied, mapped, photographed, and then covered up again. Terrible conflict around the turn of the twenty-first century has virtually obliterated the surface remains of the old community, but the knowledge of its significance as an archaeological window into the past remains.

The medieval era has yielded hardly more sites that have been found and excavated by archaeologists. The galleries cut in the bedrock under St. Andrews Castle in Scotland are the best known of the era, but a remnant of a gallery has been found at Bungay. In the Middle East a small remnant of a gallery has been identified at Montfort. Other than the amazing finds of galleries and shoring timbers at Limerick (King John's) Castle, no significant mining sites of the period from the seventeenth to the late eighteenth centuries have been found or explored, although several cities have intact permanent countermine systems that date to this time period. At a handful of these sites, guided tours are permitted through a portion of the system. The countermines at Maastricht, The Netherlands, are a particularly good place to visit if one wants to get an authentic feel for what it was like to walk through galleries securely lined with masonry shoring. In North America the mine started by American besiegers of the British garrison of Ninety-Six was discovered, photographed, and covered up again. During the American Civil War, what is left of the famous Crater at Petersburg is preserved and the site of the mine at Kennesaw Mountain is marked. But elsewhere nothing has been done to find, explore, or study the extensive mines at Schweidnitz, Sebastopol, or Port Arthur.

In the area of mine remnants, as in every other way, the Western Front towers over the history of military mining, for there are many remains of shafts, galleries, and craters in France and Belgium. Most of the effort to discover those extensive remains has been undertaken by private individuals and groups, although trained archaeologists are becoming more involved as well. No one has yet catalogued the many finds except for surface remnants to a limited degree. The underground remains are the most numerous, dangerous to explore, and time consuming to study; the great effort to uncover and catalog them will continue for a long time to come.

An archaeological team headed by Birger Stichelbaut has provided ample proof of the scale of mining on the Western Front through examination of aerial photographs taken by reconnaissance flights over the trenches. The scientists studied 14,616 photographs and mapped the evidence for surface features along thirty-seven kilometers (22.99 miles) of the front, from the North Sea to the French-Belgian border. They identified more than 211,000 features, ranging from trenches to bunkers, huts, gun emplacements, ammunition dumps, and war cemeteries. Included in this array were 301 identifiable mine craters between 32 and 246 feet (9.75 and 74.98 meters) in diameter. Team members were able to date the craters to obtain a better idea of operational flow over time. They linked 214 of them to "16 mine hot spots" at the Ypres Salient. They also identified 19 craters associated with the famous Messines Ridge operation, noting that they were among the biggest found and covered a combined area of cratering 18 acres (7.28 hectares) large. This was nearly "equal to the total surface area of all 236 previous mines together," which was 18.2 acres. The craters identified as occurring after the Battle of Messines Ridge were mostly 26–65 feet (7.92–19.81 meters) in diameter and tended to be tank traps along roads in front of and to the rear of trench lines.[9]

When team members examined an airborne laser-scanning dataset of Flanders to gauge what the landscape of the western front looked like in the early twenty-first century, they were able to learn what became of the 301 craters they had discovered on wartime air-reconnaissance photographs. Seventy of them are preserved and readily discernible to the naked eye. Another 26 are preserved "as subtle landscape" features visible with technological aids. An additional 111 of the craters have been backfilled on sites currently used as pasture or arable land. Five craters were destroyed by later mine explosions and have no remnants of any kind, while 89 have been "destroyed or built on for roads or buildings."[10]

Given the intense nature of military operations on such a narrow sliver of countryside and the great need for rehabilitating that landscape after the war, it is impressive that so many craters have survived. Nearly a quarter (23.25 percent) of them are preserved well enough to be seen and visited by Great War tourists, while another 8.63 percent are preserved well enough to be detected with modern technology. Almost one-third (31.22 percent) were destroyed by other mine explosions during the war or by postwar road and building construction, and those in the largest category, 111 (or 36.87 percent), were destroyed by farmers reclaiming their land for agricultural purposes.

Of the 70 fairly well-preserved mine craters listed by the team, a handful have become major centers of tourist attraction, commemoration, and memorialization. A collection of surface features was preserved early on at Vimy Ridge, where the Canadian government purchased ground for the Vimy Ridge National Historic Site of Canada. Initially, it was maintained in its "war-time condition," according to W. Grant Grieve and Bernard Newman. "Here is a maze of trenches, barbed-wire, and the medlied rubbish of a battlefield." Highly perishable elements such as sandbags and wooden duckboards were reconstructed in concrete. Trenches also were reconstructed in concrete because of the never-ending maintenance required for earthen scarps and parapets. Underground remnants of subways also remain, with Grange Tunnel as the highlight after 820 feet (249.93 meters) of its 2,624-foot

(799.79-meter) length were opened to visitors in the late 1980s. Hundreds of examples of soldier graffiti also are preserved on the chalk walls of those subways. Two major mines that exploded at the start of the Canadian attack on April 9, 1917, became the centerpiece of the surface remnants in the park. Known as Lichfield and Zivy, they had produced craters around which heavy fighting took place. Canadian soldiers buried many bodies in both craters, and after the war a stone wall was built around each to protect the resting place of more than one hundred soldiers. The craters were afterward filled in and planted over with grass.[11]

Two craters associated with the Somme battle have been preserved to good effect. Lochnagar remains a massive, well-formed mine crater frequently visited by tourists and providing a fascinating image when viewed from an aerial perspective. The remains of a British soldier were found in its rim in 1999. The Hawthorn Ridge Redoubt mine became the first mine explosion to be photographed in the field when it went up on July 1, 1916. In addition to the two still images by official photographer Ernest Brooks and twelve seconds of motion-picture footage by Geoffrey Malins of the blast itself, several photographs of the crater were exposed after British troops captured the ground. The crater remains well defined in the twenty-first century, but the surface is largely covered with mature trees.[12]

Much attention tends to be drawn to Messines Ridge and its famous multiple-mine attack. The ridge is a shallow topographical feature, a slight upturn across the flat countryside of Flanders with long, gentle slopes and a gently rounded top. Hill 60, the northernmost of the Messines mines, became a major tourist attraction right after the war. Local landowners capitalized on the interest with information booths, cafes, souvenir sales, private museums, and the redigging of some trenches. They also filled in some of the mine craters. Those that were left mostly filled with water to become something like agricultural ponds in the rural landscape. In fact, Spannbroekmolen's crater has become known as the Pool of Peace.[13]

The Butte de Vauquois, however, is the most impressive site of mining warfare in the world. The butte rises very prominently, with an absolutely commanding view of the countryside all around. Soon after the opposing lines settled on its top—about 250 yards (228.60 meters) wide, north to south, by about 600 yards (548.64 meters) long east to west—the medieval village of Vauquois was dismantled, and stones from its buildings were used to revet trenches. When the French and Germans began to mine and countermine, something happened that was relatively rare in mining history. Both sides became obsessed with trying to outdo the other in digging and blowing. The emotionally charged nature of tunneling developed an impulse of its own so that, by the time the war neared an end, a total of

Lochnagar Crater, 2016. One of the nineteen mines supporting the start of the British offensive at the Somme on July 1, 1916, the name derived from the designation of the trench (Lochnagar Street) from which the gallery was started more than 900 yards (822.96 meters) from the German stronghold called Schwaben Höhe. Touched off at 7:28 A.M., the charge produced a crater 69 feet (21.03 meters) deep and 330 feet (100.58 meters) wide. British troops captured it immediately, but little progress took place to right and left of the position. This unusually well-preserved crater has been developed as a tourist attraction with a pathway around the rim. Trees and brush have been cleared for a comprehensive view of the entire crater. Still from Jock Tamson, *The Lochnagar Crater, La Boisselle by Drone*, YouTube, May 22, 2016, https://www.youtube.com/watch?v=IQTI-XrxKsw.

519 mines and countermines had been blown on the surface of this small space of contested ground. The multiple explosions literally tore up the top of the butte as if a giant had been at work with a shovel digging for treasure. The site has been left largely intact except for allowing trees to regrow and sowing grass to stabilize the surface remnants. The top of the butte remains a bizarre place to visit, a monument to the power of underground charges and their ability to transform landscapes. But it is a strange monument, to be sure. Despite all those explosions, mining failed both the French and the Germans—neither side ever produced anything like a decisive edge over the opponent. The Butte de Vauquois is, in many ways, a haunting memorial to both the power and the limitations of military mining.

With all the mine craters remaining along the corridor of countryside that once was the Western Front, many of them accessible to visitors, issues of safety naturally come up. Archaeologists and geologists contribute to the understanding of those problems with modern technology that relate to what they term "geomechanics stability assessments." Much of this has been done at the Canadian Vimy Ridge National Historic Site. The process measures the likelihood of subsidence, or collapse, of a number of physical landscape features within that park, including mine craters.[14]

Most of the historic remnants of military mining history are along what is left of the Western Front in Belgium and France, although some remnants have been identified on the Gallipoli Peninsula. Seven collapsed mine galleries have been discovered there over the years, four of them dug by Australians and three by Turkish troops. Two of the galleries have been reinforced with wooden beams to stabilize them.[15]

The only remnant of mine warfare outside Europe is in the United States, where the Petersburg National Battlefield preserves what is left of the Crater, formed by the Pleasants mine of July 30, 1864. Sections of the gallery had caved in by September 1865, five months after the war ended, and in July 1866 the U.S. government exhumed at least three hundred bodies of mostly Union soldiers killed in the battle, which the Confederates had buried in the bottom of the Crater. The local landowner erected a fence around the hole and charged admission a year later, building the Crater Saloon to serve tourists and setting up a private museum to house relics picked up on the battlefield. Local Confederate veterans transformed the site into a monument to their lost cause and invited Union veterans to reunions there. Reenactments of the battle were staged in the early twentieth century even as the site passed from private ownership to federal control by 1932. Only with the further development of the national battlefield were all the commercial elements eliminated from the site and historic preservation, shorn of ideological content, applied to this battlefield. Remnants of countermines at four other locations along the Petersburg lines were also discovered by private individuals. While some attempted to make money from their discoveries, all such efforts soon collapsed, and public access to those underground remnants were closed.[16]

Visiting the remnants of tunnel mining is not the only expression of public interest in underground warfare. With the development of modern photography, it became possible to document the history of military mining. By the 1850s, the wet plate process, which involved bulky equipment and demanded a somewhat lengthy exposure time, was unsuitable for the capture of action. In other words, it would not have been possible for a photographer to take a picture of the Petersburg mine explosion as it happened, but he could certainly take photographs of the Crater right after the war ended. By the 1880s, improvement in the photographic process and the development of small, portable cameras allowed imagists to take action photographs. The first image of an actual mine explosion apparently is one taken of a practice mine during the Royal Engineers exercises at Chatham in 1907.[17]

The only well-documented still photography of an actual mine explosion in field operations was performed by Lt. Ernest Brooks, the first official photographer of the British Army. Brooks, with the authority to roam the Western Front during most of the conflict, exposed 4,400 photographs of unusual quality. He took two images of the Hawthorn Ridge Redoubt mine explosion that have become iconic of the Great War. Taken just as the dirt plume was rising and after it mostly subsided at 7:20 A.M., July 1, 1916, it signaled the start of the great Somme offensive by British forces.[18]

This same mine explosion also was captured in motion-picture images by cinematographer Geoffrey Malins. Also given a commission so he could roam along the front, Malins later claimed that he shot a mine explosion a few days before July 1 but failed to specify where or when; he also failed to report what was done with the footage. But his shot of the Hawthorn Ridge Redoubt mine on July 1 created a sensation. Incorporated as a twelve-second spot in a seventy-minute documentary called *The Battle of the Somme*, the explosion scenes contributed to the enormous general interest in the film when it was viewed throughout Great Britain a month after the start of the Somme offensive. For the first time in history, the public at large could see what an actual mine blow looked like on the surface.[19]

As far as visual representation was concerned, the Hawthorn Ridge Redoubt mine has reigned supreme as the most prominent event in mining history. Yet it was only one of some 5,000 mine and countermine explosions along the Western Front of World War I and not even the largest or most important of them.

Three feature-length films of the mid-twentieth to early twenty-first centuries recreated military mine explosions for dramatic effect, reaching literally millions of viewers over time. Cinema spread visual imagery of mine warfare far and wide, across international borders, and ostensibly forever—or at least as long as disc players and computers exist. Filmmakers were inspired by history, but they did not necessarily reproduce it. Directors, set designers, and cinematographers had their own take on how to represent mine explosions, which usually did not faithfully represent the actual events. And sometimes filmmakers deliberately entered into the realm of propaganda.

Augusto Genina, an Italian director, embarked on an effort to promote fascist values on the screen when he made *The Siege of the Alcazar*, inspired by the fighting at that place in 1936 during the Spanish Civil War. Filmed in Italy and released in August 1940, the movie was approved by the Mussolini regime for portraying the Nationalist defenders as noble exemplars of conservative values and the Republican besiegers as seedy, ignoble, and dangerous radicals. Genina streamlines the true story of the siege, positioning one

Mine Explosion, *The Siege of the Alcazar*. Italian director Augusto Genina fashioned this rightist propaganda film to glorify Francisco Franco's Nationalist forces and denigrate the pro–Republican government forces in the Spanish Civil War. It fails to accurately depict the mine explosion in the famous siege of the Alcázar, the Spanish military academy in Toledo, on September 18, 1936, in another example of promoting propaganda over truth. Still from *The Siege of the Alcazar*, directed by Augusto Genina (1940).

mine explosion that crumbles the old Alcázar building and the follow-up attack as the climax of his story, with Francisco Franco's forces saving the beleaguered garrison at the last minute. Most of what takes place before that climax focuses on the people inside the besieged complex. The film was released with Spanish dialogue and a prologue in German in October 1940.[20]

The film's depiction of mine warfare fails to capture its reality. As part of its portrayal of the Republicans and their sympathizers, it shows a civilian crowd gathering to watch the explosion, with still and moving-picture photographers ready to record it. The explosion itself looks like any blowup produced by placing explosives inside a building rather than how it would look if they were buried some distance under the building. It makes for a more spectacular explosion, of course, but gives viewers the wrong impression of how mining charges affect targets. Moreover, while the poster for the Spanish-language release shows a man about to push the plunger of a detonator, the film itself does not. In the movie a seedy looking Republican casually throws a switch nailed to the wall to make electrical connection to set off the charge.

Mine Explosion, *Cold Mountain*. The famous battle at the Petersburg Crater during the American Civil War begins the movie *Cold Mountain*. Although filmed on location in Romania, it is a typical Hollywood explosion, with fire and brimstone rather than a dirt plume like the actual Pleasants mine blast. Compare this with authentic images of actual mine explosions: the still from Geoffrey Malins's documentary film and Ernest Brooks's photograph of the Hawthorn Ridge Redoubt mine of 1916 (chapter 12) and the illustration of the Pleasants mine at Petersburg in 1864 by an eyewitness (chapter 8). Still from *Cold Mountain*, directed by Anthony Minghella (2003).

But of course, when one makes propaganda, reality and truth are un-important. The siege of the Alcázar immediately became a subject of deep interest to conservatives across Europe, who pumped it into a major event in the triumph of fascism in Spain. Genina's film has to be seen within this context. It took its place alongside the writings of many ultraconservative figures and reached millions of viewers with its warped perspective on the politics of mid-twentieth century Europe.[21]

There is no overt propaganda in *Cold Mountain*, an American film re-leased in 2003. But director Anthony Minghella and his team certainly do put a slant on the battle at the Petersburg Crater that opens this film. Based on the 1997 novel of the same name and filmed in Romania, the movie fails to do a better job of portraying mine warfare than does Genina's effort. The designer planned and built a pretty elaborate setting but in this also failed to replicate trench warfare at Petersburg in a historically accurate way. The trenches are too wide and open, and Confederate artillery is positioned en barbette so that gun crews have no protection when serving the pieces. The explosion, when it comes, looks like a fireball created by petroleum in-stead of a dirt plume. Again, it is very dramatic but has nothing to do with an actual mine explosion. In the brief depiction of the battle that follows, Minghella tends to portray the Confederates as heroic defenders, ending the battle sequence with a high-angle view of Union soldiers fighting in the Crater as if they are worms squirming in a hole. Unfortunately, *Cold Mountain* fails to offer the discerning viewer a convincing visual depiction of historical reality.[22]

In contrast, *Beneath Hill 60*, an Australian film by director Jeremy Sims released in 2010, manages to portray the Western Front more accurately than *Cold Mountain* portrays the Crater battle but with significant limitations. Inspired by the war career of Oliver H. Woodward of the First Australian Tunnelling Company, the screenwriter took great liberties with the plot and characters to construct a typical cinema approach to the subject. Filmed in Australia, the fine dust of Down Under could not substitute for the mud and clay of Flanders, and the relatively authentic use of World War I–period equipment and material tends to be overwhelmed by the cooked-up dramaturgy of the movie.[23]

The surface expressions of military mining are seen in the explosions and the craters formed by them, and filmmakers tend to concentrate on those levels. But no one from Genina to Minghella to Sims has accurately replicated an authentic mine explosion. In *Beneath Hill 60* the blast looks like Dante's inferno—another petroleum explosion, big and noisy, rather than the dirt plume of Malins's documentary film. The demands of cinematic expectations on the part of both moviemaker and audience shove reality to the side. Sims has been the only one who tried to replicate the interior of the tunnels that made those explosions possible. Still, the galleries for *Beneath Hill 60* were built on a soundstage and are far bigger in dimension than the actual attack galleries on the Western Front (or anywhere else in military mining history).

The unseen part of military mining—that which takes place underground—is the least represented in visual art of any kind. There are surprisingly few images of true attack galleries and countermine galleries, probably because of the difficulty of exposing still photographs in the confined and dark spaces of those tunnels. A rare photograph taken by Lt. Col. Robert David Perceval-Maxwell of a sapper at the face of the Hawthorn Ridge Redoubt gallery is the only authentic picture I know of to date. There are a number of postwar photographs of galleries dug during the American Civil War and on the Western Front, but these tend to document the deterioration of tunnels over time. The World War I galleries included in this category also tend to be those dug through chalk and other medium that did not need timbering. Filmmakers understandably had to recreate galleries more spaciously to provide room for camera equipment and crew.

In many ways, it is difficult to obtain a good understanding of the look and feel of a military tunnel in the twenty-first century, but a team of

Mine Galleries, *Beneath Hill 60*. Director Jeremy Sims filmed the only
movie centered on tunnelers and their work in this 2010 release. Inspired
by the story of Oliver H. Woodward of the 1st Australian Tunnelling
Company and filmed in Australia, set designers fashioned this gallery
on a sound stage, depicting it bigger and more heavily shored than
was typical for miners serving with the British Expeditionary Force.
Compare this view with the photograph of a miner at the face of the
Hawthorn Ridge Redoubt gallery (chapter 11). Still from beneathhill60,
Beneath Hill 60 (B-Rolls)—The Tunnels, YouTube, February 1, 2010, https://
www.youtube.com/watch?v=w3BIu4OM25k.

technicians at Virginia Polytechnic and State University have come up with
an interesting solution. By 2020, the team had put together an exhibit called
Visualizing History: Tunnels of Vauquois. Members mapped and digitized
the historic galleries at the butte and created a tour of them by combining
virtual reality with physical objects to replicate the feel of a gallery. The
team calls it "a hybridized physical and virtual exhibit that maps the tunnels
both literally and metaphorically using ground truth data obtained through
laser scanning and photogrammetry." They erected panels to replicate the
sides of tunnels, adding spray foam and painting for texture, and coordinated a walking tour through the exhibit by using sounds, explanations, and
introductions to describe the use of geophones, lanterns, and other props.
The tour included a replication of a mine explosion, with "virtual dust and
debris" falling on the visitor. "Plausibility illusion is the degree to which
you believe that what you are seeing is real," report the team members.
The key is to combine physical touch with the images and information fed
to the visitor through the headset he or she wears while walking through
the exhibit. It has become a hit with visitors, especially school children, and
promises much in the effort to understand in a visceral way what military
miners experienced in the past.[24]

The last traditional mining attack took place at Dien Bien Phu in 1954. It is uncertain whether any more will ever take place. During Russia's on-going war in Ukraine, which had settled into static field defenses by 2023, there have been limited reports of Russian use of offensive tunnels to cross part of no- man's-land so as to avoid the deadly use of drones by Ukrainian defenders. The invaders apparently have been using those tunnels to emerge in places closer to Ukrainian defenses rather than aiming to blow up the de-fending works.[25] That tactic is a modification of Vaubanian principles with precedents throughout the history of military mining.

The enormous mining efforts of World War I tend to overshadow the 2,000 years of military mining that preceded the Great War. But that should not obscure the fact that offensive and defensive tunneling has a very long and multilayered history, involving technology, civilian mining, military tactics and operations, memory, culture, and more. Public fascination with the tunnelers of the Western Front has ensured that the memory of un-derground warfare will not soon disappear, even if the modern battlefield offers fewer and fewer opportunities for the military miner to work at his unusual craft.

NOTES

Introduction

1. On the spelling of "fuse" versus "fuze" for mining purposes, "fuze" is more appropriate. The word derives from *fuzee* (a tube filled with combustibles), while "fuse" derives from *fusus* (to melt) and is more appropriately applied to electricity. Hess, *Civil War Field Artillery*, xiii.

2. West, *Innovation*, 12, 18.

3. Bradbury, *Medieval Siege*, 271; Kaufmann and Kaufmann, *Medieval Fortress*, 60; Youngblood, *Development of Mine Warfare*, 1; Leonard, *Beneath the Killing Fields*, 25–26.

4. Wiggins, *Siege Mines*, 10; Grieve and Newman, *Tunnellers*.

5. *Appian's Roman History*, 4:429–431, 435, 437.

1. The Near East and Greece

1. Berger, "Mining History," 2, 4–5.

2. Sherwood et al., *Greek and Roman Technology*, 185–187, 194, 204; Healy, *Mining and Metallurgy*, 70.

3. Healy, *Mining and Metallurgy*, 77–78, 80–81; Berger, "Mining History," 5.

4. Healy, *Mining and Metallurgy*, 82–85, 132–133.

5. Healy, 83–84. Similar development of civilian gallery mining can be seen in Britain during the period about 2330–1740 BCE. West, *Innovation*, 18–19.

6. West, *Innovation*, 19–21.

7. Yadin, *Art of Warfare in Biblical Lands*, 33–34, 146–147, 316–317; Brice, *Stronghold*, 35; Ferrill, *Origins of War*, 74–76; Schneck, "Origins of Military Mines: Part I," 50; Wiggins, *Siege Mines and Underground Warfare*, 9. For the wall-panel relief showing mining, see British Museum, Museum No. 124554, www.britishmuseum.org /collection/object/W_1849-1222-23. For the wall-panel relief showing sapping. see British Museum, Museum No. 124931, www.britishmuseum.org/collection /object/W_1856-0909-17-18_2.

8. Winter, *Greek Fortifications*, 296; Herodotus, *The History*, 354.

9. Kern, *Ancient Siege Warfare*, 60–61; Lawrence, *Greek Aims*, 56; Davies, "Roman Offensive Siege Works," 253; Maier and von Wartburg, "Reconstruction of a Siege," 8.

10. Davies, "Roman Offensive Siege Works," 257.

11. Anglim et al., *Fighting Techniques*, 180–181; Ober, "Hoplites and Obstacles," 183.

12. Thucydides, *Peloponnesian War*, 87–88; Brice, *Stronghold*, 45–46; Lawrence, *Greek Aims*, 41–42.

13. Xenophon, *History of My Times*, 97.

14. Lawrence, *Greek Aims*, 49–50, 52; Winter, *Greek Fortifications*, 307, 310; *Diodorus of Sicily*, 5:289, 291, 295.

15. *Aineias the Tactician*, 8, 12, 91–94.

16. *Arrian*, 1:217.

17. Winter, *Greek Fortifications*, 272–273.

18. *Diodorus of Sicily*, 9:203, 205; Warry, *Warfare in the Classical World*, 91.

19. Lawrence, *Greek Aims*, 103.

20. Polybius, *The Histories*, 3:11, 13, 239, 241.

21. Polybius, 4:93, 95, 97; Lawrence, *Greek Aims*, 64; Toy, *History of Fortification*, 27.

22. Polybius, *The Histories*, 5:23. Philip's siege of Lamia in 191 BCE is noted by Polybius as having involved another aborted mining effort. The Macedonians encountered deposits of flint that proved to be "almost unworkable with iron," as Livy put it. This delayed the mining so much that events elsewhere forced Philip to call off the siege. *Livy*, 10:233, 235; Kern, *Ancient Siege Warfare*, 274.

23. Polybius, *The Histories*, 5:67, 69, 71, 73, 77.

24. Lawrence, *Greek Aims*, 53–66.

2. Rome

1. Healy, *Mining and Metallurgy*, 86, 90–91, 135; Timberlake, "Archaeology of Mining," 38.

2. West, *Innovation*, 22; Timberlake, "Archaeology of Mining," 54; Healy, *Mining and Metallurgy*, 93–95.

3. Healy, *Mining and Metallurgy*, 91–92.

4. Healy, 86; *Livy*, 3:67, 73, 75.

5. Polybius, *The Histories*, 1:116–117, 119, 129, 135, 137; *Livy*, 6:61.

6. Polybius, *The Histories*, 1:299.

7. Polybius, 1:299, 302, 305.

8. *Appian's Roman History*, 2:303, 305.

9. *Appian's Roman History*, 2:309, 311.

10. *Appian's Roman History*, 2:387; Davies, "'Dig for Victory'!," 46.

11. Caesar, *Conquest of Gaul*, 58, 83, 162, 166–167.

12. Caesar, 220; Davies, "Roman Offensive Siege Works," 250, 403–404.

13. Josephus, *The Jewish War*, 2:163, 165; Brice, *Forts and Fortresses*, 26.

14. Josephus, *The Jewish War*, 2:493, 495; Brice, *Forts and Fortresses*, 26–27.

15. Josephus, *The Jewish War*, 3:347.

16. Josephus, 3:285, 397; Brice, *Forts and Fortresses*, 27, 30, 32; Kern, *Ancient Siege Warfare*, 318–320.

17. Baird, *Dura-Europos*, 18–19, 23, 29, 31, 35–36, 112–113; Butcher, *Roman Syria*, 259, 261; Hopkins, "Siege of Dura," 251–253.

18. Baird, *Dura-Europos*, 113–114; Anglim et al., *Fighting Techniques*, 217; James, "Stratagems, Combat, and 'Chemical Warfare,'" 69.

19. Anglim et al., *Fighting Techniques*, 217; Hopkins, "Siege of Dura," 257–258.

20. Hopkins, "Siege of Dura," 256–257.

21. James, "Stratagems, Combat, and 'Chemical Warfare,'" 70.

22. James, 72, 76–79; Davies, "Roman Offensive Siege Works," 255 (quote), 257.

23. James, "Stratagems, Combat, and 'Chemical Warfare,'" 79–82, 88.

24. James, 91.

25. James, 92–96.

26. James, 74, 100; Butcher, *Roman Syria*, 57; Coulston, "Archaeology of Roman Conflict," 38, 41.

27. Elton, *Warfare in Roman Europe*, 82–86.

28. *Ammianus Marcellinus*, 2:145, 437, 441, 443; Zosimus, *Historia Nova*, 119–122.

29. Vegetius, *Epitome*, 122, 132, 135–136.

30. Procopius, *History of the Wars*, 1:379.

31. Procopius, 1:381.

32. Procopius, 1:53, 55, 57, 59.

33. Procopius, 1:503, 505, 507, 515.

34. Procopius, 1:539, 541, 543.

35. Procopius, 5:151, 153, 161.

3. Medieval Mining

1. Merdinger, "Tunnels through the Ages," [pt. 1], 33; West, *Innovation*, 23, 58–59.

2. Timberlake, "Archaeology of Mining," 38; Purton, *Medieval Military Engineer*, 190.

3. Bachrach, *Early Carolingian Warfare*, 103, 231; Purton, *Medieval Military Engineer*, 143; Bradbury, *Medieval Siege*, 37, 41–42.

4. Bradbury, *Medieval Siege*, 273; Purton, *Medieval Military Engineer*, 143–144; Oman, *Art of War in the Middle Ages*, 2:50; Prestwich, *Armies and Warfare*, 281–304.

5. Prestwich, *Armies and Warfare*, 293–294; Kaufmann and Kaufmann, *Medieval Fortress*, 25; Kenyon, *Medieval Fortifications*, 198; Hogg, *History of Fortification*, 31.

6. Oman, *Art of War in the Middle Ages*, 2:50; Seymour, *Great Sieges of History*, 6; Warner, *Sieges of the Middle Ages*, 23, 26; Bradbury, *Medieval Siege*, 273; Schmale, *Fontes Italici*, 112–113; Rogers, *Wars of Edward III*, 192–193; Comnena, *Alexiad*, 329.

7. Wiggins, *Siege Mines*, 16–17; France, *Western Warfare*, 116–117; Oman, *Art of War in the Middle Ages*, 1:134.

8. McGeer, "Byzantine Siege Warfare," 125–128.

9. Marshall, *Warfare in the Latin East*, 241, 246–247; Rogers, *Latin Siege Warfare*, 19, 22, 25, 41, 87–88.

10. Rogers, *Latin Siege Warfare*, 114–115.

11. Marshall, *Warfare in the Latin East*, 229–230; Bradbury, *Medieval Siege*, 117.

12. Purton, *Medieval Military Engineer*, 15, 144–145, 148, 191, 193, 223–225.

13. Rogers, *Latin Siege Warfare*, 103–104; France, *Western Warfare*, 116; Marshall, *Warfare in the Latin East*, 238.

14. Marshall, *Warfare in the Latin East*, 233; Wiggins, *Siege Mines*, 18; Oman, *Art of War in the Middle Ages*, 2:51; Hitti, *Arab-Syrian Gentleman and Warrior*, 102–105.

15. Purton, *Medieval Military Engineer*, 142; Clermont-Ganneau, "Taking of Jerusalem," 37n.

16. David, *De Expugnatione Lyxbonensi*, 40, 142–143; Rogers, *Latin Siege Warfare*,

182–186. Some have asserted that this is the first reliable evidence of a medieval gallery mine on record. See historiographical note in David, *De Expugnatione Lyxbonensi*, 143n.

17. David, *De Expugnatione Lyxbonensi*, 142–143, 145, 161, 165, 178.

18. Rogers, *Latin Siege Warfare*, 134–135; Bradbury, *Medieval Siege*, 149, 208–209, 272; Oman, *Art of War in the Middle Ages*, 2:52; *Chronicle of James I*, 1:147–149.

19. Fulton, "Siege of Montfort," 714–716.

20. Marvin, "War in the South," 375, 380, 383–385.

21. Marvin, 391–392.

22. Oman, *Art of War in the Middle Ages*, 2:50–51.

23. Oman, 2:34–36.

24. Rogers, *Wars of Edward III*, 117; Rogers, *War Cruel and Sharp*, 344, 406.

25. Rogers, *Wars of Edward III*, 192–193; Dunster, *Chronicles of England*, 132–133; Froissart, *Chronicles*, 2:66–67.

26. Porter, *History of the Corps of Royal Engineers*, 1:16–18; Purton, *Medieval Military Engineer*, 224; Bradbury, *Medieval Siege*, 164, 166; Halle, *Union of the Two Noble Families*, xxv.

27. Purton, *Medieval Military Engineer*, 224; Shakespeare, *Life of King Henry the Fifth*, 44.

28. Oman, *Art of War in the Middle Ages*, 1:134; Potter, *Gesta Stephani*, 35.

29. Braun, *Bungay Castle*, 16, 24–26, 37.

30. France, *Western Warfare*, 116; Purton, *Medieval Military Engineer*, 191; Rogers, *Soldiers' Lives*, 121.

31. Amt, "Besieging Bedford," 113–114.

32. Freeman, "Wall-Breakers," 8; Caple, *Excavations at Dryslwyn Castle*, 1, 188–190, 197; Prestwich, *Armies and Warfare*, 287; Rogers, *War Cruel and Sharp*, 335.

33. Bradbury, *Medieval Siege*, 222, 224–225, 273.

34. Sawyer, *Fire and Water*, 23, 191.

35. Sawyer, 33–35, 191–192.

36. Sawyer, 63, 323, 356–357; Wallacker, "Studies in Medieval Chinese Siegecraft: The Siege of Chien-K'ang," 48, 50–52.

37. Sawyer, *Fire and Water*, 349–350; Wallacker, "Studies in Medieval Chinese Siegecraft: The Siege of Yü-pi," 796–798.

38. Wallacker, "Studies in Medieval Chinese Siegecraft: The Siege of Yü-pi," 798.

39. Sawyer, *Fire and Water*, 119–120, 191, 193; Wallacker, "Studies in Medieval Chinese Siegecraft: The Siege of Fengtian," 190–191.

40. Sawyer, *Fire and Water*, 193.

41. Smith, *Warfare and Diplomacy*, 110–111.

42. Rogers et al., "Investigating the Outcome of Sieges," 26–27, 32, 36 Rogers, *Soldiers' Lives*, 121.

4. Gunpowder

1. Bury, "Early History of the Explosive Mine," 23–24; Wiggins, *Siege Mines*, 20; Duffy, *Siege Warfare*, 11.

2. Bury, "Early History of the Explosive Mine," 24–25; Pepper and Adams, *Firearms & Fortifications*, 17.

3. Duffy, *Siege Warfare*, 11; Bury, "Early History of the Explosive Mine," 27; Gille, *Engineers of the Renaissance*, 151, 213–214.

4. Smith and DeVries, *Rhodes Besieged*, 9–10, 95, 106, 110–111, 114–117, 119, 128–129.

5. Duffy, *Siege Warfare*, 11–12; Pepper, "Underground Siege," 31; Pepper and Adams, *Firearms & Fortifications*, 105, 112–113.

6. Bury, "Early History of the Explosive Mine," 23, 26; Pepper, "Underground Siege," 31–32.

7. Bury, "Early History of the Explosive Mine," 26; Taylor, *Art of War in Italy*, 143.

8. Pepper, "Underground Siege," 32, 35; Wiggins, *Siege Mines*, 21–22.

9. Pepper, "Underground Siege," 32; Wiggins, *Siege Mines*, 22.

10. Pepper, "Underground Siege," 32.

11. Pepper, 33; Taylor, *Art of War in Italy*, 146.

12. Hogg, *History of Fortification*, 99–100; Pepper and Adams, *Firearms & Fortifications*, 20–22; Bury, "Early History of the Explosive Mine," 27; Pepper, "Underground Siege," 34.

13. Smith and DeVries, *Rhodes Besieged*, 111; Pepper, "Underground Siege," 34; Pepper and Adams, *Firearms & Fortifications*, 110.

14. Wiggins, *Siege Mines*, 26; Toy, *Castles*, 150. For color photographs of the gallery remains at Saint Andrews Castle, see Tabraham, *Scottish Castles*, 17–18.

15. Wiggins, *Siege Mines*, 23; Pepper, "Underground Siege," 33.

16. Pepper, "Underground Siege," 33; Pepper and Adams, *Firearms & Fortifications*, 24.

17. Wild, "Black Powder in Mining," 203–204, 208.

18. Gille, *Engineers of the Renaissance*, 200.

19. Herbert Clark Hoover and Lou Henry Hoover, introduction to Agricola, *De Re Metallica*, xvii.

20. Agricola, 102, 104–105, 124–125, 129–130, 200, 203, 207.

21. Brice, *Forts and Fortresses*, 81.

22. Gommans, *Mughal Warfare*, 142; Gascoigne, *Great Moghuls*, 88–91; Searles, "Oriental Siege," 397–399; Beveridge, *Akbarnāmā of Abu-L-Fazl*, 2:464, 467–469.

23. Gascoigne, *Great Moghuls*, 88–91; Beveridge, *Akbarnāmā of Abu-L-Fazl*, 2:471.

24. Duffy, *Siege Warfare*, 169, 173, 238.

25. Gommans, *Mughal Warfare*, 134.

5. The Maturation of Mine Warfare

1. Wiggins, *Siege Mines*, 32–33; Harrington, *English Civil War Archaeology*, 71, 94–95.

2. Harrington, *English Civil War Archaeology*, 95.

3. Roberts, *Pontefract Castle*, 95–96, 111–112, 414–415, 425; Harrington, *English Civil War Archaeology*, 53.

4. Wiggins, *Anatomy of a Siege*, 10.

5. Toal, *North Kerry Archaeological Survey*, 36–37; Wiggins, *Anatomy of a Siege*, 10–11.

6. Wiggins, *Anatomy of a Siege*, 1–4, 13–14, 36, 38.

7. Wiggins, 45–46, 73, 79, 83, 95, 99, 102–103.

8. Wiggins, 105, 107, 113–114, 124–125.

9. Wiggins, 115, 126–127, 135–136.

10. Wiggins, 136, 138–140, 143–144, 152.

11. Wiggins, 157–159, 162, 166–168.

12. Wiggins, 169, 175.

13. Wiggins, 189–190, 192, 208–209.

14. Wiggins, 214, 216, 218–219.

15. Wiggins, 181–182, 220; Wiggins, *Place of Great Consequence*, 218, 220–221, 226, 231, 234, 236–237, 240.

16. Wiggins, *Anatomy of a Siege*, 236–238, 240–241. Christopher Duffy acknowledges English inferiority in most aspects of siege warfare but believes they might have excelled in mining techniques beyond the general capabilities of their Continental counterparts. He believes their use of mineral and coal miners from both England and Scotland gave them an edge in this regard over Continental military miners. This is not a convincing argument; English sieges cannot provide any clear evidence that mining was significantly more successful or sophisticated than those seen in Europe, and Continental armies also employed civilian miners at times. Duffy, *Siege Warfare*, 147, 159.

17. Langins, *Conserving the Enlightenment*, 108–109.

18. Lynn, *Wars of Louis XIV*, 369–370; Lynn, *Giant of the Grand Siècle*, 531–532.

19. Vauban, *Manual of Siegecraft and Fortification*, 86, 89–90.

20. Vauban, 87.

21. Vauban, 107, 109–112.

22. Vauban, 109. On the use of "fuze" rather than "fuse," see introduction, note 1.

23. Vauban, 109.

24. Schneck, "Origins of Military Mines: Part I," 51.

25. Duffy, *Siege Warfare*, 214–215.

26. Stoye, *Siege of Vienna*, 157, 159.

27. Stoye, 168–169, 171–173, 239–240.

28. Stoye, 241, 257, 259.

29. Lenihan, "Namur Citadel," 282.

30. Porter, *History of the Corps of Royal Engineers*, 1:134–135.

31. Chandler, *Journal of Marlborough's Campaigns*, 85, 87.

32. Browning, *War of the Austrian Succession*.

33. Browning, 259–260.

34. Drinker, *Tunneling*, 45; Vernon, *Treatise on the Science of War*, 2:266; Day and McNeil, *Biographical Dictionary*, 90; Duffy, *Fortress in the Age of Vauban*, 127; "Mining, Military," 400.

35. "Mining, Military," 400–401.

36. "Mining, Military," 401.

37. "Mining, Military," 401; translation of essay by Belidor in Manningham, *Complete Treatise of Mines*, 167–168.

38. Vernon, *Treatise on the Science of War*, 2:267–268; Drinker, *Tunneling*, 45; "Mining, Military," 401.

39. "Mining, Military," 402; Drinker, *Tunneling*, 45.

40. "1762—Siege of Schweidnitz—First Phase of Mining Operations," www .kronoskaf.com/syw/index.php?title=1762_-_Siege_of_Schweidnitz_-_First _phase_of_mining_operations.

41. "1762—Siege of Schweidnitz—First Phase of Mining Operations."

42. "1762—Siege of Schweidnitz—First Phase of Mining Operations"; Mahan, *Elementary Course*, 171.

43. Courtney, "Archaeology of the Early-Modern Siege," 107; "Great Underground Passages of the Citadel of Namur," accessed July 21, 2024, https://www .namurtourisme.be/en/a-voir-a-faire/visites-guidees/grands-souterrains-de -citadelle-de-namur/; Rorive, *La guerre de siège sous Louis XIV*, 160–162, 164–165.

44. Duffy, *Fortress in the Age of Vauban*, 52–56; Fitzpatrick, "Explore Turin's Underground Secrets," https://www.italy-villas.com/to-italy/2015/northwest-italy /piedmont/turin-underground-tour; Rorive, *La guerre de siège sous Louis XIV*, 163; "Fort Manoel Countermines," Wikipedia, https://en.wikipedia.org/wiki/Fort _Manoel#Construction; Castillo de San Fernando, Figueres, Spain, http://www .starforts.com/figueres.html.

45. Šeguljev, "Warfare with Landmines in Permanent Fortification and Countermines System in Peterwardein Fortress," accessed July 21, 2024, https://www .academia.edu/45520850/Warfare_with_Landmines_in_Permanent_Forti fication_and_Countermines_System_in_Peterwardein_Fortress; "Fortress Josefov," accessed July 21, 2024, https://www.josefov.com/josefov_english.htm; "Klodzko Fortress, the Passages of the Mine-Laying Labyrinth," Polish Tourism Organization, accessed July 21, 2024, https://www.polen.travel/no/tourist -attractions/klodzko-fortress-the-passages-of-the-mine-laying-labyrinth (page removed); "Fortress of Terezín," https://whc.unesco.org/en/tentativelists /1561/.

46. Lepage, *Dutch Fortifications*, 144; Morreau, *Bolwerk de Nederlanden*, 286–315; Wouters, *Fortress Maastricht*, 56–57; "Maastricht Casemates," https://www .exploremaastricht.nl/en/casemates.

6. The New World and the Napoleonic Era

1. Lee, "Fortify, Fight, or Flee," 738.

2. Lee, 740, 743; Barnwell, "Second Tuscarora Expedition," 34, 37, 38.

3. Larrabee, "Archaeological Research at the Fortress of Louisbourg," 8–10.

4. Larrabee, 32–33.

5. Hart, *Siege of Havana*, 17–19, 23, 30–31; Anderson, *Crucible of War*, 499–501; "Chief Engineer's Journal of the Siege of the Moro Fort," 463–464; Lydenberg, *Archibald Robertson*, 60–61.

6. "Chief Engineer's Journal of the Siege of the Moro Fort," 464–465; Lydenberg, *Archibald Robertson*, 62; Porter, *History of the Corps of Royal Engineers*, 1:195–197.

7. Hart, *Siege of Havana*, 33–35, 44, 48–49, 53; Anderson, *Crucible of War*, 499–501.

8. Faragher, *Daniel Boone*, 194–195, 198.

9. Lumpkin, *From Savannah to Yorktown*, 191.

10. Pancake, *This Destructive War*, 209–215; Dunkerly and Williams, *Old Ninety Six*, 36–38; Conrad, *Papers of General Nathanael Greene*, 8:374–375, 391, 419; Cann, "War in the Backcountry," 9; Haiman, *Kosciuszko in the American Revolution*, 113–114; Collins, Doering, and Gonzalez, "Terrestrial and Airborne LiDAR Digital Documentation of Kosciuszko Mine," 10, 15.

11. Dunkerly and Williams, *Old Ninety Six*, 36–38; Collins, Doering, and Gonzalez, "Terrestrial and Airborne LiDAR Digital Documentation of Kosciuszko Mine," 28, 48.

12. Schneck, "Origins of Military Mines: Part I," 51; Schneid, *Napoleon's Italian Campaigns*.

13. Rothenburg, *Art of Warfare in the Age of Napoleon*, 219–220.

14. Porter, *History of the Corps of Royal Engineers*, 1:318, 322; Jones, *Journals of Sieges*, 1:276, 288–290, 292, 294, 335.

15. Jones, *Journals of Sieges*, 1:295–297.

16. Jones, 1:298, 301–305.

17. Jones, 1:305–306; Porter, *History of the Corps of Royal Engineers*, 1:324–327.

18. Porter, *History of the Corps of Royal Engineers*, 1:336–339; Jones, *Journals of Sieges*, 2:41–42.

19. Porter, *History of the Corps of Royal Engineers*, 1:342–347; Jones, *Journals of Sieges*, 2:71–72, 74.

20. Smith, "Military Training Earthworks in Crowthorne Wood," 422–423, 425–426.

21. Jones, *Journals of Sieges*, 1:xv.

22. Vernon, *Treatise on the Science of War*, 2:262, 270–279.

23. Vernon, 2:282–283, 296.

24. Vernon, 2:284, 286–289.

25. Vernon, 2:280–281, 291–293, 300, 302.

26. Halleck, *Elements of Military Art and Science*, 373–374; *Aide-Mémoire to the Military Sciences*, vol. 1 (London: 1853), 282; Straith, *Introductory Essay to the Study of Fortification*, 62, 68–69.

27. Straith, *Introductory Essay to the Study of Fortification*, 61–62; "Safety Fuse," Wikipedia, https://enwikipedia.org/wiki/Safety_fuse; Drinker, *Tunneling*, 110.

7. Sebastopol, India, and Electrical Detonation

1. Baumgart, *Crimean War*, 115–124.

2. Lloyd, "Mine Warfare at Sebastopol," 93–94; Kulikov, Kulikov, and Robins, "Mining and Counter-Mining in the Crimean War," 339–340. Lloyd's article is tightly based on Neil's published report of French mining, while Kulikov, Kulikov, and Robins's article is tightly based on Totleben's published report. For British practice mining, see Barton, Doyle, and Vandewalle, *Beneath Flanders Fields*, 43; and Smith, "Chatham Siege Operations in 1907," 91.

3. Lloyd, "Mine Warfare at Sebastopol," 94; Kulikov, Kulikov, and Robins, "Mining and Counter-Mining in the Crimean War," 340, 342.

4. Lloyd, "Mine Warfare at Sebastopol," 94; Kulikov, Kulikov, and Robins, "Mining and Counter-Mining in the Crimean War," 340.

5. Lloyd, "Mine Warfare at Sebastopol," 95–97; Kulikov, Kulikov, and Robins, "Mining and Counter-Mining in the Crimean War," 343.

6. Lloyd, "Mine Warfare at Sebastopol," 96–98; Kulikov, Kulikov, and Robins, "Mining and Counter-Mining in the Crimean War," 343–344.

7. Lloyd, "Mine Warfare at Sebastopol," 98; Kulikov, Kulikov, and Robins, "Mining and Counter-Mining in the Crimean War," 344.

8. Lloyd, "Mine Warfare at Sebastopol," 98; Kulikov, Kulikov, and Robins, "Mining and Counter-Mining in the Crimean War," 345; McClellan, "Report upon the Operations in the Crimea," 13.

9. Lloyd, "Mine Warfare at Sebastopol," 99–100; Kulikov, Kulikov, and Robins, "Mining and Counter-Mining in the Crimean War," 345, 347.

10. Lloyd, "Mine Warfare at Sebastopol," 100–101; Kulikov, Kulikov, and Robins, "Mining and Counter-Mining in the Crimean War," 347–349; Baumgart, *Crimean War*, 160–162.

11. Kulikov, Kulikov, and Robins, "Mining and Counter-Mining in the Crimean War," 354; Mahan, *Elementary Course of Military Engineering*, 172–173.

12. Kulikov, Kulikov, and Robins, "Mining and Counter-Mining in the Crimean War," 349.

13. Kulikov, Kulikov, and Robins, 349.

14. Kulikov, Kulikov, and Robins, 353–354.

15. Kulikov, Kulikov, and Robins, 353–354.

16. Lloyd, "Mine Warfare at Sebastopol," 101; Kulikov, Kulikov, and Robins, "Mining and Counter-Mining in the Crimean War," 351; Mahan, *Elementary Course of Military Engineering*, 173, 176.

17. Lloyd, "Mine Warfare at Sebastopol," 107–110; Kulikov, Kulikov, and Robins, "Mining and Counter-Mining in the Crimean War," 351, 354.

18. Lloyd, "Mine Warfare at Sebastopol," 102–103; Mahan, *Elementary Course of Military Engineering*, 172.

19. Lloyd, "Mine Warfare at Sebastopol," 104–105.

20. Lloyd, 104.

21. Lloyd, 102.

22. Brooks, *Military Mining*, 6.

23. Delafield, *Report on the Art of War in Europe*, 110–111; Mordecai, *Military Commission to Europe*, 78.

24. Cannon-Brookes, "FSG Study Tour to the Crimea—May 2002," 11.

25. Delafield, *Report on the Art of War in Europe*, 215; Lendy, *Elements of Fortification*, 163.

26. Lake, *Journals of the Sieges of the Madras Army*, 209–210, 218–219, 233n.

27. Lake, 232.

28. Lake, 232–233, 241n.

29. Lake, 181–183.

30. Lake, 183–184, 187, 189–191.

31. Lake, 191–194, 229.

32. Sandes, *Military Engineer in India*, 1:203.

33. Lake, *Journals of the Sieges of the Madras Army*, 222–223.

34. Sandes, *Military Engineer in India*, 1:264–265.

35. Seymour, *Great Sieges of History*, 139, 145, 147, 149–150; Sandes, *Military Engineer in India*, 1:352.

36. Sandes, *Military Engineer in India*, 1:353–354.

37. Sandes, 1:354–355; Porter, *History of the Corps of Royal Engineers*, 1:487.

38. Seymour, *Great Sieges of History*, 150.

39. Schiffer, *Power Struggles*, 13–14, 19, 75–79; "Voltaic Pile," Wikipedia, https://en.wikipedia.org/wiki/Voltaic_pile; Lockwood, *Electricity, Magnetism, and Electric Telegraphy*, 24–26, 28.

40. Schiffer, *Power Struggles*, 22, 27–28, 49, 51–52, 191; Schellen, *Magneto-Electric and Dynamo-Electric Machines*, 1:15–16; "Charles Wheatstone," Wikipedia, https://en.wikipedia.org/wiki/Charles_Wheatstone.

41. Schiffer, *Power Struggles*, 120–121; King, *Torpedoes*, 77; Lockwood, *Electricity, Magnetism, and Electric Telegraphy*, 61.

42. Ilsley and Hooker, *Electric Shot-Firing*, 3–4.

43. Ilsley and Hooker, 2–3.

44. Straith, *Introductory Essay to the Study of Fortification*, 71; Macaulay, *Treatise on Field Fortification*, 215–216, 224–230.

45. "Nitrocellulose [Guncotton]," Wikipedia, https://en.wikipedia.org/wiki/Nitrocellulose.

46. Macaulay, *Treatise on Field Fortification*, 214, 230. In addition to Macaulay, another fine discussion of the principles and practice of underground warfare is "Mining, Military." This article was authored by several English engineer officers and published in *Aide-Mémoire to the Military Sciences*.

47. West, *Innovation*, 29–35, 37–39.

48. West, 11–13, 129, 135–137.

8. The American Civil War

1. Duane, *Manual for Engineer Troops*, 207.

2. Duane, 217, 233–234.

3. Duane, 231, 236.

4. Duane, 228–231.

5. *War of the Rebellion*, ser. 1, 24(2):172, 202 (hereafter *OR*; all citations to ser. 1 unless otherwise indicated); Hickenlooper, "Vicksburg Mine," 540; Hickenlooper diary, June 22, 1863 Andrew Hickenlooper Collection, Cincinnati Museum Center (hereafter CMC); Henry Otis Dwight to W. P. Gault, September 26, 1901, 20th Ohio Folder, Vicksburg National Military Park (hereafter VNMP).

6. *OR*, 24(2):202; Hickenlooper, "Vicksburg Mine," 541; Hickenlooper to Comstock, November 17, 1863, Chief Engineer Letters Received, National Archives

and Records Administration (hereafter NARA); "Scene at Vicksburg at Night," correspondence of *Cleveland Herald* in *Indianapolis Daily Journal*, June 20, 1863.

7. Hickenlooper, "Reminiscences of General Andrew Hickenlooper," 27, *Civil War Times Illustration* Collection, U.S. Army Military History Institute (hereafter CWTI-USAMHI); D. Wintter to Lockett, June 23, 1863, vol. 4, folder 124, Samuel H. Lockett Papers, University of North Carolina (hereafter UNC); *OR*, 24(2):333; L. T. Camp account, in Yeary, *Reminiscences of the Boys in Gray*, 115; account attached to Christopher C. Scott to William T. Rigby, January 30, 1904, 3rd Louisiana Folder, VNMP.

8. Hickenlooper, "Vicksburg Mine," 541–542; Hickenlooper, "Reminiscences of General Andrew Hickenlooper," 27, CWTI-USAMHI; *OR*, 24(2):202; Hickenlooper to Comstock, November 17, 1863, Chief Engineer Letters Received, NARA.

9. *OR*, 24(2):202, 372; T. C. Murphy, "Blowing Up Fort Hill: Exciting Hours in the Crater at Vicksburg," *National Tribune*, January 3, 1907; Hickenlooper, "Vicksburg Mine," 542; Hickenlooper diary, June 25, 1863, Andrew Hickenlooper Collection, CMC; Christopher C. Scott, "The Storming of the 3rd La. Redan at Vicksburg June 25th 1863," 96, Camden, Arkansas, Civil War Record Book, University of Arkansas; account attached to Christopher C. Scott to William T. Rigby, January 30, 1904, 3rd Louisiana Folder, VNMP.

10. *OR*, 24(2):372; Hickenlooper, "Reminiscences of General Andrew Hickenlooper," 27, CWTI-USAMHI; Richard R. M. Springer to J. A. Edmiston, March 18, 1902, 20th Illinois Folder, VNMP.

11. *OR*, 24(2):294, 313, 333, 415–416; account attached to Christopher C. Scott to William T. Rigby, January 30, 1904, 3rd Louisiana Folder, VNMP; Wilbur F. Crummer to William T. Rigby, October 21, 1902, 45th Illinois Folder, VNMP.

12. *OR*, 24(2):373; Bearss, *Unvexed to the Sea*, 924; John T. Wiesman, "Useless Sacrifice: Folly of the Fighting in the Crater at Vicksburg," *National Tribune*, December 13, 1894.

13. Hickenlooper, "Vicksburg Mine," 542; Hickenlooper to Comstock, November 17, 1863, Chief Engineer Letters Received, NARA; Bearss, *Unvexed to the Sea*, 925.

14. *OR*, 24(2):333.

15. Hickenlooper to Comstock, November 17, 1863, Chief Engineer Letters Received, NARA; *OR*, 24(2):179.

16. Thomas T. Taylor to William T. Rigby, March 19, 1903, 47th Ohio Folder, VNMP; *OR*, 24(2):173, 334, 368, 416; Lockett, "Defense of Vicksburg," 491.

17. *OR*, 24(2):189–190; Shoup, "Vicksburg," 173.

18. *OR*, 24(2):172, 190, 333, 409; D. Wintter to Lockett, June 23, 1863, vol. 4, folder 124, Samuel H. Lockett Papers, UNC.

19. *OR*, 24(2):190.

20. *OR*, 24(2):191–192.

21. *OR*, 24(2):191–192; Bearss, *Unvexed to the Sea*, 900.

22. *OR*, 24(2):192, 421; Bearss, *Unvexed to the Sea*, 904.

23. Bearss, *Unvexed the Sea*, 936–937; *OR*, 24(2):185, 334; James H. Dean to William T. Rigby, March 14, 1901, 23rd Iowa Folder, VNMP.

24. D. Wintter to Lockett, June 20, 28, July 2, 1863, vol. 4, folder 124, Samuel H. Lockett Papers, UNC; *OR*, 24(2):185; Hiram J. Lewis to friend, July 3, 1863, Lewis and Moulton Family Papers, Wisconsin Historical Society; Peter C. Hains to Walter B. Scates, July 2, 1863, E. O. C. Ord Papers, University of California, Berkeley; Bearss, *Unvexed to the Sea*, 889–891.

25. Grant, *Personal Memoirs*, 1:370; Bearss, *Unvexed to the Sea*, 953.

26. *OR*, 24(2):175, 334; Mahan, *Elementary Course of Military Engineering*, 180; Lockett, "Defense of Vicksburg," 491.

27. *OR*, 24(2):177.

28. Dodd, "Recollections of Vicksburg," 7; Morris, Hartwell, and Kuykendall, *31st Regiment Illinois*, 74; *OR*, 24(2):367.

29. Arthur L. Perkins to William T. Rigby, September 22, 1904, Company A, 1st Mississippi Light Artillery Folder, VNMP; Harry Beard to mother, July 13, 1863, Daniel Carter Beard Family Papers, Library of Congress (hereafter LC); Wells, *Siege of Vicksburg*, 97; Howard, *124th Regiment Illinois*, 130–131; Bearss, *Unvexed to the Sea*, 918n.

30. Palfrey, "Port Hudson," 60; *OR*, 26(1):48, 74.

31. *OR*, 26(1):48, 146; Palfrey, "Port Hudson," 60.

32. *OR*, 26(1):46, 48, 75; John W. Millett, "At Port Hudson: A Boy of the 24th Me. Has Some Exciting Adventures," *National Tribune*, August 19, 1909: Palfrey, "Port Hudson," 60.

33. Hess, *Kennesaw Mountain*, 182–183; Stormont, *Fifty-Eighth Regiment of Indiana*, 335.

34. Hess, *Trench Warfare under Grant & Lee*, 171–172, 176–177; Turtle, "History of the Engineer Battalion," 8.

35. Hess, *Into the Crater*, 1–3, 5–6, 11; Henry Pleasants testimony, January 13, 1865, *Report of the Committee on the Conduct of the War on the Attack on Petersburg*, 126–127 (hereafter *RCCW*); *OR*, 40(1):556–557, 40(2):484; James C. Fitzpatrick dispatch, *New York Herald*, August 1, 1864.

36. Hess, *Into the Crater*, 12–14; Inners, "Colonel Henry Pleasants and the Military Geology of the Petersburg Mine," 3–5, 7; *OR*, 40(1):557, 40(2):590–591.

37. Hess, *Into the Crater*, 2, 14–15; Henry Pleasants testimony, January 13, 1865, *RCCW*, 126; *OR*, 40(1): 557–558, 40(2):396–397; James C. Fitzpatrick dispatch, *New York Herald*, August 1, 1864.

38. Hess, *Into the Crater*, 17–20, 35, 75; *OR*, 40(3):791–792, 795, 797, 801, 806–808, 813, 816; newspaper clipping, E. N. Wise to editor, November 1905, in Stewart, "Charge of the Crater," Museum of the Confederacy; Hugh T. Douglas, "Confederate Countermining in Front of Petersburg," Petersburg Crater Collection, University of Virginia (hereafter UVA).

39. *OR*, 40(1):524, 557–558, 40(3):283, 290, 300–310, 336–337, 354; Henry Pleasants testimony, January 13, 1865, *RCCW*, 127; Hess, *Into the Crater*, 21–22, 24.

40. *OR*, 40(1):137, 557, 561, 40(3):526–530, 565–566; Ambrose E. Burnside testimony,

December 17, 1864, and Henry Pleasants testimony, January 13, 1865, *RCCW*, 18, 23–24, 127–129; Hess, *Into the Crater*, 51–53.

41. Hess, *Into the Crater*, 15, 28.

42. Hess, 55–56, 58–59, 62.

43. Hess, 77, 79–83; Henry Reese account in Guthrie, *Camp-Fires of the Afro-Americans*, 529; Bryant, *Diary*, 176; Cunningham, *Three Years with the Adirondack Regiment*, 139.

44. Hess, *Into the Crater*, 84–85; newspaper clipping, E. N. Wise to editor, November 1905, in Stewart, "Charge of the Crater," Museum of the Confederacy.

45. Hess, *Into the Crater*, 87, 92, 95, 104, 111.

46. Hess, *Into the Crater*, 117, 123, 135, 152, 157, 160–167, 173, 175–177, 184, 187–188, 192, 200.

47. Hess, *Into the Crater*, 197–198, 227–229; *OR*, 42(1):896–897; *OR*, 42(2):58, 1160; Douglas, "Confederate Countermining in Front of Petersburg," Petersburg Crater Collection, UVA; Blackford, *War Years with Jeb Stuart*, 265–267; Fleet and Fuller, *Green Mount*, 337; Andrews, *Sketch of Company A*, 24–25; Bernard, *War Talks*, 334; Fagan, "Petersburg Crater," in Fortin, "Colonel Hilary A. Herbert's History of the Eighth Alabama," 151; Graham, "Fifty-Sixth Regiment," 375.

48. Wiley, *Norfolk Blues*, 141.

49. Hess, *In the Trenches at Petersburg*, 112–113, 121; Venable, "In the Trenches at Petersburg," 59–60.

50. Hess, *In the Trenches at Petersburg*, 120–121, 358n28; Blackford, *War Years with Jeb Stuart*, 262–263; Venable, "In the Trenches at Petersburg," 59–60.

51. Hess, *In the Trenches at Petersburg*, 114–115, 154, 184–185; *OR*, 42(2):551–552, 1155, 1159–1160, 1163, 1287; Venable, *Eighty Years After*, 21, 43–44, 47.

52. Hess, *In the Trenches at Petersburg*, 121–122; Jackson, *First Regiment Engineer Troops*, 87, 90, 92–93.

53. Hess, *In the Trenches at Petersburg*, 118.

54. Hess, 118–119; *OR*, 42(2):1155, 1158, 1162–1163; Douglas, "Confederate Countermining in Front of Petersburg," Petersburg Crater Collection, UVA.

55. Hess, *In the Trenches at Petersburg*, 119–120, 342n; *OR*, 42(1):792–793, 885, 42(2):52–53, 59–60, 71, 79, 1163.

56. Hess, *In the Trenches at Petersburg*, 115, 122, 206–207, 358n; *OR*, 42(1):176–179, 669, 42(3):516.

57. Hess, *In the Trenches at Petersburg*, 115, 117, 206; *OR*, 42(1):183, 42(2):19, 26, 42(3):857; Thompson, *Engineer Battalion*, 79–80; Gilbert Thompson Journal, August 5, 1864, LC.

58. Wise, *Gate of Hell*, 195–203; Hearn, *Mobile Bay and the Mobile Campaign*, 159–201; *Supplement to the Official Records*, pt. 1, 7:961–962.

59. Orlando B. Willcox testimony, December 20, 1864, *RCCW*, 91.

60. Cilella, *Correspondence of Major General Emory Upton*, 190.

61. Dennis Hart Mahan to Sherman, September 8, 1865, William T. Sherman Papers, LC.

9. Before the Great War

1. Croll, *History of Landmines*, 9; "Safety Fuse," Wikipedia, https://enwikipedia.org/wiki/Safety_fuse.

2. Kochan and Wideman, *Civil War Torpedoes*, 334, 343; Wolters, "Electric Torpedoes," 777.

3. Schiffer, *Power Struggles*, 88–89, 167; *Index of Patents*, 6, 8; Kochan and Wideman, *Civil War Torpedoes*, 299, 343, 346; Plum, *Military Telegraph*, 2:88–98; Delafield, *Art of War in Europe*, 111–112; King, *Torpedoes*, 53–54, 56–58; "New Infernal Machine," *New York Herald*, July 2, 1865; Hess, *Civil War Torpedoes*, 156–157.

4. Schiffer, *Power Struggles*, 232; Simon, *Papers of Ulysses S. Grant*, 14:72n; "Taliaferro Preston Shaffner," Wikipedia, https://en.wikipedia.org/wiki/Taliaferro_Preston_Shaffner; Hess, *Civil War Torpedoes*, 157.

5. Simon, *Papers of Ulysses S. Grant*, 14:70–71, 71n-73n; *OR*, 46(1):172; Hess, *In the Trenches at Petersburg*, 254–279; Hess, *Civil War Torpedoes*, 158.

6. Beach, "Sieges of the Franco-German War," 363, 365–366, 433–435.

7. Ernst, *Manual of Practical Military Engineering*, 15–18, 22–23, 37–41, 48.

8. Ernst, 42–44, 47–48.

9. Mahan, *Elementary Course of Military Engineering*, 159.

10. Ernst, *Manual of Practical Military Engineering*, 44.

11. Mercur, *Attack of Fortified Places*, 119–120, 138–139–160, 170–171.

12. Mercur, 131–133, 135.

13. Mercur, 163.

14. Ilsley and Hooker, *Electric Shot-Firing*, 6.

15. Mercur, *Attack of Fortified Places*, 169–173.

16. Mercur, 169.

17. Drinker, *Tunneling*, 28, 31, 676; Simms, *Practical Tunnelling*, 9–23, 191–193, 332, 536; West, *Innovation*, 101, 110–115, 175–176, 179.

18. Kerr, *Practical Coal Mining*, 31–32, 38, 40; West, *Innovation*, 160–161.

19. Kerr, *Practical Coal Mining*, 72–75, 80–82, 85; West, *Innovation*, 60–63.

20. Kerr, *Practical Coal Mining*, 92, 97–99; West, *Innovation*, 85, 87, 226–247, 270–280.

21. Kerr, *Practical Coal Mining*, 296–297, 350.

22. Barnatt, *Archaeology of Underground Mines and Quarries*, 53–54, 64, 69, 95–96, 100, 102–103; Farrenkopf, "Accidents and Mining," 196–198.

23. Thomson, *Chitral Campaign*, vii-viii, 44, 48, 56, 76; "Chitral Expedition," Wikipedia, http://en.wikipedia.org/wiki/Chitral_Expedition.

24. Thomson, *Chitral Campaign*, 79–84.

25. Thomson, 82–84.

26. Hogg, *History of Fortification*, 185–186; James, *Siege of Port Arthur*, 62–63.

27. James, *Siege of Port Arthur*, 106, 109, 113, 139; Livermore, "Field and Siege Operations," 438.

28. James, *Siege of Port Arthur*, 145, 168–169.

29. James, 171.

30. James, 171–172.

31. James, 165–166, 176–177.

32. James, 178, 180.

33. James, 173, 175, 214–219, 221.

34. James, 206, 225–230.

35. James, 273.

36. Brooks, *Military Mining*, 8.

37. Brooks, 8; Saunders, *Trench Warfare*, 94.

38. Willig, "German Military Geology and Military Mining," 136.

39. Barton, Doyle, and Vandewalle, *Beneath Flanders Fields*, 43; Smith, "Chatham Siege Operations in 1907," 91.

40. Barton, Doyle, and Vandewalle, *Beneath Flanders Fields*, 44; Smith, "Chatham Siege Operations in 1907," 91–92.

41. Smith, "Chatham Siege Operations in 1907," 89, 90, 92.

10. Underground Warfare on the Western Front

1. Harvey, "British Military Mining," 509; Jones, *Underground Warfare*, 253.

2. Jones, *Underground Warfare*, 29.

3. Jones, 30, 32–36, 39–40.

4. Rhyne, "War in the Lithosphere," 61; Barton, Doyle, and Vandewalle, *Beneath Flanders Fields*, 53.

5. Norton-Griffiths, "Origin of Tunnelling Companies," 87–88.

6. Norton-Griffiths, 88–89; R. Napier Harvey, foreword to Grieve and Newman, *Tunnellers*, 11; *Work of the Royal Engineers*, 2; Logan, "Difficulties and Dangers of Mine-Rescue Work," 198.

7. Graham, *Life of a Tunnelling Company*, 14, 37; Grieve and Newman, *Tunnellers*, 316.

8. Jones, *Underground Warfare*, 165–166; Woodward, "Notes on the Work of an Australian Tunnelling Company," 1–3; MacLeod, "Phantom Soldiers," 32, 34.

9. Coulthard, "Tunnelling at the Front," 445; Pascas, "Clay-kickers of Flanders Fields," 3–4; Morley, "Tunneling on the Western Front," 344; Neill, *New Zealand Tunnelling Company*, 3, 8, 15, 21.

10. Harvey, "Military Mining in the Great War," 538; *Work of the Royal Engineers*, 2–3.

11. MacLeod, "Phantom Soldiers," 40.

12. Jones, *Underground Warfare*, 165.

13. *History of the 27th Engineers*, 1, 4–5, 18; Mayo, "Military Tunneling," 276; Trounce, "Military Mining on the Western Front," 81; Brooks, *Military Mining*, 5.

14. Jones, *Underground Warfare*, 93–96; *Work of the Royal Engineers*, 3, 5, 13; Harvey, "British Military Mining," 513.

15. *Work of the Royal Engineers*, 104.

16. *Work of the Royal Engineers*, 3, 135, 137; Stichelbaut et al., "First World War from Above and Below," 70; Ball, "Work of the Miner," 241.

17. Davis, "Tunneling Reminiscences," 476; *Work of the Royal Engineers*, 22; Jones, *Underground Warfare*, 76–79, 81–82, 84–85.

18. Jones, *Underground Warfare*, 45, 47–48.

19. Jones, 48–53.

20. Barton, Doyle, and Vandewalle, *Beneath Flanders Fields*, 97; Murray, "Tunnelling in the Ypres Salient," 116; Coulthard, "Tunnelling at the Front," 447.

21. Jones, *Underground Warfare*, 127–129.

22. Brooks, *Military Mining*, 29.

23. Brooks, 25.

24. Brooks, 25–27.

25. Brooks, 27.

26. Trounce, *Fighting the Boche Underground*, 78.

27. Graham, *Life of a Tunnelling Company*, 25–26.

28. Graham, 29–30, 34.

29. Ball, "Work of the Miner," 196; Harvey, "Military Mining in the Great War," 538–540.

30. Jones, *Underground Warfare*, 53, 55–56; *Mine Warfare by Captain Cussenot*, 5.

31. *Work of the Royal Engineers*, 103.

32. *Work of the Royal Engineers*, 102–103, 105–106.

33. Morley, "Tunnelling on the Western Front," 342.

34. Reynolds, "Mining in Chalk," 466–467.

35. Morley, "Tunneling on the Western Front," 342.

36. Neill, *New Zealand Tunnelling Company*, 48.

37. Trounce, *Fighting the Boche Underground*, 12–14, 189, 195–196, 215; Trounce, *Notes on Military Mining*, 1, 62.

11. The Materials and Methods of Mining in World War I

1. Brooks, *Military Mining*, 20; *Mine Warfare by Captain Cussenot*, 6.

2. *Work of the Royal Engineers*, 105.

3. Brooks, *Military Mining*, 22, 29–30; *Work of the Royal Engineers*, 105.

4. *Work of the Royal Engineers*, 25; Coulthard, "Tunnelling at the Front," 460.

5. Brooks, *Military Mining*, 29–30; Barton, Doyle, and Vandewalle, *Beneath Flanders Fields*, 80–82; Coulthard, "Tunnelling at the Front," 451; Doyle, "Examples of the Influence of Groundwater," 81; Jones, *Underground Warfare*, 170.

6. Coulthard, "Tunnelling at the Front," 452.

7. Grieve and Newman, *Tunnellers*, 252.

8. Brooks, *Military Mining*, 27.

9. Brooks, 22–23; *Mine Warfare by Captain Cussenot*, 9.

10. Coulthard, "Tunnelling at the Front," 457.

11. Reynolds, "Mining in Chalk," 468.

12. *Work of the Royal Engineers*, 112.

13. *Work of the Royal Engineers*, 110.

14. Brooks, *Military Mining*, 21; *Work of the Royal Engineers*, 106, 112; Neill, *New Zealand Tunnelling Company*, 43.

15. *Work of the Royal Engineers*, 22, 112–113; Grieve and Newman, *Tunnellers*, 220; Coulthard, "Tunnelling at the Front," 453–454; Morley, "Tunnelling on the

Western Front," 344; Reynolds, "Mining in Chalk," 469; Davis, "Tunnelling Reminiscences," 480.

16. Graham, *Life of a Tunnelling Company*, 92–93.

17. Reynolds, "Mining in Chalk," 469; Barton, Doyle, and Vandewalle, *Beneath Flanders Fields*, 111.

18. Morley, "Tunnelling on the Western Front," 344–345; Coulthard, "Tunnelling at the Front," 460–461; *Work of the Royal Engineers*, 107; Neill, *New Zealand Tunnelling Company*, 48.

19. Reynolds, "Mining in Chalk," 469; *Work of the Royal Engineers*, 110.

20. Brooks, *Military Mining*, 22; *Work of the Royal Engineers*, 110–111.

21. *Mine Warfare by Captain Cussenot*, 6; Brooks, *Military Mining*, 19.

22. Neill, *New Zealand Tunnelling Company*, 34.

23. Brooks, *Military Mining*, 34–35, 37.

24. Jones, *Underground Warfare*, 171, 174; Reynolds, "Mining in Chalk," 472; Grieve and Newman, *Tunnellers*, 177.

25. Barton, Doyle, and Vandewalle, *Beneath Flanders Fields*, 179–180; Barrie, *War Underground*, 196–203; West, *Innovation*, 249, 251, 253; "Whitaker Machine," 81.

26. Morley, "Tunnelling on the Western Front," 344; Rose and Rosenbaum, "British Military Geologists," 43–45, 47; Brooks, "Use of Geology," 105–106; Barton, Doyle, and Vandewalle, *Beneath Flanders Fields*, 71–72, 77.

27. Brooks, "Use of Geology," 87, 109.

28. Brooks, 108.

29. Brooks, 108.

30. Barton, Doyle, and Vandewalle, *Beneath Flanders Fields*, 13; Tatham, "Tunnelling in the Sand Dunes," 249, 251, 253.

31. Barton, Doyle, and Vandewalle, *Beneath Flanders Fields*, 13–15, 76, 79; Doyle, "Examples of the Influence of Groundwater," 74, 76, 77; Grieve and Newman, *Tunnellers*, 220; Jones, *Underground Warfare*, 92; Doyle, "Geology and the War on the Western Front," 184–186.

32. Barton, Doyle, and Vandewalle, *Beneath Flanders Fields*, 14, 75–76; Reynolds, "Mining in Chalk," 465; Doyle, "Field Guide to the Geology of the British Sector," 408.

33. Brooks, "Use of Geology," 107; Brooks, *Military Mining*, 14.

34. Brooks, "Use of Geology," 105; Ball, "Work of the Miner," 190–191.

35. Grieve and Newman, *Tunnellers*, 12; *Work of the Royal Engineers*, 113; Morse, "War Mining Notes," 480.

36. Brooks, *Military Mining*, 23.

37. Reynolds, "Mining in Chalk," 471–472.

38. Barton, Doyle, and Vandewalle, *Beneath Flanders Fields*, 118; Reynolds, "Mining in Chalk," 472.

39. Marshall, *Dictionary of Explosives*, 5; Norton-Griffiths, "Origin of Tunnelling Companies," 89; Trounce, "Military Mining," 60; Barton, Doyle, and Vandewalle, *Beneath Flanders Fields*, 119; *Manual of Field Works (All Arms)*, 177; Reynolds, "Mining in Chalk," 470; "TNT," Wikipedia, https://en.wikipedia.org/wiki/TNT.

40. Neill, *New Zealand Tunnelling Company*, 47; *Work of the Royal Engineers*, 113–114.

41. *Manual of Field Works (All Arms)*, 176; Ball, "Work of the Miner," 194; Trounce, "Military Mining," 60; Trounce, *Notes on Military Mining*, 67.

42. Ball, "Work of the Miner," 194; Trounce, *Notes on Military Mining*, 66–69; Trounce, "Military Mining," 60–61.

43. Trounce, *Notes on Military Mining*, 70; "Cheddite," Wikipedia, https://en .wikipedia.org/wiki/Cheddite; Ball, "Work of the Miner," 195.

44. Jones, *Underground Warfare*, 184; Trounce, *Fighting the Boche Underground*, 3.

45. Trounce, *Notes on Military Mining*, 78.

46. Trounce, 79–80.

47. Trounce, 80–81; *Manual of Field Works (All Arms)*, 177–178, 182–183.

48. Trounce, *Notes on Military Mining*, 80, 82, 85; *Manual of Field Works (All Arms)*, 183–184; Coulthard, "Tunnelling at the Front," 458–459.

49. Coulthard, "Tunnelling at the Front," 458.

50. Trounce, *Notes on Military Mining*, 82; *Manual of Field Works (All Arms)*, 183.

51. Trounce, *Notes on Military Mining*, 83; *Work of the Royal Engineers*, 21; Neill, *New Zealand Tunnelling Company*, 47–48; Coulthard, "Tunnelling at the Front," 458; Reynolds, "Mining in Chalk," 472; Grieve and Newman, *Tunnellers*, 53; Trounce, *Notes on Military Mining*, 85.

52. Reynolds, "Mining in Chalk,"472; Neill, *New Zealand Tunnelling Company*, 48; Coulthard, "Tunnelling at the Front," 459; Grieve and Newman, *Tunnellers*, 43n; *Work of the Royal Engineers*, 21; Trounce, *Notes on Military Mining*, 84; Pascas, "Clay-Kickers of Flanders Fields," 16; Brooks, *Military Mining*, 31.

53. Reynolds "Mining in Chalk," 472; Brooks, *Military Mining*, 25.

54. Jones, *Underground Warfare*, 185–189.

55. Jones, 212, 214; Grieve and Newman, *Tunnellers*, 169–171.

56. Davis, "Tunnelling Reminiscences," 479; Coulthard, "Tunnelling at the Front," 455–456; Trounce, *Notes on Military Mining*, 75.

57. Graham, *Life of a Tunnelling Company*, 115–116; Stichelbaut et al., "First World War from Above and Below," 70.

58. Grieve and Newman, *Tunnellers*, 12; Davis, "Tunnelling Reminiscences," 478; Barton, Doyle, and Vandewalle, *Beneath Flanders Fields*, 99; Barrie, *War Underground*, 53.

59. Malins, *How I Filmed the War*, 97–98.

60. Barton, Doyle, and Vandewalle, *Beneath Flanders Fields*, 101; Brooks, *Military Mining*, 39; Ball, "Work of the Miner," 207; Trounce, "Military Mining," 59.

61. Trounce, "Military Mining," 58.

62. Trounce, 58; Woodward, "Notes on the Work of an Australian Tunnelling Company," 39; Murray, "Tunnelling in the Ypres Salient," 115.

63. Ball, "Work of the Miner," 204, 206.

64. Murray, "Tunnelling in the Ypres Salient," 115; Brooks, *Military Mining*, 39–40; *Work of the Royal Engineers*, 129.

65. Murray, "Tunnelling in the Ypres Salient," 115–116.

66. Brooks, *Military Mining*, 14; Jones, *Underground Warfare*, 178.

67. Graham, *Life of a Tunnelling Company*, 13–14; Barton, Doyle, and Vandewalle, *Beneath Flanders Fields*, 105.

68. Morley, "Tunnelling on the Western Front," 344–345; Davis, "Tunnelling Reminiscences," 479; Reynolds, "Mining in Chalk," 462, 464–465; Ball, "Work of the Miner," 207.

69. Trounce, *Fighting the Boche Underground*, 74.

70. *Work of the Royal Engineers*, 29–30.

71. Logan, "Difficulties and Dangers of Mine-Rescue Work," 201–204; Barton, Doyle, and Vandewalle, *Beneath Flanders Fields*, 126.

72. Barton, Doyle, and Vandewalle, *Beneath Flanders Fields*, 125; *Work of the Royal Engineers*, 59; Logan, "Difficulties and Dangers of Mine-Rescue Work," 205.

73. Reynolds, "Mining in Chalk," 470; Barton, Doyle, and Vandewalle, *Beneath Flanders Fields*, 127–129, 134; *Work of the Royal Engineers*, 61–63, 65; Logan, "Difficulties and Dangers of Mine-Rescue Work," 206; Trounce, "Mine Rescue Work," 549–564.

74. Logan "Difficulties and Dangers of Mine-Rescue Work," 208, 217.

75. Logan, 216.

76. *Work of the Royal Engineers*, 74–75, 78–79.

77. Logan, "Difficulties and Dangers of Mine-Rescue Work," 208–209.

78. *Work of the Royal Engineers*, 3, 73, 135, 137; Logan, "Difficulties and Dangers of Mine-Rescue Work," 210–211.

79. Logan, "Difficulties and Dangers of Mine-Rescue Work," 212.

80. *Work of the Royal Engineers*, 12, 83; Reynolds "Mining in Chalk," 470.

81. Logan, "Difficulties and Dangers of Mine-Rescue Work," 203–204.

82. *Work of the Royal Engineers*, 67.

83. Reynolds, "Mining in Chalk," 469–470; *Work of the Royal Engineers*, 68.

84. Reynolds, "Mining in Chalk," 470; "Western Transept," accessed July 28, 2024, https://www.snwm.org/tour/?view=station_5; Norton-Griffiths, "Origin of Tunnelling Companies," 92.

12. 1916

1. Jones, *Underground Warfare*, 98–99; Barrie, *War Underground*, 146–149, 157.

2. Jones, *Underground Warfare*, 132; Graham, *Life of a Tunnelling Company*, 28–29.

3. Jones, *Underground Warfare*, 114.

4. *Work of the Royal Engineers*, 33–34; Ball, "Work of the Miner," 216.

5. Jones, *Underground Warfare*, 115.

6. Malins, *How I Filmed the War*, 101–102, 105–106.

7. Malins, 124, 161–163.

8. Reeves, "Cinema, Spectatorship, and Propaganda," 12–16; Jones, *Underground Warfare*, 118–120.

9. Jones, *Underground Warfare*, 121–126.

10. Jones, 130.

11. *Work of the Royal Engineers*, 34–35; Jones, *Underground Warfare*, 199–202, 204, 206–208.

12. *Work of the Royal Engineers*, 35; Jones, *Underground Warfare*, 131.

13. Jones, *Underground Warfare*, 131–132.

14. Alexander Johnston diary, in Cave, *Arras, Vimy Ridge*, 68.

15. Jones, *Underground Warfare*, 98; Alexander Johnston diary, in Cave, *Arras, Vimy Ridge*, 69, 72, 79–80.

16. Grieve and Newman, *Tunnellers*, 149–150, 152, 155.

17. Trounce, *Fighting the Boche Underground*, 180–182, 184; Trounce, "Military Mining Operations," 190.

18. Coulthard, "Tunnelling at the Front," 460; Graham, *Life of a Tunnelling Company*, 26–27.

19. Ball, "Work of the Miner," 190; Harvey, "Military Mining in the Great War," 548; Trounce, *Notes on Military Mining*, 2; MacLeod, "Phantom Soldiers," 36.

20. Logan, "Difficulties and Dangers of Mine-Rescue Work," 201.

21. *Work of the Royal Engineers*, 32.

22. Davis, "Tunnelling Reminiscences," 479.

24. Grieve and Newman, *Tunnellers*, 106–107.

24. Barton, Doyle, and Vandewalle, *Beneath Flanders Fields*, 141; Reynolds, "Mining in Chalk," 473.

25. Grieve and Newman, *Tunnellers*, 316; Barrie, *War Underground*, 56–57, 126–127.

26. Grieve and Newman, *Tunnellers*, 91–93; Trounce, *Fighting the Boche Underground*, 47–52; Ball, "Work of the Miner," 225.

27. Trounce, *Fighting the Boche Underground*, 54; Grieve and Newman, *Tunnellers*, 93; Ball, "Work of the Miner," 225–226.

28. Ball, "Work of the Miner," 226; Grieve and Newman, *Tunnellers*, 94.

29. Grieve and Newman, *Tunnellers*, 94.

30. For other examples of underground combat, see *Work of the Royal Engineers*, 24–26; Grieve and Newman, *Tunnellers*, 180; and Jones, *Underground Warfare*, 82, 84, 139–140, 142–143.

31. Barrie, *War Underground*, 227–229, 231–239; Barton, Doyle, and Vandewalle, *Beneath Flanders Fields*, 139–141.

32. Barton, "Second Canadian Tunnellers at Mount Sorrel," https://cmehistory .wordpress.com/2013/05/25/the-second-canadian-tunnellers-at-mount-sorrell/; Barton, Doyle, and Vandewalle, *Beneath Flanders Fields*, 141.

33. Barton, "Second Canadian Tunnellers at Mount Sorrel"; Barton, Doyle, and Vandewalle, *Beneath Flanders Fields*, 141.

34. Barton, Doyle, and Vandewalle, *Beneath Flanders Fields*, 174–175; Grieve and Newman, *Tunnellers*, 220–221.

35. Logan, "Difficulties and Dangers of Mine-Rescue Work," 199–200; Davis, "Tunnelling Reminiscences," 480; Barton, Doyle, and Vandewalle, *Beneath Flanders Fields*, 175.

13. 1917

1. Jones, *Underground Warfare*, 215, 222, 224.

2. Neill, *New Zealand Tunnelling Company*, 64; *Work of the Royal Engineers*, 38; Jones, *Underground Warfare*, 230.

3. *Work of the Royal Engineers*, 36–37, 115; Graham, *Life of a Tunnelling Company*, 61–62; Ball, "Work of the Miner," 217; Jones, *Underground Warfare*, 225. The Grange Subway is now partly open to visitors in the Vimy Ridge National Historic Site of Canada.

4. Brooks, *Military Mining*, 41, 43; *Work of the Royal Engineers*, 36; Jones, *Underground War*, 226.

5. Grieve and Newman, *Tunnellers*, 158; Ball, "Work of the Miner," 218; Jones, *Underground Warfare*, 230–231.

6. Jones, *Underground Warfare*, 228, 230.

7. Boire, "Underground War," 20–21; Jones, *Underground Warfare*, 136, 226–227.

8. Norton-Griffiths, "Origin of Tunneling Companies," 90.

9. Ball, "Work of the Miner," 222–224.

10. Ball, 222; Barrie, *War Underground*, 247; Woodward, "Notes on the Work of an Australian Tunnelling Company," 32; Grieve and Newman, *Tunnellers*, 211–212.

11. Barrie, *War Underground*, 248; Cowley, "Tunnels of Hill 60," 68; Ball, "Work of the Miner," 220.

12. Murray, "Tunnelling in the Ypres Salient," 122–123; Barrie, *War Underground*, 251; Ball, "Work of the Miner," 222.

13. Barrie, *War Underground*, 252; Ball, "Work of the Miner," 222.

14. Ball, "Work of the Miner," 222.

15. Grieve and Newman, *Tunnellers*, 219.

16. Barrie, *War Underground*, 252; Ball, "Work of the Miner," 222.

17. Ball, "Work of the Miner," 223.

18. Ball, 223; Grieve and Newman, *Tunnellers*, 227–228.

19. Ball, "Work of the Miner," 223; Barton, Doyle, and Vandewalle, *Beneath Flanders Fields*, 178.

20. Ball, "Work of the Miner," 223; Barrie, *War Underground*, 253–254.

21. Grieve and Newman, *Tunnellers*, 234; Jones, *Underground Warfare*, 155.

22. Barrie, *War Underground*, 254; Ball, "Work of the Miner," 223.

23. Ball, "Work of the Miner," 223; Barrie, *War Underground*, 254.

24. Barrie, *War Underground*, 254; Ball, "Work of the Miner," 223. For a good map of the Messines Ridge mine array, see Barton, Doyle, and Vandewalle, *Beneath Flanders Fields*, 163. For a detailed narrative of the digging of the Messines mines, see Grieve and Newman, *Tunnellers*, 204–247.

25. Ball, "Work of the Miner," 220.

26. Mullins, "Mines at Messines," 22–23; Grieve and Newman, *Tunnellers*, 245.

27. Barrie, *War Underground*, 256–257, 259; Harvey, "British Military Mining," 515; Woodward, "Notes on the Work of an Australian Tunnelling Company," 9.

28. Barrie, *War Underground*, 257; Ball, "Work of the Miner," 222–223.

29. Prior and Wilson, *Passchendaele*, 49–65.

30. Grieve and Newman, *Tunnellers*, 204, 245, 247; Barrie, *War Underground*, 261; Jones, *Underground Warfare*, 145, 157, 159.

31. Norton-Griffiths, "Origin of Tunnelling Companies," 90–91; Barrie, *War Underground*, 261; Grieve and Newman, *Tunnellers*, 253.

32. Stichelbaut et al., "First World War from Above and Below," 70; Grieve and Newman, *Tunnellers*, 241, 245; Ball, "Work of the Miner," 220–221.

33. Barton, Doyle, and Vandewalle, *Beneath Flanders Fields*, 191–193.

34. Barton, Doyle, and Vandewalle, 195–196.

35. Jones, *Underground Warfare*, 238–239.

36. Jones, 240–241, 243.

37. Jones, 43–44.

38. Jones, 56–57.

39. Leonard, *Beneath the Killing Fields*, 167; Jones, *Underground Warfare*, 58–59.

40. Leonard, *Beneath the Killing Fields*, 167; Jones, *Underground Warfare*, 59–60.

41. Leonard, *Beneath the Killing Fields*, 167; Jones, *Underground Warfare*, 60–61.

42. Leonard, *Beneath the Killing Fields*, 171; Mosier, *Myth of the Great War*, 138–139; Jones, *Underground Warfare*, 61–62.

43. Jones, *Underground Warfare*, 64, 66.

44. Jones, 64, 66.

45. Jones, 67–68, 70–71, 73.

46. Leonard, *Beneath the Killing Fields*, 166; Mosier, *Myth of the Great War*, 138–139; Jones, *Underground Warfare*, 73. For more details, diagrams, and photographs, see Buchner, *Der Minenkrieg auf Vauquois*.

47. Association des Amis de Vauquois et de sa Région, Butte de Vauquois, https://butte-vauquois.fr/en/.

48. Jones, *Underground Warfare*, 74.

49. Ball, "Work of the Miner," 215; *Work of the Royal Engineers*, 18.

50. Grieve and Newman, *Tunnellers*, 252–253; Neill, *New Zealand Tunnelling Company*, 38–40; Jones, *Underground Warfare*, 168.

51. Barton, Doyle, and Vandewalle, *Beneath Flanders Fields*, 66–70; Grieve and Newman, *Tunnellers*, 251–252.

14. Other Fronts of World War I

1. Doyle, "Geology as an Interpreter," 243.

2. Bean, *Story of Anzac*, 98–99; Jones, *Underground Warfare*, 190.

3. Jones, *Underground Warfare*, 191–192.

4. Jones, 192–194; Stevenson, *To Win the Battle*, 126, 128–129.

5. Grieve and Newman, *Tunnellers*, 80; Jones, *Underground Warfare*, 86, 88.

6. Jones, *Underground Warfare*, 90.

7. Grieve and Newman, *Tunnellers*, 78; East, *Gallipoli Diary of Sergeant Lawrence*, 153–154.

8. East, *Gallipoli Diary of Sergeant Lawrence*, 29, 42, 156–159.

9. East, 24, 31.

10. East, 25–26.

11. East, 31, 38, 52.

12. East, 39, 41, 45.

13. East, 81.

14. Willig, "German Military Geology," 131.

15. Willig, 131–133.

16. "Mines on the Italian Front (World War I)," Wikipedia, https://en.wikipedia.org/wiki/Mines_on_the_Italian_front_(World_War_I).

17. "Mines on the Italian Front (World War I)."

18. "Mines on the Italian Front (World War I)."

19. "Mines on the Italian Front (World War I)."

20. Lang, "Some Notes on the Turkish Trenches," 311.

15. After the Great War

1. Grieve and Newman, *Tunnellers*, 318–319.

2. Cron, "Military Geology in Time of War," 85; Trounce, "Military Mining Operations," 187–190; *Engineer Field Manual: Professional Papers of the Corps of Engineers*, 393–423.

3. Brooks, *Military Mining*, 10.

4. Harvey, "British Military Mining," 518.

5. Mitchell, *Army Engineering*, 220; *Fortification: Underground Water and Its Relation to Fieldworks*, 8; *Engineer Field Manual*, vol. 2, *Military Engineering (Tentative)*, pt. 2, *Defensive Measures*, 148–173, 241–244.

6. Harvey, "British Military Mining," 518.

7. *Engineer Field Manual: Field Fortifications*, 251–258; Brooks, *Military Mining*, 10.

8. Seymour, *Great Sieges of History*, 238, 241, 245–246, 248, 250–251, 254–256.

9. Barton, Doyle, and Vandewalle, *Beneath Flanders Fields*, 290.

10. Willig, "German Military Geology," 136; Nemchinsky, Yuriev, and Galkin, "Underground Warfare," 190–191.

11. Nemchinsky, Yuriev, and Galkin, "Underground Warfare," 191–192.

12. Gibby, "Battle of Shangganling," 61–62, 64, 69–70, 72, 74,76–78, 80, 84–85.

13. Mayo, "Military Tunneling," 276.

14. Mayo, 277.

15. Mayo, 277.

16. Bermudez, *North Korean Special Forces*, 169–172.

17. Sarantakes, "Quiet War," 439–457.

18. Boylan, "No 'Technical Knockout,'" 1381–1382.

19. Boylan, 1357, 1382.

20. Codó, "Subterranean Warfare," 96; Forman, *Report from Red China*, 142–144; Petchey, "Second World War Japanese Defences on Watom Island," 29–51; Deacon, "Cave Warfare on Biak," 4–6; Mushynsky, McKinnon, and Camacho, "Archaeology of World War II Karst Defences," 198–222; Price and Knecht, "Peleliu 1944," 5–48; Bartley, *Iwo Jima*, 15; Floyd, "Cave Warfare on Okinawa," 6–9.

21. Gibby, "Battle of Shangganling," 53–89; Levine, "Headfirst into Underground Battle," 42–48; Lehrer, "Viet Cong Tunnels," 243–247; Trainor, "Tunnel Destruction in Vietnam," 341–342.

22. Grau and Jalali, "Underground Combat," 20–23; Marcus, "Learning 'under Fire,'" 344–370.

Conclusion

1. Taylor, *Art of War in Italy*, 138.

2. Brooks, *Military Mining*, 5, 10; Trounce, *Notes on Military Mining*, 49.

3. Harvey, "British Military Mining," 518.

4. Harvey, "Military Mining in the Great War," 548; Brooks, *Military Mining*, 17; Rhyne, "War in the Lithosphere," 64, 68; MacLeod, "Phantom Soldiers," 40.

5. Morley, "Tunnelling on the Western Front," 343.

6. Barton, "Second Canadian Tunnellers at Mount Sorrel," https://cmehistory .wordpress.com/2013/05/25/the-second-canadian-tunnellers-at-mount-sorrell/.

7. Barrie, *War Underground*, ix; Banks, "Digging in the Dark," 165–167.

8. Barrie, *War Underground*, 247; Barton, Doyle, and Vandewalle, *Beneath Flanders Fields*, 198–199; Leonard, *Beneath the Killing Fields*, 70; Jones, *Underground Warfare*, 135.

9. Stichelbaut et al., "First World War from Above and Below," 68–70.

10. Stichelbaut et al., "First World War from Above and Below," 70; Stichelbaut et al., "Non-Invasive Research of Tunneling Heritage," 110.

11. Grieve and Newman, *Tunnellers*, 161–162; Doyle, "Field Guide to the Geology of the British Sector," 397–398; Leonard, *Beneath the Killing Fields*, 42, 44, 120; Rosenbaum, "Geological Influences on Tunnelling," 135, 140.

12. Doyle, "Field Guide to the Geology of the British Sector," 404, 408.

13. Cowley, "Tunnels of Hill 60," 75; Wiggins, *Siege Mines*, 51; Stichelbaut et al., "First World War from Above and Below," 70.

14. Hutchinson et al., "Geomechanics Stability Assessment of World War I Military Excavations," 76.

15. Sagona et al., "ANZAC [Ariburnu] Battlefield," 328, 330.

16. Hess, *Into the Crater*, 239–241, 243–246; Hess, *In the Trenches at Petersburg*, 293–294, 342.

17. Smith, "Chatham Siege Operations in 1907," 89.

18. "Ernest Brooks (Photographer)," Wikipedia, https://en.wikipedia.org /wiki/Ernest_Brooks_(photographer). Franky Bostyn includes a German photograph of a British mine explosion in 1916 showing the dirt plume as a shaft in form rather than a wave as in the Brooks photograph. See Bostyn, "Zero Hour," 230.

19. Malins, *How I Filmed the War*, 101–102, 105–106, 124, 161–163.

20. "The Siege of the Alcazar," Wikipedia, https://en.wikipedia.org/wiki/The _Siege_of_the_Alcazar

21. Schue, "Remember the Alcazar!," 132–133.

22. "Cold Mountain (Film)," Wikipedia, https://en.wikipedia.org/wiki/Cold _Mountain_(film).

23. "Beneath Hill 60," Wikipedia, https://en.wikipedia.org/wiki/Beneath_Hill _60.

24. Duer et al., "Making the Invisible Visible," 40–44.

25. Ankel, "Russian Soldiers Are Digging Tunnels to Sneak Up on Ukrainian Positions," www.businessinsider.com/russian-soldiers-digging-tunnels-to-sneak -up-ukraine-positions-army-2023-10.

BIBLIOGRAPHY

Archives
Cincinnati Museum Center, Cincinnati, Ohio
    Andrew Hickenlooper Collection
Library of Congress, Manuscript Division, Washington, DC
    Daniel Carter Beard Family Papers
    William T Sherman Papers
    Gilbert Thompson Journal
Museum of the Confederacy, Richmond, Virginia
    William H. Stewart, comp., "The Charge of the Crater: Personal Recollections
        of Participants in the Charge of the Crater at Petersburg, Va., July 30th,
        1864"
National Archives and Records Administration, Washington, DC
    Chief Engineer Letters Received (G1879), Record Group 77
U.S. Army Military History Institute, Carlisle, Pennsylvania
    Civil War Times Illustrated Collection
        Gordon Hickenlooper, ed., "The Reminiscences of General Andrew Hick-
            enlooper, 1861–1865"
University of Arkansas, Special Collections, Fayetteville
    Camden, Arkansas, Civil War Record Book
University of California, Bancroft Library, Berkeley
    E. O. C. Ord Papers
University of North Carolina, Southern Historical Collection, Chapel Hill
    Samuel Henry Lockett Papers
University of Virginia, Special Collections, Charlottesville
    Petersburg Crater Collection
        Hugh T. Douglas, "Confederate Countermining in Front of Petersburg:
            Experience and Recollection"
        ———, "The Petersburg Crater"
Vicksburg National Military Park, Vicksburg, Mississippi
    20th Illinois Folder
    45th Illinois Folder
    23rd Iowa Folder
    3rd Louisiana Folder
    Company A, 1st Mississippi Light Artillery Folder
    20th Ohio Folder
    47th Ohio Folder

Wisconsin Historical Society, Madison
    Lewis and Moulton Family Papers

Internet Sources

"1762—Siege of Schweidnitz—First Phase of Mining Operations." In *Beyträge zur Kriegs-Kunst und Geschichte des Krieges von 1756 bis 1763*, by J. G. Tielke, 4:151–359. Freiberg: Barthel, 1781. Abridged English translation accessed at Kronoskaf: The Virtual Time Machine. www.kronoskaf.com/syw/index.php?title=1762 _-_Siege_of_Schweidnitz_-_First_phase_of_mining_operations.

Ankel, Sophia. "Russian Soldiers Are Digging Tunnels to Sneak Up on Ukrainian Positions without Being Seen by Their Lethal Drones, Kyiv Says." *Business Insider*, October 22, 2023. https://www.businessinsider.com/russian-soldiers -digging-tunnels-to-sneak-up-ukraine-positions-army-2023-10.

Association des Amis de Vauquois et de sa Région. Butte de Vauquois. https:// butte-vauquois.fr/en/.

Barton, Peter. "The Second Canadian Tunnellers at Mount Sorrel." *CME History* (blog), May 25, 2013. https://cmehistory.wordpress.com/2013/05/25/the-sec ond-canadian-tunnellers-at-mount-sorrel/.

British Museum. Museum Number 124554, Wall Panel Relief of Mining. https:// www.britishmuseum.org/collection/object/W_1849-1222-23.

———. Museum Number 124931, Wall Panel Relief of Sapping at Hamanu. https://www.britishmuseum.org/collection/object/W_1856-0909-17-18_2.

Castillo de San Fernando, Figueres, Spain. Star Forts. www.starforts.com/figueres .html.

Collins, Lori, Travis Doering, and Jorge Gonzalez. "Terrestrial and Airborne Li-DAR Digital Documentation of Kosciuszko Mine, Ninety Six National His-toric Site." 2015. Digital Heritage and Humanities Collections, Faculty and Staff Publications 10. https://scholarcommons.usf.edu/dhhc_facpub/10.

Fitzpatrick, Colette. "Explore Turin's Underground Secrets." Easy Reserve, December 16, 2015. https://www.italy-villas.com/to-italy/2015/northwest-italy/piedmont/turin-underground-tour.

"Fortress Josefov." Pevnost Josefov. https://www.josefov.com/josefov_english.htm.

"The Fortress of Terezín." UNESCO World Heritage Convention, July 6, 2001. https://whc.unesco.org/en/tentativelists/1561/.

"The Great Underground Passages of the Citadel of Namur." Visit Namur. https://www.namurtourisme.be/en/a-voir-a-faire/visites-guidees/grands -souterrains-de-citadelle-de-namur/.

"Maastricht Casemates." Maastricht Underground. https://www.exploremaa stricht.nl/en/casemates.

Šeguljev, Nenad. "Warfare with Landmines in Permanent Fortification and Countermines System in Peterwardein Fortress." *Material for the Study of the Cultural Monuments of Vojvodina 33* (2020). Academia, https://www.academia .edu/45520850/Warfare_with_Landmines_in_Permanent_Fortification_and _Countermines_System_in_Peterwardein_Fortress.

"The Western Transept." Scottish National War Memorial. https://www.snwm
.org/tour/?view=station_5.
Wikipedia. https://en.wikipedia.org/.

Articles and Books

Agricola, Georgius. *De Re Metallica*. Translated by Herbert Clark Hoover and Lou Henry Hoover. New York: Dover, 1950.

Aide-Mémoire to the Military Sciences, Framed from Contributions of Officers of the Different Services, and Edited by a Committee of the Corps of Royal Engineers, 1853. 2nd ed. corr. Vol. 1 (London: J. Weale, 1853).

Aineias the Tactician: How to Survive under Siege. Translated by David Whitehead. New York: Oxford University Press, 1990.

Ammianus Marcellinus. 3 vols. Translated by John C. Rolfe. Cambridge, MA: Harvard University Press, 1956–1958.

Amt, Emilie. "Besieging Bedford: Military Logistics in 1224." *Journal of Medieval Military History* 1 (2001): 101–124.

Anderson, Fred. *Crucible of War: The Seven Years' War and the Fate of Empire in British North America, 1754–1766*. New York: Knopf, 2001.

Andrews, W. J. *Sketch of Company K, 23rd South Carolina Volunteers, in the Civil War, from 1862–1865*. Richmond, VA: Whittet and Shepperson, n.d.

Anglim, Simon, Phyllis G. Jestice, Rob S. Price, Scott M. Rusch, and John Serrati. *Fighting Techniques of the Ancient World, 3,000 BC–500 AD: Equipment, Combat Skills, and Tactics*. New York: St. Martin's, 2002.

Appian's Roman History. 4 vols. Translated by Horace White. Cambridge, MA: Harvard University Press, 1955.

Arrian. 2 vols. Translated by P. A. Brunt and E. Iliff Robson. Cambridge, MA: Harvard University Press, 1966–1976.

Bachrach, Bernard S. *Early Carolingian Warfare: Prelude to Empire*. Philadelphia: University of Pennsylvania Press, 2001.

Baird, J. A. *Dura-Europos*. London: Bloomsbury Academic, 2018.

Ball, H. Standish. "The Work of the Miner on the Western Front." *Transactions of the Institution of Mining and Metallurgy* 28 (1918–1919): 189–285.

Banks, Iain. "Digging in the Dark: The Underground War on the Western Front in World War I." *Journal of Conflict Archaeology* 9, no. 3 (September 2014): 156–176.

Barnatt, John. *The Archaeology of Underground Mines and Quarries in England*. Swindon, Eng.: Historic England, 2019.

Barnwell, Joseph W. "The Second Tuscarora Expedition." *South Carolina Historical and Genealogical Magazine* 10, no. 1 (January 1909): 33–48.

Barrie, Alexander. *War Underground: The Tunnellers of the Great War*. Staplehurst, Eng.: Spellmount, 2000.

Bartley, Whitman S. *Iowa Jima: Amphibious Epic*. Washington, DC: Government Printing Office, 1954.

Barton, Peter, Peter Doyle, and Johan Vandewalle. *Beneath Flanders Fields: The Tunneller's War, 1914–1918*. Staplehurst, Eng.: Spellmount, 2004.

Baumgart, Winfried. *The Crimean War, 1853–1856*. London: Bloomsbury, 1999.

Beach, Lansing H. "The Sieges of the Franco-German War." *Military Engineer* 14 (1922): 363–366.

Bean, C. E. W. *The Story of Anzac from 4 May, 1915, to the Evacuation of the Gallipoli Peninsula*. 11th ed. Sydney, Australia: Angus and Robertson, 1941.

Bearss, Edwin C. *Unvexed to the Sea: The Campaign for Vicksburg*. Vol. 3. Dayton, OH: Morningside, 1986.

Berger, Stefan. "Mining History: Sub-fields and Agendas." In *Making Sense of Mining History: Themes and Agendas*, edited by Stefan Berger and Peter Alexander, 1–24. Oxfordshire, Eng.: Routledge, 2020.

Bermudez, Joseph S., Jr., *North Korean Special Forces*. Coulsdon, Surrey, Eng.: Jane's, 1988.

Bernard, George S., ed. *War Talks of Confederate Veterans*. Petersburg, VA: Fenn and Owen, 1892.

Beveridge, H., trans. *The Akbarnāmā of Abu-L-Fazl: History of the Reign of Akbar including an Account of His Predecessors*. 3 vols. 1907. Reprint, Calcutta, India: Asiatic Society, 2000.

Blackford, W. W. *War Years with Jeb Stuart*. New York: Charles Scribner's Sons, 1945.

Boire, Michael. "The Underground War: Military Mining Operations in Support of the Attack on Vimy Ridge, 9 April 1917." *Canadian Military History* 1, no. 1 (1992): 15–24.

Bostyn, Franky. "Zero Hour: Historical Notes on the British Underground War in Flanders, 1915–1917." In *Fields of Battle: Terrain in Military History*, edited by Peter Doyle and Matthew R. Bennett, 225–236. Cham, Switz.: Springer, 2002.

Boylan, Kevin M. "No 'Technical Knockout': Giap's Artillery at Dien Bien Phu." *Journal of Military History* 78, no. 4 (October 2014): 1349–1383.

Bradbury, Jim. *The Medieval Siege*. Woodbridge, Eng.: Boydell, 1992.

Braun, Hugh. *Bungay Castle: Historical Notes and Account of the Excavations*. Bungay, Eng.: Bungay Castle Trust, 1991.

Brice, Martin H. *Forts and Fortresses*. New York: Facts on File, 1990.

———. *Stronghold: A History of Military Architecture*. New York: Schocken Books, 1985.

Brooks, Alfred H. *Military Mining*. Washington, DC: Government Printing Office, 1920.

———. "The Use of Geology on the Western Front." *United States Geological Survey Professional Papers* 128-D (1920): 85–124.

Browning, Reed. *The War of the Austrian Succession*. New York: St. Martin's, 1993.

Bryant, Elias A. *The Diary of Elias A. Bryant of Francestown, N.H.* Concord, NH: Rumford, n.d.

Buchner, Adolf. *Der Minenkrieg auf Vauquois*. Karlsfeld, W. Ger.: W. Goertz, 1982.

Buel, Clarence Clough, and Robert Underwood Johnson, eds. *Battles and Leaders of the Civil War*. 4 vols. New York: Thomas Yoseloff, 1956.

Bury, J. B. "The Early History of the Explosive Mine." *Fort* 10 (1982): 23–29.

Butcher, Kevin. *Roman Syria*. London: British Museum, 2003.

Caesar, Julius. *The Conquest of Gaul*. Translated by S. A. Handford. New York: Penguin Books, 1982.

Cann, Marvin L. "War in the Backcountry: The Siege of Ninety Six, May 22-June 19, 1781." *South Carolina Historical Magazine* 72, no. 1 (January 1971): 1–14.

Cannon-Brookes, Stephen. "FSG Study Tour to the Crimea—May 2002." *Casemate* 66 (January 2003): 9–16.

Caple, Chris. *Excavations at Dryslwyn Castle, 1980–95*. London: Routledge, 2007.

Cave, Nigel. *Arras, Vimy Ridge*. London: Leo Cooper, 1996.

Chandler, D. G., ed. *A Journal of Marlborough's Campaigns during the War of the Spanish Succession, 1704–1711*. Society for Army Historical Research, 1984.

"The Chief Engineer's Journal of the Siege of the Moro Fort, and the Havannah." *Gentleman's Magazine and Historical Chronicle* 32 (1762): 463–466.

The Chronicle of James I, King of Aragon, Surnamed the Conqueror (Written by Himself). 2 vols. Translated by John Forster. London: Chapman and Hall, 1883.

Cilella, Salvatore G., Jr., ed. *Correspondence of Major General Emory Upton*. Vol. 1, *1857–1875*. Knoxville: University of Tennessee Press, 2017.

Clermont-Ganneau, Charles. "The Taking of Jerusalem by the Persians, A.D. 614." In *Palestine Exploration Fund, Quarterly Statement for 1898*, 36–54. London: Palestine Exploration Fund, n.d.

Codó, Enrique Martinez. "Subterranean Warfare." *Military Review* 48 (February 1968): 92–96.

Comnena, Anna. *The Alexiad of the Princess Anna Comnena*. Translated by Elizabeth A. S. Dawes. London: Kegan Paul, Trench, Trubner, 1928.

Conrad, Dennis M., ed. *The Papers of General Nathanael Greene*. 10 vols. Chapel Hill: University of North Carolina Press, 1976–1998.

Coulston, Jon. "The Archaeology of Roman Conflict." In *Fields of Conflict: Progress and Prospect in Battlefield Archaeology*, edited by P. W. M. Freeman and A. Pollard, 23–49. BAR International Series 958. Oxford, Eng.: Archaeopress, 2001.

Coulthard, R. W. "Tunneling at the Front." *Transactions of the Canadian Mining Institute* 22 (1919): 444–461.

Courtney, Paul. "The Archaeology of the Early-Modern Siege." In *Fields of Conflict: Progress and Conflict in Battlefield Archaeology*, edited by P. W. M. Freeman and A. Pollard, 105–115. BAR International Series 958. Oxford, Eng.: Archaeopress, 2001.

Cowley, Robert. "The Tunnels of Hill 60." *Military History Quarterly* 1 (Autumn 1988): 66–75.

Croll, Mike. *The History of Landmines*. Barnsley, Eng.: Leo Cooper, 1998.

Cron, Robert E., Jr. "Military Geology in Time of War." *Military Engineer* 25 (1933): 84–88.

Cunningham, John L. *Three Years with the Adirondack Regiment: 118th New York Volunteers Infantry*. Privately published, 1920.

David, Charles Wendell, ed. *De Expugnatione Lyxbonensi: The Conquest of Lisbon*. New York: Columbia University Press, 1936.

Davies, Gwyn. "'Dig for Victory!': Competitive Fieldwork in Classical Siege Operations." In *The Art of Siege Warfare and Military Architecture from the Classical World to the Middle Ages*, edited by Michael Eisenberg and Rabei G. Khamisy, 45–53. Havertown, PA: Oxbow, 2021.

———. "Roman Offensive Siege Works." Ph.D. diss., University of London, 2001.

Davis, A. W. "Tunnelling Reminiscences." *Transactions of the Canadian Mining Institute* 22 (1919): 475–481.

Day, Lance, and Ian McNeil. *Biographical Dictionary of the History of Technology.* London: Routledge, 1996.

Deacon, Kenneth J. "Cave Warfare on Biak." *Military Engineer* 52 (1962): 4–6.

Delafield, Richard. *Report on the Art of War in Europe, in 1854, 1855, and 1856.* Washington, DC: George W. Bowman, 1860.

Diodorus of Sicily. 12 vols. Translated by C. H. Oldfather, Charles L. Sherman, C. Bradford Welles, Russell M. Geer, and Francis R. Walton. Cambridge, MA: Harvard University Press, 1956–1967.

Dodd, W. O. "Recollections of Vicksburg during the Siege." *Southern Bivouac* 1, no. 1 (September 1882): 2–11.

Doyle, Peter. "Examples of the Influence of Groundwater on British Military Mining in Flanders, 1914–1917." In *Military Aspects of Hydrogeology*, edited by E. P. F. Rose and J. D. Mather, 73–83. Bath, Eng.: Geological Society, 2012.

———. "A Field Guide to the Geology of the British Sector of the Western Front, 1914–18." In *Geology and Warfare: Examples of the Influence of Terrain and Geologists on Military Operations*, edited by Edward P. F. Rose and C. Paul Nathanail, 381–412. Bath, Eng.: Geological Society, 2000:.

———. "Geology and the War on the Western Front, 1914–1918." *Geology Today* 30, no. 5 (September–October 2014): 183–191.

———. "Geology as an Interpreter of Great War Battle Sites." In *Fields of Conflict: Progress and Prospect in Battlefield Archaeology*, edited by P. W. M. Freeman and A. Pollard, 237–252. BAR International Series 958. Oxford, Eng.: Archaeopress, 2001:.

Drinker, Henry S. *Tunneling, Explosive Compounds, and Rock Drills.* 2nd ed. New York: John Wiley and Sons, 1882.

Duane, J. C. *Manual for Engineer Troops.* New York: D. Van Nostrand, 1862.

Duer, Zach, Todd Ogle, David Hicks, Scott Fralin, Thomas Tucker, and Run Yu. "Making the Invisible Visible: Illuminating the Hidden Histories of the World War I Tunnels at Vauquois through a Hybridized Virtual Reality Exhibition." *Digital Object Identifier* 10 (July–August 2020): 39–50.

Duffy, Christopher. *The Fortress in the Age of Vauban and Frederick the Great, 1660–1789.* London: Routledge and Kegan Paul, 1985.

———. *Siege Warfare: The Fortress in the Early Modern World, 1494–1660.* London: Routledge and Kegan Paul, 1979.

Dunkerly, Robert M., and Eric K. Williams. *Old Ninety Six: A History and Guide.* Charleston, SC: History Press, 2006.

Dunster, H. P., [ed.]. *The Chronicles of England, France, and Spain by Sir John Froissart.* New York: E. P. Dutton, 1961.

East, Ronald, ed. *The Gallipoli Diary of Sergeant Lawrence of the Australian Engineers, 1st A.I.F. 1915*. Carlton, Australia: Melbourne University Press, 1981.

Elliott, F. Haws. *Trench Fighting*. Boston: Houghton Mifflin, 1917.

Elton, Hugh. *Warfare in Roman Europe, A.D. 350–425*. Oxford, Eng.: Oxford University Press, 1997.

Engineer Field Manual. Vol. 2, *Military Engineering (Tentative)*. Pt. 2, *Defensive Measures*. Washington, DC: Government Printing Office, 1932.

Engineer Field Manual: Field Fortifications. FM 5-15. Washington, DC: Government Printing Office, 1940.

Engineer Field Manual: Professional Papers of the Corps of Engineers, U.S. Army, No. 29. 5th ed. New York: Military Publishing, [1918].

Ernst, O. H. *A Manual of Practical Military Engineering, Prepared for the Use of the Cadets of the U.S. Military Academy, and for Engineer Troops*. New York: D. Van Nostrand, 1873.

Faragher, John Mack. *Daniel Boone: The Life and Legend of an American Pioneer*. New York: Henry Holt, 1993.

Farrenkopf, Michael. "Accidents and Mining: The Problem of the Risk of Explosion in Industrial Coal Mining in Global Perspective." In *Making Sense of Mining History: Themes and Agendas*, edited by Stefan Berger and Peter Alexander, 193–211. Oxfordshire, Eng.: Routledge, 2020.

Ferrill, Arthur. *The Origins of War from the Stone Age to Alexander the Great*. Boulder, CO: Westview, 1997.

Fleet, Betsy, and John D. P. Fuller, eds. *Green Mount: A Virginia Plantation Family during the Civil War: Being the Journal of Benjamin Robert Fleet and Letters of His Family*. Lexington: University Press of Kentucky, 1962.

Floyd, Dale E. "Cave Warfare on Okinawa." *Army History* 34 (Spring/Summer 1995): 6–9.

Forman, Harrison. *Report from Red China*. London: Robert Hale, 1946.

Fortification: Underground Water and Its Relation to Fieldworks. Technical Regulations 1195-65. Washington, DC: Government Printing Office, 1926.

Fortin, Maurice S., ed. "Colonel Hilary A. Herbert's History of the Eighth Alabama Volunteer Regiment, C.S.A." *Alabama Historical Quarterly* 39, nos. 1–4 (1977): 5–321.

France, John. *Western Warfare in the Age of the Crusades, 1000–1300*. Ithaca, NY: Cornell University Press, 1999.

Freeman, A. Z. "Wall-Breakers and River-Bridgers: Military Engineers in the Scottish Wars of Edward I." *Journal of British Studies* 10, no. 2 (May 1971): 1–16.

Froissart, John. *Sir John Froissart's Chronicles of England, France, and the Adjoining Countries, from the Latter Part of the Reign of Edward II to the Coronation of Henry IV*. 2 vols. Translated by Thomas Johnes. Hafod, 1804.

Fronsperger, Leonhardt. *Kriegsbuch*. Frankfurt Am Main, 1573.

Fulton, Michael S. "The Siege of Montfort and Mamluk Artillery Technology in 1271: Integrating the Archaeology and Topography with the Narrative Sources." *Journal of Military History* 83, no. 3 (July 2019): 689–717.

Gascoigne, Bamber. *The Great Moghuls*. London: Constable, 1998.

"The German Shelters on the Somme." *Professional Memoirs, Corps of Engineers, United States Army, and Engineer Department at Large* 9 (1917): 399–406.

Gibby, Bryan R. "The Battle of Shangganling, Korea, October–November 1952." *Journal of Chinese Military History* 6 (2017): 53–89.

Gille, Bertrand. *Engineers of the Renaissance*. Cambridge, MA: MIT Press, 1966.

Gommans, Jos. *Mughal Warfare: Indian Frontiers and High Roads to Empire, 1500–1700*. London: Routledge, 2002.

Graham, H. W. *The Life of a Tunnelling Company, Being an Intimate Story of the Life of the 185th Tunnelling Company, Royal Engineers, in France, during the Great War, 1914–1918*. Hexham, Eng.: J. Catherall, 1927.

Graham, Robert D. "Fifty-Sixth Regiment." In *Histories of the Several Regiments and Battalions from North Carolina in the Great War, 1861–'65*, edited by Walter Clark, 3:313–404. Goldsboro, NC: Nash Brothers, 1901.

Grant, Ulysses S. *Personal Memoirs of U.S. Grant*. 2 vols. in 1. New York: Viking, 1990.

Grau, Lester W., and Ali Ahmad Jalali. "Underground Combat: Stereophonic Blasting, Tunnel Rats, and the Soviet-Afghan War." *Engineer* 28 (November 1998): 20–23.

Grieve, W. Grant, and Bernard Newman. *Tunnellers: The Story of the Tunnelling Companies, Royal Engineers, during the World War*. London: Herbert Jenkins, 1936.

Guthrie, James M. *Camp-Fires of the Afro-American; or, The Colored Man as a Patriot*. Philadelphia: Afro-American Publishing, 1899. Reprint, New York: Johnson Reprint, 1970.

Haiman, Miecislaus. *Kosciuszko in the American Revolution*. New York: Polish Institute of Arts and Sciences in America, 1943.

Halle, Edward. *The Union of the Two Noble Families of Lancaster and York*. Menston, Eng.: Scolar, 1970.

Halleck, H. Wager. *Elements of Military Art and Science; or, Course of Instruction in Strategy, Fortification, Tactics of Battles, &C., Embracing the Duties of Staff, Infantry, Cavalry, Artillery, and Engineers, Adapted to the Use of Volunteers and Militia*. 3rd ed. New York: D. Appleton, 1862.

Harrington, Peter. *English Civil War Archaeology*. London: B. T. Batsford, 2004.

Hart, Francis Russell. *The Siege of Havana, 1762*. Boston: Houghton Mifflin, 1931.

Harvey, R. N. "British Military Mining, 1915–1917." *Military Engineer* 23 (1931): 509–518.

———. "Military Mining in the Great War." *Royal Engineers Journal* 43 (December 1929): 537–548.

Healy, John F. *Mining and Metallurgy in the Greek and Roman World*. London: Thames and Hudson, 1978.

Hearn, Chester G. *Mobile Bay and the Mobile Campaign: The Last Great Battles of the Civil War*. Jefferson, NC: McFarland, 1993.

Herodotus. *The History*. Translated by David Grene. Chicago: University of Chicago Press, 1987.

Hess, Earl J. *Civil War Field Artillery: Promise and Performance on the Battlefield.* Baton Rouge: Louisiana State University Press, 2023.

———. *Civil War Torpedoes and the Global Development of Landmine Warfare.* Lanham, MD: Rowman and Littlefield, 2023.

———. *In the Trenches at Petersburg: Field Fortifications and Confederate Defeat.* Chapel Hill: University of North Carolina Press, 2009.

———. *Into the Crater: The Mine Attack at Petersburg.* Columbia: University of South Carolina Press, 2010.

———. *Kennesaw Mountain: Sherman, Johnston, and the Atlanta Campaign.* Chapel Hill: University of North Carolina Press, 2013.

———. *Trench Warfare under Grant & Lee: Field Fortifications in the Overland Campaign.* Chapel Hill: University of North Carolina Press, 2007.

Hickenlooper, Andrew. "The Vicksburg Mine." In Buell and Johnson, *Battles and Leaders of the Civil War,* 3:539–542.

History of the 27th Engineers, U.S.A., 1917–1919. New York: Association of the 27th Engineers, 1920.

Hitti, Philip K., trans. *An Arab-Syrian Gentleman and Warrior in the Period of the Crusades: Memoirs of Usāmah Ibn-Munqidh.* New York: Columbia University Press, 1929.

Hogg, Ian. *The History of Fortification.* New York: St. Martin's, 1981.

Hopkins, Clark. "The Siege of Dura." *Classical Journal* 42, no. 5 (February 1947): 251–259.

Howard, R. L. *History of the 124th Regiment Illinois Infantry Volunteers.* Springfield, IL: H. W. Rokker, 1880.

Hutchinson, D. Jean, Mark Diederichs, Peeter Pehme, Peter Sawyer, Phillip Robinson, Al Puxley, and Hélène Robichaud. "Geomechanics Stability Assessment of World War I Military Excavations at the Canadian National Vimy Memorial Site, France." *International Journal of Rock Mechanics & Mining Sciences* 45 (2008): 59–77.

Index of Patents Relating to Electricity, Granted by the United States Prior to July 1, 1881, with an Appendix Embracing Patents Granted from July 1, 1881, to June 30, 1882. Washington, DC: Government Printing Office, 1882.

Inners, John D. "Colonel Henry Pleasants and the Military Geology of the Petersburg Mine—June–July, 1864." *Pennsylvania Geology* 40 (October 1989): 3–10.

Ilsley, L. C., and A. B. Hooker. *Electric Shot-Firing in Mines, Quarries, and Tunnels.* Washington, DC: Government Printing Office, 1926.

Jackson, Harry L. *First Regiment Engineer Troops, P.A.C.S.: Robert E. Lee's Combat Engineers.* Louisa, VA: R. A. E. Design and Publishing, 1998.

James, David H. *The Siege of Port Arthur: Records of an Eye-Witness.* London: T. Fisher Unwin, 1905.

James, Simon. "Stratagems, Combat, and 'Chemical Warfare' in the Siege Mines of Dura-Europos." *American Journal of Archaeology* 115, no. 1 (January 2011): 69–101.

Jones, John T. *Journals of Sieges Carried On by the Army under the Duke of Wellington in Spain, during the Years 1811 to 1814.* 3 vols. London: John Weale, 1846.

Jones, Simon. *Underground Warfare, 1914–1918*. Barnsley, Eng.: Pen & Sword, 2010.

Josephus. *The Jewish War*. 3 vols. Translated by H. St. J. Thackeray. Cambridge, MA: Harvard University Press, 1957.

Kaufmann, J. E., and H. W. Kaufmann. *The Medieval Fortress: Castles, Forts, and Walled Cities of the Middle Ages*. Conshohocken, PA: Combined Publishing, 2001.

Kenyon, John R. *Medieval Fortifications*. Leicester, Eng.: Leicester University Press, 1991.

Kern, Paul Bentley. *Ancient Siege Warfare*. Bloomington: Indiana University Press, 1999.

Kerr, George L. *Practical Coal Mining: A Manual for Managers, Under-Managers, Colliery Engineers, and Others*. 4th ed. London: Charles Griffin, 1905.

King, W. R. *Torpedoes: Their Invention and Use, from the First Application to the Art of War to the Present Time, for the Use of the Officers of the Corps of Engineers*. Washington, DC: Government Printing Office, 1866.

Kochan, Michael P., and John C. Wideman. *Civil War Torpedoes: A History of Improvised Explosive Devices in the War between the States*. DVD. By the authors, 2012.

Kulikov, Yuri, Elena Kulikov, and Colin Robins. "Mining and Counter-Mining in the Crimean War." *Journal of the Society for Army Historical Research* 97 (2019): 339–354.

Lake, Edward. *Journals of the Sieges of the Madras Army in the Years 1817, 1818, and 1819*. London: Kingsbury, Parbury, and Allen, 1825.

Lang, W. H. "Some Notes on the Turkish Trenches at Sannayait." *Professional Memoirs, Corps of Engineers, United States Army and Engineer Department at Large* 10 (1918): 309–314.

Langins, Janis. *Conserving the Enlightenment: French Military Engineering from Vauban to the Revolution*. Cambridge, MA: MIT Press, 2004.

Larrabee, Edward McM. "Archaeological Research at the Fortress of Louisbourg, 1961–1965." *Canadian Historic Sites: Occasional Papers in Archaeology and History*. No. 2. Ottawa, Can.: National Historic Sites Service, 1971.

Lawrence, A. W. *Greek Aims in Fortification*. Oxford, Eng.: Oxford University Press, 1979.

Lee, Wayne E. "Fortify, Fight, or Flee: Tuscarora and Cherokee Defensive Warfare and Military Culture Adaptation." *Journal of Military History* 68, no. 3 (July 2004): 713–770.

Lehrer, Glenn H. "Viet Cong Tunnels." *Military Engineer* 60 (July–August 1968): 243–247.

Lendy, A. F. *Elements of Fortification: Field and Permanent, for the Use of Students, Civilian and Military*. London: John W. Parker and Son, 1857.

Lenihan, Pádraig. "Namur Citadel, 1695: A Case Study in Allied Siege Tactics." *War in History* 18, no. 3 (2011): 282–303.

Leonard, Matthew. *Beneath the Killing Fields: Exploring the Subterranean Landscapes of the Western Front*. Barnsley, Eng.: Pen & Sword, 2016.

Lepage, Jean-Denis G. G. *Dutch Fortifications: An Illustrated History from the Roman Era to the Cold War*. Jefferson, NC: McFarland, 2021.

Levine, Beth. "Headfirst into Underground Battle." *Military History* (February 1987): 42–48.

Livermore, William R. "Field and Siege Operations in the Far East." *Journal of the Military Service Institute of the United States* 36, no. 135 (May–June 1905): 421–441.

Livy. 14 vols. Cambridge, MA: Harvard University Press, 1919–1959.

Lloyd, E. M. "The Mine Warfare at Sebastopol." *Professional Papers of the Corps of Royal Engineers* 1 (1877): 93–111.

Lockett, S. H. "The Defense of Vicksburg." In Buel and Johnson, *Battles and Leaders of the Civil War*, 3:482–492.

Lockwood, Thomas D. *Electricity, Magnetism, and Electric Telegraphy: A Practical Guide and Hand-Book*. New York: D. Van Nostrand, 1883.

Logan, D. Dale. "The Difficulties and Dangers of Mine-Rescue Work on the Western Front, and Mining Operations Carried Out by Men Wearing Rescue-Apparatus." *Transactions of the Institution of Mining Engineers* 56 (1918–1919): 197–222.

Lumpkin, Henry. *From Savannah to Yorktown: The American Revolution in the South*. New York: Paragon, 1987.

Lydenberg, Harry Miller, ed. *Archibald Robertson, Lieutenant-General Royal Engineers: His Diaries and Sketches in America, 1762–1780*. New York: New York Public Library, 1930.

Lynn, John A. *Giant of the Grand Siècle: The French Army, 1610–1715*. New York: Cambridge University Press, 1997.

———. *The Wars of Louis XIV, 1667–1714*. London: Longman, 1999.

Macaulay, J. S. *A Treatise on Field Fortification, the Attack of Fortresses, Military Mining, and Reconnoitring*. London: Bosworth and Harrison, 1860.

MacLeod, Roy. "Phantom Soldiers: Australian Tunnellers on the Western Front, 1916–18." *Journal of the Australian War Memorial* (October 1988): 31–41.

Mahan, D. H. *An Elementary Course of Military Engineering, Part I, Comprising Field Fortification, Military Mining, and Siege Operations*. New York: John Wiley and Son, 1867.

Maier, Franz Georg, and Marie-Louise von Wartburg. "Reconstruction of a Siege: The Persians at Paphos." In *Ancient Cyprus in the British Museum: Essays in Honour of Dr. Veronica Tatton-Brown*, edited by Thomas Kiely, 7–20. London: British Museum, 2009.

Malins, Geoffrey H. *How I Filmed the War*. London: Herbert Jenkins, 1920.

Manningham, Henry. *A Complete Treatise of Mines: Extracted from the Memoires d'Artillerie*. London: A. Millar, 1756.

Manual of Field Works (All Arms). London: His Majesty's Stationery Office, 1921.

Marcus, Raphael D. "Learning 'under Fire': Israel's Improvised Military Adaptation to Hamas Tunnel Warfare." *Journal of Strategic Studies* 42, nos. 3–4 (2019): 344–370.

Marshall, Arthur. *Dictionary of Explosives*. Philadelphia: P. Blakiston's Son, 1920.

Marshall, Christopher. *Warfare in the Latin East, 1192–1291*. Cambridge: Cambridge University Press, 1992.

Marvin, Laurence W. "War in the South: A First Look at Siege Warfare in the Albigensian Crusade, 1209–1218." *War in History* 8, no. 4 (2001): 373–395.

Mayo, Robert S. "Military Tunneling." *Military Engineer* 45 (1953): 276–277.

McClellan, George B. "Report upon the Operations in the Crimea." In *Report of the Secretary of War, Communicating the Report of Captain George B. McClellan (First Regiment United States Cavalry), One of the Officers Sent to the Seat of War in Europe, in 1855 and 1856*, 5–24. Washington, DC: A. O. P. Nicholson, 1857.

McGeer, Eric. "Byzantine Siege Warfare in Theory and Practice." In *The Medieval City under Siege*, edited by Ivy A. Corfis and Michael Wolfe, 123–130. Woodbridge, Eng.: Boydell, 1995.

Mercur, James. *Attack of Fortified Places including Siege-Works, Mining, and Demolitions*. New York: John Wiley, 1894.

Merdinger, C. J. "Tunnels through the Ages." [Pt. 1]. *Military Engineer* 50 (1958): 32–38.

Mine Warfare by Captain Cussenot. Nancy, Fr.: Berger-Levrault, [1917].

"Mining, Military." In *Aide-Mémoire to the Military Sciences*, 2:361–424. London: Lockwood, 1860.

Mitchell, William A. *Army Engineering*. 5th ed. Washington, DC: Society of American Military Engineers, 1938.

Mordecai, Alfred. *Military Commission to Europe, in 1855 and 1856*. Washington, DC: George W. Bowman, 1860.

Morley, George H. "Tunnelling on the Western Front." *Canadian Mining Journal* 11 (May 14, 1919): 341–345.

Morreau, L. J. *Bolwerk de Nederlanden*. Assen, Neth.: Van Gorcum, 1979.

Morris, W. S., L. D. Hartwell, and J. B. Kuykendall. *History 31st Regiment Illinois Volunteers, Organized by John A. Logan*. Carbondale: Southern Illinois Press, 1998.

Morse, R. V. "War Mining Notes from the Western Front." *Transactions of the Institution of Engineers, Australia* 1 (1920): 480.

Mosier, John. *The Myth of the Great War: A New Military History of World War I*. New York: Harper Collins, 2001.

Mullins, Lawrence E. "The Mines at Messines." *Military Review* 45, no. 4 (April 1965): 18–24.

Murray, R. R. "Tunnelling in the Ypres Salient." *Journal of the United Services Institute of Nova Scotia* 2 (1929): 114–123.

Mushynsky, Julie, Jennifer McKinnon, and Fred Camacho. "The Archaeology of World War II Karst Defences in the Pacific." *Journal of Conflict Archaeology* 13, no. 3 (2018): 198–222.

Neill, J. C., ed. *The New Zealand Tunnelling Company, 1915–1919*. Auckland, N.Z.: Whitcombe and Tombs, 1922.

Nemchinsky, N., V. Yuriev, and Y. Galkin. "Underground Warfare." *Military Engineer* 36, no. 224 (June 1944): 190–192.

Norton-Griffiths, John. "The Origin of Tunnelling Companies, R.E." *Royal Engineers Journal* (March 1928): 87–92.

Ober, Josiah. "Hoplites and Obstacles." In *Hoplites: The Classical Greek Battle Experience*, edited by Victor Davis Hanson, 173–196. London: Routledge, 1994.

Oman, Charles. *A History of the Art of War in the Middle Ages*. 2nd ed. 2 vols. New York: Burt Franklin, 1924.

Palfrey, John C. "Port Hudson." In *The Mississippi Valley, Tennessee, Georgia, Alabama, 1861–1864: Papers of the Military Historical Society of Massachusetts*, 8: 23–63. Boston: Military Historical Society of Massachusetts, 1910.

Pancake, John S. *This Destructive War: The British Campaign in the Carolinas, 1780–1782*. Tuscaloosa: University of Alabama Press, 1985.

Pascas, Brian. "Clay-Kickers of Flanders Fields: Canadian Tunnellers at Messine Ridge, 1916–1917." *Canadian Military History* 2, no. 2 (2018): 1–31.

Pepper, Simon. "The Underground Siege." *Fort* 10 (1982): 31–38.

Pepper, Simon, and Nicholas Adams. *Firearms & Fortifications: Military Architecture and Siege Warfare in Sixteenth-Century Siena*. Chicago: University of Chicago Press, 1986.

Petchey, Peter. "Second World War Japanese Defences on Watom Island, Papua New Guinea." *Journal of Conflict Archaeology* 10, no. 1 (2015): 29–51.

Plum, William R. *The Military Telegraph during the Civil War in the United States*. 2 vols. Chicago: Jansen, McClurg, 1882.

Polybius. *The Histories*. 6 vols. Translated by W. R. Paton. New York: G. P. Putnam's Sons, 1922.

Porter, Whitworth. *History of the Corps of Royal Engineers*. 2 vols. London: Longmans, Green, 1889.

Potter, K. R., ed. *Gesta Stephani*. Oxford, Eng.: Clarendon, 1976.

Prestwich, Michael. *Armies and Warfare in the Middle Ages: The English Experience*. New Haven, CT: Yale University Press, 1999.

Price, Neil, and Rick Knecht. "Peleliu 1944: The Archaeology of a South Pacific D-Day." *Journal of Conflict Archaeology* 7, no. 1 (January 2012): 5–48.

Prior, Robin, and Trevor Wilson. *Passchendaele: The Untold Story*. New Haven, CT: Yale University Press, 1996.

Procopius. *History of the Wars*. 5 vols. Translated by H. B. Dewing. Cambridge, MA: Harvard University Press, 1953.

Purton, Peter. *The Medieval Military Engineer from the Roman Empire to the Sixteenth Century*. Woodbridge, Eng.: Boydell, 2018.

Reeves, Nicholas. "Cinema, Spectatorship, and Propaganda: 'Battle of the Somme' (1916) and Its Contemporary Audience." *Historical Journal of Film, Radio and Television* 17, no. 1 (1997): 5–28.

Report of the Committee on the Conduct of the War on the Attack on Petersburg on the 30th Day of July, 1864. Washington, DC: Government Printing Office, 1865.

Reynolds, L. B. "Mining in Chalk on the Western Front, with Some Notes on the Explosion of Large Charges of High Explosives." *Transactions of the Canadian Mining Institute* 22 (1919): 462–474.

Rhyne, David W. "War in the Lithosphere." *Army* 36 (August 1986): 60–64, 68.

Roberts, Ian. *Pontefract Castle Archaeological Excavations, 1982–86*. Morley, Eng.: West Yorkshire Archaeology Service, 2002.

Rogers, Clifford J. *Soldiers' Lives through History: The Middle Ages*. New York: Greenwood, 2007.

———. *War Cruel and Sharp: English Strategy under Edward III, 1327–1360*. Woodbridge, Eng.: Boydell, 2000.

———, ed. *The Wars of Edward III: Sources and Interpretations*. Woodbridge, Eng.: Boydell, 1999.

Rogers, Clifford J., Daniel Berardino, Ryne Hicks, Liam Kane, and Zachary Watters. "Investigating the Outcome of Sieges during the Era of the Hundred Years' War: A Quantitative Reconnaissance." *Medieval Warfare* 11, no. 6 (2022): 24–36.

Rogers, R. *Latin Siege Warfare in the Twelfth Century*. Oxford, Eng.: Oxford University Press, 1997.

Rorive, Jean-Pierre. *La guerre de siège sous Louis XIV en Europe et à Huy*. Brussels, Belg.: Editions Racine, 1998.

Rose, E. P. F., and M. S. Rosenbaum. "British Military Geologists: The Formative Years to the End of the First World War." *Geologist's Associations Proceedings* 104 (1993): 41–49.

Rosenbaum, M. S. "Geological Influences on Tunnelling under the Western Front at Vimy Ridge." *Proceedings of the Geologist's Association* 100 (1989): 135–140.

Rothenburg, Gunther E. *The Art of Warfare in the Age of Napoleon*. Bloomington: Indiana University Press, 1980.

Sagona, Antonio, Mithat Atabay, Richard Reid, Ian McGibbon, Chris Mackie, Muhammet Erat, and Jessie Birkett-Rees. "The ANZAC [Ariburnu] Battlefield: New Perspectives and Methodologies in History and Archaeology." *Australian Historical Studies* 42, no. 3 (2011): 313–336.

Sandes, E. W. C. *The Military Engineer in India*. 2 vols. Chatham, Eng.: Institution of Royal Engineers, 1933.

Sarantakes, Nicholas Evan. "The Quiet War: Combat Operations along the Korean Demilitarized Zone, 1966–1969." *Journal of Military History* 64, no. 2 (April 2000): 439–457.

Saunders, Anthony. *Trench Warfare, 1850–1950*. Barnsley, Eng.: Pen & Sword, 2010.

Sawyer, Ralph D. *Fire and Water: The Art of Incendiary and Aquatic Warfare in China*. Boulder, CO: Westview, 2004.

Schellen, H. *Magneto-Electric and Dynamo-Electric Machines: Their Construction and Practical Application to Electric Lighting and the Transmission of Power*. Vol. 1. New York: D. Van Nostrand, 1884.

Schiffer, Michael Brian. *Power Struggles: Scientific Authority and the Creation of Practical Electricity before Edison*. Cambridge, MA: MIT Press, 2008.

Schmale, Franz-Joseph, ed. and trans. *Fontes Italici de rebus a Frederico I. imperatore in Italia gestis et epistola de eiusdem expeditione sacra*. Darmstadt, W. Ger.: Wissenschaftliche Buchgesselschaft, 1986.

Schneck, William C. "The Origins of Military Mines: Part I." *Engineer* 28 (July 1998): 49–55.

Schneid, Frederick C. *Napoleon's Italian Campaigns, 1805–1815.* Westport, CT: Praeger, 2005.

Schue, Paul. "Remember the Alcazar! The Creation of Nationalist Myths in the Spanish Civil War: The Writings of Robert Brasillach." *National Identities* 10, no. 2 (2008): 131–147.

Searles, P. J. "An Oriental Siege." *Military Engineer* 34 (1942): 397–401.

Seymour, William. *Great Sieges of History.* Washington, DC: Brassey's, 1991.

Shakespeare, William. *The Life of King Henry the Fifth.* New York: Washington Square, 1960.

Sherwood, Andrew N., Milorad Nikolie, John W. Humphrey, and John P. Oleson. *Greek and Roman Technology: A Sourcebook of Translated Greek and Roman Texts.* 2nd ed. London: Routledge, 2019.

Shoup, Francis A. "Vicksburg: Some New History in the Experience of Gen. Francis A. Shoup." *Confederate Veteran* 2 (1894): 172–174.

Simms, Frederick Walter. *Practical Tunnelling.* 4th ed. New York: D. Van Nostrand, 1896.

Simon, John Y., ed. *The Papers of Ulysses S. Grant.* 28 vols. Carbondale: Southern Illinois University Press, 1967–2005.

Smith, Nicola. "Military Training Earthworks in Crowthorne Wood, Berkshire: A Survey by the Royal Commission on the Historical Monuments of England." *Archaeological Journal* 152, no. 1 (1995): 422–440.

Smith, Robert Douglas, and Kelly DeVries. *Rhodes Besieged: A New History.* Stroud, Eng.: History Press, 2011.

Smith, Robert S. *Warfare and Diplomacy in Pre-Colonial West Africa.* London: James Currey, 1989.

Smith, Victor T. C. "Chatham Siege Operations in 1907." *Fort* 5 (Spring 1978): 87–93.

Stevenson, Robert E. *To Win the Battle: The 1st Australian Division in the Great War, 1914–18.* Cambridge: Cambridge University Press, 2013.

Stichelbaut, Birger, Nicolas Note, Timothy Saey, Daan Hanssens, Hanne Van den Berghe, Jean Bourgeois, Marc Van Meirvenne, Veele Van Eetvelde, and Wouter Gheyle. "Non-Invasive Research of Tunneling Heritage in the Ypres Salient (1914–1918)—Research of the Tor Top Tunnel System." *Journal of Cultural Heritage* 26 (2017): 109–117.

Stichelbaut, Birger, Wouter Gheyle, Timothy Saey, Veerle Van Eetvelde, Marc Van Meirvene, Nicholas Note, Hanne Van den Berghe, and Jean Bourgeois. "The First World War from Above and Below: Historical Aerial Photographs and Mine Craters in the Ypres Salient." *Applied Geography* 66 (2016): 64–72.

Stormont, Gilbert R., comp. *History of the Fifty-Eighth Regiment of Indiana Volunteer Infantry, Its Organization, Campaigns, and Battles from 1861 to 1865.* Princeton, IN: Clarion, 1895.

Stoye, John. *The Siege of Vienna.* New York: Holt, Rinehart, and Winston, 1964.

Straith, Hector. *Introductory Essay to the Study of Fortification, for Young Officers of the Army.* London: Parker, Furnivall, and Parker, 1849.

Supplement to the Official Records of the Union and Confederate Armies. 100 vols. Wilmington, NC: Broadfoot, 1993–2000.

Tabraham, Christopher. *Scottish Castles and Fortifications.* Edinburgh, Scot.: Crown, 1986.

Tatham, H. "Tunnelling in the Sand Dunes of the Belgian Coast." *Transactions of the Institution of Mining and Metallurgy* 28 (1918–1919): 249–260.

Taylor, F. L. *The Art of War in Italy, 1494–1529.* Cambridge: Cambridge University Press, 1921.

Thompson, Gilbert. *The Engineer Battalion in the Civil War.* Washington, DC: Press of the Engineer School, 1910.

Thomson, H. C. *The Chitral Campaign: A Narrative of Events in Chitral, Swat, and Bajour.* London: William Heineman, 1895.

Thucydides. *The Peloponnesian War.* Translated by Walter Blanco. New York: W. W. Norton, 1998.

Timberlake, Simon. "Archaeology of Mining in the Pre-Industrial Age: The Recognition and Interpretation of Ancient Mines." In *Making Sense of Mining History: Themes and Agendas*, edited by Stefan Berger and Peter Alexander, 25–64. Oxfordshire, Eng.: Routledge, 2020.

Toal, Caroline. *North Kerry Archaeological Survey.* Dingle, Ire.: Brandon, 1995.

Toy, Sidney. *Castles: Their Construction and History.* New York: Dover, 1985.

———. *A History of Fortification from 3000 BC to AD 1700.* London: Heinemann, 1966.

Trainor, Francis E. "Tunnel Destruction in Vietnam." *Military Engineer* 60 (September–October 1968): 341–342.

Trounce, H. D. *Fighting the Boche Underground.* New York: Charles Scribner's Sons, 1918.

———. "Military Mining." *Military Engineer* 30 (1938): 58–61.

———. "Military Mining on the Western Front." *Military Engineer* 29 (1937): 81–85.

———. "Military Mining Operations." *Military Engineer* 29 (1937): 187–190.

———. "Mine Rescue Work." *Professional Memoirs, Corps of Engineers, United States Army and Engineer Department at Large* 10 (1918): 549–564.

———. *Notes on Military Mining.* Washington, DC: Press of the Engineer School, 1918.

Turtle, Thomas. "History of the Engineer Battalion." *Printed Papers of the Essayons Club of the Corps of Engineers* 1 (1868–1872): 1–9.

Vauban, Sébastien Le Prestre de. *A Manual of Siegecraft and Fortification.* Translated by George A. Rothrock. Ann Arbor: University of Michigan Press, 1968.

Vegetius. *Epitome of Military Science.* Translated by N. P. Milner. Liverpool, Eng.: Liverpool University Press, 1996.

Venable, M. W. "In the Trenches at Petersburg." *Confederate Veteran* 14 (1906): 178–179.

Venable, Matthew Walton. *Eighty Years After; or, Grandpa's Story.* Charleston, WV: Hood-Hiserman-Brodhag, 1929.

Vernon, Simon François Baron Gay de. *A Treatise on the Science of War and*

Fortification. 2 vols. Translated by John Michael O'Connor. New York: J. Seymour, 1817.

Wallacker, Benjamin E. "Studies in Medieval Chinese Siegecraft: The Siege of Chien-K'ang, A.D. 548–549." *Journal of Asian History* 5, no. 1 (1971): 35–54.

———. "Studies in Medieval Chinese Siegecraft: The Siege of Fengtian, A.D. 783." *Journal of Asian History* 33, no. 2 (1999): 185–193.

———. "Studies in Medieval Chinese Siegecraft: The Siege of Yü-pi, A.D. 546." *Journal of Asian Studies* 28, no. 3 (May 1969): 789–802.

Warner, Philip. *Sieges of the Middle Ages*. 1968. Reprint, New York: Barnes and Noble, 1994.

War of the Rebellion: A Compilation of the Official Records of the Union and Confederate Armies. 70 vols. in 128 parts. Washington, DC: Government Printing Office, 1880–1901.

Warry, John. *Warfare in the Classical World: An Illustrated Encyclopedia of Weapons, Warriors, and Warfare in the Ancient Civilizations of Greece and Rome*. Norman: University of Oklahoma Press, 1995.

Wells, Seth J. *The Siege of Vicksburg*. Detroit: William H. Rowe, 1915.

West, Graham. *Innovation and the Rise of the Tunnelling Industry*. Cambridge: Cambridge University Press, 1988.

"The Whitaker Machine." *Quarterly Journal of Engineering Geology* 25 (1992): 81–82.

Wiggins, Kenneth. *Anatomy of a Siege: King John's Castle, Limerick, 1642*. Woodbridge, Eng.: Boydell, 2001.

———. *A Place of Great Consequence: Archaeological Excavations at King John's Castle, Limerick, 1990–8*. Dublin, Ire.: Wordwell, 2016.

———. *Siege Mines and Underground Warfare*. Princes Risborough, Eng.: Shire, 2003.

Wild, Heinz Walter. "Black Powder in Mining—Its Introduction, Early Use, and Diffusion over Europe." In *Gunpowder: The History of an International Technology*, edited by Brenda J. Buchanan, 203–217. Bath, Eng.: Bath University Press, 1996.

Wiley, Kenneth, ed. *Norfolk Blues: The Civil War Diary of the Norfolk Light Artillery Blues*. Shippensburg, PA: Burd Street, 1997.

Willig, Dierk. "German Military Geology and Military Mining on the Eastern Front in World War I." In *Military Aspects of Geology: Fortification, Excavation and Terrain Evaluation*, edited by E. P. F. Rose, J. Ehlen, and U. L. Lawrence, 131–150. London: Geological Society, 2019.

Winter, Frederick E. *A History of Greek Fortifications*. Toronto, Can.: University of Toronto Press, 1971.

Wise, Stephen R. *Gate of Hell: Campaign for Charleston Harbor, 1863*. Columbia: University of South Carolina Press, 1994.

Wolters, Timothy S. "Electric Torpedoes in the Confederacy: Reconciling Conflicting Histories." *Journal of Military History* 72, no. 3 (July 2008): 755–783.

Woodward, O. H. "Notes on the Work of an Australian Tunnelling Company in France." *Proceedings of the Australasian Institute of Mining and Metallurgy* 37 (1920): 1–54.

The Work of the Royal Engineers in the European War, 1914–19: Military Mining. Chatham, Eng.: W. & J. Mackay, 1922.

Wouters, H. H. E. *The Fortress Maastricht.* Heerlen, Neth.: DSM, n.d.

Xenophon. *History of My Times (Hellenica).* Translated by Rex Warner. Baltimore: Penguin Books, 1966.

Yadin, Yigael. *The Art of Warfare in Biblical Lands in the Light of Archaeological Discovery.* London: Weidenfeld and Nicolson, 1963.

Yeary, Mamie, comp. *Reminiscences of the Boys in Gray, 1861–1865.* Dallas, TX: Smith-Lamar, 1912.

Youngblood, Norman. *The Development of Mine Warfare: A Most Murderous and Barbarous Conduct.* Westport, CT: Praeger, 2006.

Zosimus. *Historia Nova: The Decline of Rome.* Translated by James J. Buchanan and Harold T. Davis. San Antonio, TX: Trinity University Press, 1967.

INDEX

Abel, Frederick Augustus, 128
Abeokuta, mining at siege of, 55
Abydos, mining at siege of, 19
Acre, mining at sieges of, 44, 98
Afghanistan, mining during Soviet
  occupation, 284
Africa, 6, 12, 55–56
Agricola, Georgius (Georg Bauer), 66
Aineas the Tactician, 17
Akbar, 67
Alcázar, mining at siege of the, 277
Aleppo, 43, 45
Alexander the Great, 17–18
Alicante, mining at siege of, 83–84
Amasis of Persia, 15
Ambracia, mining at siege of, 23–24, 32
Amida, mining at siege of, 35–36
Ammianus Marcellinus, 33, 34
Ammonal, 181, 189, 207, 208, 220, 248
Amulneir, mining at siege of, 120
Anderson, John, 123
Andrade, Tonio, 11
Antonius, Lucius, 9
Appian, 9, 24, 25
Aquileia, mining at siege of, 33
Arras, mining at, 244–246, 264, 281, 282
Artois, Second Battle of, 181
Assyrians, 15
Asti, mining at siege of, 84
Athens, mining at siege of, 24
Augusta, mining at siege of, 97
Avaricum, mining at siege of, 26

Ball, H. Standish, 185
Ballyshannon Castle, mining at siege of, 72
Banks, Nathaniel P., 142
Barca, mining at siege of, 15, 17
Barnwell, John, 94
Baron Hugh Bigod, 50
Barrie, Alexander, 255, 291
Barton, Peter, 292

Beardslee, George W., 159, 162
Bedford Castle, mining at siege of, 51
Bedson, William, 241–242
Beirut, mining at siege of, 43
Belgrade, mining at siege of, 58, 63
Belidor, Bernard Forest de, 85, 86, 89, 90,
  93, 104
Bellucci, G. B., 60, 62
Beneath Hill 60, 301–302
Bere-Ferrers, 40
Bergamo, mining at siege of, 41
Berwick, mining at siege of, 44, 52
Besuchis, mining at siege of, 34
Bharatpore, mining at siege of, 122
Bickford, William, 105
Bickford fuze, 115, 118, 125, 127, 132, 148,
  159, 209
Biringuccio, Vannoccio, 60, 62
Black, Jeremy, 11
Black, Thomas, 238
Boire, Michael, 245
Bone, Phillis, 223
Boonesborough, mining at siege of, 97
Bostyn, Franky, 328n18
Boulogne, mining at siege of, 63
Brescia, mining at siege of, 46
Brisco, Richard, 238, 240
Britain, 6, 56
Brock, Kevin, 11
Brooks, Alfred H., 118, 179, 183, 184, 193,
  197, 200, 203, 205, 212, 275, 289, 290
Brooks, Ernest, 227, 229, 230, 295, 298, 300
Brussels, mining at siege of, 84
Buisson, Robert du Mesnil du, 28–30, 32, 34
Bunel, Marc Isambard, 128
Bungay Castle, mining at siege of, 50, 293
Bunsen, Robert Wilhelm, 161
Burgos, mining at siege of, 99–101
Burnside, Ambrose E., 144, 148, 150
Bury, J. B., 60
Busca, Gabriello, 61

mining, military (*cont.*)

gallery lengths, 3–5, 14, 16, 19, 28, 29, 41, 42, 46, 50, 52, 64, 65, 74, 75, 81, 85–89, 92, 93, 95, 97–101, 107, 108, 110, 113, 114, 121, 123–125, 131, 134, 136, 138, 142–144, 145, 147, 152, 155, 162, 167, 172, 188, 213, 224, 226, 227, 229, 231, 234, 237, 244–253, 256, 257, 259, 261, 266, 268, 269, 271, 272, 277–278, 281, 284, 287, 289–290

gas attack in, 184, 199

gases, 5, 13, 219–222

geology, 10, 28–29, 35, 64, 72, 83, 96, 100, 107, 108, 113, 121, 124, 131, 133, 140, 142–145, 151, 152, 166, 168, 195, 202–206, 224, 244, 259, 262, 266, 267, 271, 272, 277, 279–280, 306n22

geophone, 215, 216, 280

globe of compression, 4, 86, 87, 89, 93, 101, 162, 189, 206–207, 273, 288

historical periods of, 3–4, 10, 57, 106–107, 117, 243, 275, 282, 286–288, 303

illumination, 14, 22, 100, 105, 108, 124, 144, 151, 166, 172, 199, 245, 260, 279, 280

leftover mine explosions, 292

line of least resistance, 62, 85–87, 101, 120, 131, 132

listening underground, 172, 183, 188, 214–218, 234, 280

losses in, 115, 134, 136, 139, 142, 149, 150, 178, 183, 219, 256, 258, 260, 271

manuals, 4, 17–18, 35, 43, 52, 54, 58, 60–63, 66, 103, 127, 129, 131, 132, 140, 141, 161–163, 275–278

mined shelters, 243–244

miners, 6–8, 13–14, 16, 19, 21, 26, 40, 43, 44, 46, 50–52, 61, 66–67, 73, 77, 79–81, 86, 100, 107, 108, 123, 133, 139, 142, 144, 146, 151, 176–178, 191, 194, 199–201, 236, 264, 291

mine schools in theater, 180, 218, 221

navigation underground, 4, 22, 60–61, 81, 121, 198

pottery found in mines, 268

practice mines, 102, 108, 297, 312n2

preservation of craters, 293–297

professionalization of, 70, 81

prop mining, 2, 59–60

radius of rupture, 162

rescue, 5, 172, 185, 193, 218–222, 241, 251, 260, 269

seismo-microphone, 216–217

shaft depths, 66, 72, 87, 102, 110, 113, 114, 121, 124, 131, 134, 146, 151, 154, 181, 188, 197, 226, 227, 231, 251, 278

shaft dimensions, 13, 72, 133, 154, 196, 227

shafts, 3, 7, 13, 21, 165, 194–196, 259

shoring, 13, 26, 29, 30, 45, 66, 73, 75–78, 107, 124, 136, 138, 144, 151–152, 172, 195, 197, 239, 244, 256, 263, 267, 272, 293

simultaneous underground approaches, 117, 124, 129, 175, 289

spoil, 82, 114, 133, 138, 143, 144, 152, 154, 184, 194, 200, 248, 251, 259, 260, 268

success rate, 20

subterranean combat, 23–25, 29–33, 35, 42, 44, 117, 124, 237–242, 291, 292, 324n29

surface attacks, 3, 20, 22, 25, 45, 46, 48, 49, 57–59, 81–83, 95, 96, 101, 102, 123, 124, 135, 140, 141, 148, 150, 155, 156, 171, 176, 180–182, 186, 242, 225–226, 229, 231, 237, 245, 246, 253–255, 258, 261, 266–267, 276, 277, 286, 290

tamping, 4, 62, 81, 100, 101, 114, 121, 132, 134, 148, 151–153, 212, 220, 239, 248, 260, 268

technology, 7–8, 125–129, 163, 224

tempo of mine warfare, 87–90, 116–117, 162, 272–273, 288

test explosions, 84–87, 287

tools, 13, 22, 40, 63, 72, 100, 133, 143–145, 151, 172

transit tunnels, 7, 14, 23, 28, 34–35, 53–54, 65, 128, 164, 244–245

tubbing, 165, 193, 195, 204

tunnels and insurgent warfare, 8, 282–285

ventilation of, 3–5, 13–14, 22, 29, 64, 66, 100, 103, 105, 107, 113, 116, 121, 124, 131, 133, 138, 142, 143, 145, 146, 151, 152, 162, 172, 181, 183, 198, 199, 259, 266, 268, 272, 280

venting a gunpowder blast, 63–64, 133, 134

virtual reality tour of mines, 302

visual depiction of mine explosions, 299–302, 328n18

zone of destruction, 4, 86, 189

Mobile, mining at siege of, 156
Montalcino, mining at siege of, 59, 63
Mont Cenis Tunnel, 128
Mont Cornillet, mining at, 257
Monte Visio, 65–66
Montfort, mining at siege of, 47, 293
Montfort, Simon de, 47
Moore, James, 95
Morgan, Garfield, 238
Morley, George H., 188, 200, 202, 291
Mosul, 45
mud forts, 119–120
Mukden, mining site near, 171
Murray, Robert R., 217

Namur, permanent countermine at, 91
Namur Citadel, mining at siege of, 83
Napoleon III, 26
Napoleonic Wars, 94, 98
Native Americans, 94
Navarro, Pedro, 58, 59
Near East, 6, 12, 14, 39
Niel, Adolphe, 107, 112, 113, 312n2
Neoheroka, mining at siege of, 95
Newcastle upon Tyne, mining at siege of,
    71–72
Newman, Bernard, 238, 255, 275, 294
Nicaea, mining at siege of, 51
Ninety-Six, mining at siege of, 97–99, 293
nitroglycerin, 128
Nivelle, Robert, 186
North America, 6
Norton-Griffiths, John, 176–178, 187, 189,
    207, 214, 246, 255, 263
Nowa, mining at siege of, 120–122

Ober, Josiah, 16
O'Connor, John Michael, 103
Octavian, 9
Oersted, Hans Christian, 126
Oman, Charles, 41, 42, 49, 50
Ontario Farm, mining at, 247, 252, 254, 255
Orense, mining at siege of, 58
Otones of Persia, 16
Outram, Sir James, 124

Padua, mining at siege of, 60, 63
Palfrey, John C., 142

Palma de Majorca, 46
Palus, mining at siege of, 18
Paphos, mining at siege of, 15–16
Peckham Farm, mining at, 247, 251
Peninsula War, 98–103
Penman, Lieutenant, 221
penthouse (also called sow, cat, tortoise,
    testudo), 2, 71, 72
Pepper, Simon, 61
Percival-Maxwell, Robert David, 194, 301, 302
Persia, 15
Perugia, mining at siege of, 9
Petersburg, mining at, 10, 68, 117, 144–155,
    160, 190, 193, 293, 297, 300
Peterwaradin (modern Petrovaradin),
    permanent countermine at, 91
Petit Bois, mining at, 247, 250
Petra, mining at siege of, 36–37
Philip Augustus, 43, 48
Philip V of Macedon, 18–20, 306n22
Philo of Byzantium, 18
Phoenicians, 16
Pien-ching, mining at siege of, 54
Piraeus, mining at siege of, 24–25
Pisa, mining at siege of, 58
Platea, mining at siege of, 16–17
Pleasants, Henry, 144–148, 150, 155
Polyaenus, 16
Polybius, 19–20, 23, 306n22
Polyperchon, 18
Pont-Audemer, 49
Pontefract Castle, mining at siege of, 72
Port Arthur, mining at siege of, 158,
    167–170, 172, 206, 273, 289, 293
Port Hudson, mining at siege of, 142, 145
Prestwich, Michael, 41
Prime, Frederick E., 141
Prinassus, mining at siege of, 19
Prince Eugene of Savoy, 84
Princess Anna Comnena, 42
Procopius, 35–37
Publius Licinius Crassus, 25

Rees, Albert, 238
Reese, Henry, 144, 149
Remi, mining at siege of, 25
Reynolds, L. B., 188, 198, 205, 206, 211, 212,
    222, 223, 238

Ukraine War, 303
United States, 130–131, 156, 179
Upton, Emory, 156
Uxellodunum, mining at siege of, 26, 292

Valle, Battista Della, 60
Vandewalle, Johan, 292
Vauban, Sébastien Le Prestre, 2, 3–4, 71,
    78–82, 84, 85, 91–93, 129, 131, 133, 155,
    157, 158, 171, 181, 189, 192, 258, 281, 282,
    287, 303
Vegetius, 35
Veii, mining at siege of, 22–23
Venable, Matthew, 151
Verdun, permanent countermine at, 104
Vernon, Simon François Baron Gay de,
    103–104
Vicksburg, mining at siege of, 117, 129,
    132–142, 145, 155, 157, 268
Vienna, mining at siege of, 82–83
Vietnam, American war in, 283–284
Vimy Ridge, mining at, 185, 199, 225–227,
    243, 245–246, 292, 294–296
Viscount Raymond-Roger Trencavel, 47
Volta, Alessandro, 126
Vrano, John, 58, 63

Walters, John, 151
War of the Austrian Succession, 84
Watson, Samuel J., 11
Wellesley, Arthur, 99, 101, 102
Westacott, John, 240–241, 291
Western Front, 175, 190–191, 192, 223, 225,
    243, 265, 274, 277, 289, 293, 303

Western Hemisphere, 95–95, 98, 130, 132, 155
Wheatstone, Charles, 126, 160
Whitaker, Douglas, 202
Wiggins, Kenneth, 6, 11, 72, 76
Willcox, Orlando B., 156
Wilmot, Sir Charles, 72
Woodward, Oliver H., 301, 302
World War I, 3, 4–5, 7–10, 16, 76, 79, 83,
    104, 107, 116–118, 130, 157, 158, 164–166,
    170–173, 273, 282, 287
    Australian mining at Gallipoli, 266,
        269–270
    British mining in, 174, 176, 178, 179,
        183–184, 187, 189, 191, 193, 198, 201, 202,
        203, 208, 213, 215, 217, 219, 224, 231, 236,
        242, 243, 244, 255, 256, 262–264
    French mining in, 175, 178, 181, 183–184,
        186, 189, 200, 201, 203, 209, 215, 216,
        224, 243, 244, 257–264
    German mining in, 175, 178, 181, 183–184,
        186, 189, 193, 195, 197, 198, 201, 203, 209,
        215, 217, 220, 224, 231, 236, 239, 243, 244,
        255–264
    number of mine explosions in, 197,
        223–224, 236, 259, 261, 262, 264, 271, 298
    number of miners in, 236
    Turkish mining in, 269–270
World War II, 278–279, 283–285
Wright, Lieutenant, 239

Yü-Pi, mining at siege of, 53–54

Zealots, 27
Zosimus, 34